130

Cambridge Studies in Biological and Evolutionary Anthropology 47

The First Boat People

The First Boat People concerns how people travelled across the world to Australia in the Pleistocene. It traces movement from Africa to Australia offering a new view of population growth at that time, challenging current ideas and underscoring problems with the 'Out of Africa' theory of how modern humans emerged. The variety of routes, strategies and opportunities that could have been used by those first migrants is proposed against the very different regional geography that existed at that time. Steve Webb shows the impact of human entry into Australia on the megafauna using fresh evidence from his work in Central Australia, including a description of palaeoenvironmental conditions existing there during the last two glaciations. He argues for an early human arrival and describes in detail the skeletal evidence for the first Australians. This is a stimulating account for students and researchers in biological anthropology, human evolution and archaeology.

STEVE WEBB is Professor of Australian Studies at Bond University, in Queensland, Australia. He has previously carried out a pioneering palaeopathological study of Aboriginal health patterns prior to European colonisation, and has previously published *Palaeopathology of Aboriginal Australians* (1995). His research now concentrates on Australian regional human evolution, reasons for the extinction of Australia's megafauna, Upper Pleistocene migration and the earliest human settlement of the continent. His particular focus is on palaeoenvironmental change accompanying the last two glaciations in Central Australia in order to understand more fully megafaunal extinction in the region and the timing of the first human entry into Australia.

Cambridge Studies in Biological and Evolutionary Anthropology

Series editors

HUMAN ECOLOGY
C. G. Nicholas Mascie-Taylor, University of Cambridge
Michael A. Little, State University of New York, Binghamton

GENETICS
Kenneth M. Weiss, Pennsylvania State University

HUMAN EVOLUTION
Robert A. Foley, University of Cambridge
Nina G. Jablonski, California Academy of Science

PRIMATOLOGY
Karen B. Strier, University of Wisconsin, Madison

Also available in the series

30 *Human Biology of Pastoral Populations* William R. Leonard & Michael H. Crawford (eds.) 0 521 78016 0
31 *Paleodemography* Robert D. Hoppa & James W. Vaupel (eds.) 0 521 80063 3
32 *Primate Dentition* Daris R. Swindler 0 521 65289 8
33 *The Primate Fossil Record* Walter C. Hartwig (ed.) 0 521 66315 6
34 *Gorilla Biology* Andrea B. Taylor & Michele L. Goldsmith (eds.) 0 521 79281 9
35 *Human Biologists in the Archives* D. Ann Herring & Alan C. Swedlund (eds.) 0 521 80104 4
36 *Human Senescence – Evolutionary and Biocultural Perspectives* Douglas E. Crews 0 521 57173 1
37 *Patterns of Growth and Development in the Genus* Homo. Jennifer L. Thompson, Gail E. Krovitz & Andrew J. Nelson (eds.) 0 521 82272 6
38 *Neanderthals and Modern Humans – An Ecological and Evolutionary Perspective* Clive Finlayson 0 521 82087 1
39 *Methods in Human Growth Research* Roland C. Hauspie, Noel Cameron & Luciano Molinari (eds.) 0 521 82050 2
40 *Shaping Primate Evolution* Fred Anapol, Rebecca L. German & Nina G. Jablonski (eds.) 0 521 81107 4
41 *Macaque Societies – A Model for the Study of Social Organization* Bernard Thierry, Mewa Singh & Werner Kaumanns (eds.) 0 521 81847 8
42 *Simulating Human Origins and Evolution* Ken Wessen 0 521 84399 5
43 *Bioarchaeology of Southeast Asia* Marc Oxenham and Nancy Tayles (eds.) 0 521 82580 6
44 *Seasonality in Primates* Diane K. Brockman and Carel P. van Schaik 0 521 82069 3
45 *Human Biology of Afro-Caribbean Populations* Lorena Madrigal 0 521 81931 8
46 *Primate and Human Evolution* Susan Cachel 0 521 82942 9

The First Boat People

S. G. WEBB
Bond University, Queensland

CAMBRIDGE UNIVERSITY PRESS
Cambridge, New York, Melbourne, Madrid, Cape Town, Singapore,
São Paulo

CAMBRIDGE UNIVERSITY PRESS
The Edinburgh Building, Cambridge CB2 2RU, UK
Published in the United States of America by Cambridge University Press,
New York

www.cambridge.org
Information on this title: www.cambridge.org/9780521856560

© Cambridge University Press 2006

This publication is in copyright. Subject to statutory exception
and to the provisions of relevant collective licensing agreements,
no reproduction of any part may take place without
the written permission of Cambridge University Press.

First published 2006

Printed in the United Kingdom at the University Press, Cambridge

A catalogue record for this publication is available from the British Library

ISBN-13 978-0-521-85656-0 hardback
ISBN-10 0-521-85656-6 hardback

Cambridge University Press has no responsibility for the persistence or
accuracy of URLs for external or third-party internet websites referred to in
this book, and does not guarantee that any content on such websites is, or
will remain, accurate or appropriate.

This book is dedicated to the memories of Rhys Jones and Peter Clark, two great friends who were immersed in the story of Australia's human beginnings.

Contents

List of plates	*page* viii
List of figures	x
List of maps	xi
List of tables	xii
Preface	xv
Introduction	1
1 Going to Sunda: Lower Pleistocene transcontinental migration	5
2 Pleistocene population growth	40
3 From Sunda to Sahul: transequatorial migration in the Upper Pleistocene	73
4 Upper Pleistocene migration patterns on Sahul	112
5 Palaeoenvironments, megafauna and the Upper Pleistocene settlement of Central Australia	134
6 Upper Pleistocene Australians: the Willandra people	183
7 Origins: a morphological puzzle	233
8 Migratory time frames and Upper Pleistocene environmental sequences in Australia	252
9 An incomplete jigsaw puzzle	271
Appendix 1	277
Appendix 2	279
Appendix 3	281
References	287
Index	304

Plates

5.1–5.2	Human cranial vault from Lake Eyre basin	page 162–163
5.3–5.5	Various views of human right metacarpal	168–169
5.6–5.7	Burnt megafauna bone	170–171
6.1	The cremated cranium of WLH1	189
6.2	The cranium of WLH3	190
6.3A,B,C	WLH50, A – Lateral view, B – Frontal view, C – Superior view	192–193
6.4	Cross-section of three cranial vaults WLH22 (top), WLH28 (middle, cremated) and WLH63 (bottom)	195
6.5	Cranial thickness of WLH18 composed almost entirely of spongy bone	196
6.6	X-ray of WLH1 showing uniformly thin cranial vault structure	196
6.7	Cranial thickness along the sagittal suture of right parietal of WLH68 with calcination (white) from cremation	197
6.8	X-ray showing uniformly thick cranial vault structure of WLH50 with additional thickening at the superior occipital protuberance and at the prebregmatic region of the frontal bone	198
6.9	A section of the WLH50 vault (parietal) showing how it is almost entirely diploeic (spongy bone) with extraordinary thin cranial tables	199
6.10	Close up of the diploeic cranial vault bone of WLH50	200
6.11	Brow profile of WLH68 showing a bulbous forehead, a lack of any brow development, the arc of the left eye socket (bottom) and discolouration typical of cremated bone	203
6.12	Zygomatic trigones of WLH18 (top) and WLH69 (bottom)	206

6.13	Comparison of the malar bones of WLH1 (left) and WLH2 (right) with its prominent malar tuberosity and overall greater rugosity than that of WLJ1	211
6.14	Frontal sinus of WLH50, at top of picture	213
6.15	Thick humeral cortex in WLH110	218
6.16	Comparison of WLH7 tibia (centre) with a modern example (left)	220
6.17	Curved cracking, calcination and colour changes on tubular bones of WLH115, all typical indicators of the bone having been cremated in a high temperature fire	222
6.18	Cremated cranial sections of WLH68. At left is a view of the frontal and supraorbital region	223
6.19	Heavily charred pieces of the cranium of WLH28 (top)	224
6.20	WLH3 mandible showing both canine teeth missing, resorption of the alveolar bone, and a compensatory lean of adjacent teeth towards the gap	227
6.21	Parallel grooves on the lower, first molar of WLH3	228
6.22	Nanwoon occipital bone showing a hole of indeterminate origin on the right side	232

Figures

1.1	Evolution of hominid cranial capacity	*page* 24
2.1	Standard world population growth trend over 1My	42
2.2	Proposed world population growth during the Pleistocene	56
3.1	Human migrations in the Upper Pleistocene	77
3.2	Increase in world reproduction rates in the Upper Pleistocene	78
3.3	Sea levels during the Upper Pleistocene	82
5.1	Upper Pleistocene oxygen isotope stages and equivalent palaeoenvironments in the Lake Eyre basin	138
5.2	Fluorine analysis of fossil and modern bone samples from the Lake Eyre region	167
5.3	Timing and process of the Australian megafaunal extinctions	177
5.4	Final process of megafaunal extinctions	178
6.1	Supraorbital modules for Willandra, Ngandong and Choukoutien fossils groups	204
6.2	Linear regression correlations between supraorbital module and cranial vault thickness	207
6.3	Linear regression correlations between malar size/length and robusticity modules	209
6.4A	Malar size/length module	210
6.4B	Malar robusticity module	210
6.5	Linear regression correlation between malar robusticity and cranial vault thickness	211
8.1	Upper Pleistocene megalake phases of Lake Eyre (hatched) compared to sea level change	253

Maps

1.1	Erectine migration from Africa to Asia showing two possible routes to East Asia and a third into central Russia	*page* 11
2.1	Population growth centres in China, Africa and India shown with Upper Pleistocene migrations	63
2.2	A genetic pressure barrier on the Indian/Myanmar (Burma) border	68
3.1	Exposed shelf of Sunda at −145m sea level (After Vorkis 2000, Field Museum, Chicago, 2000)	83
3.2	Boundary of ash fan from the Mt Toba eruption	90
3.3	Migration patterns and genetic mixing through island Indonesia	102
3.4	Major palaeodrainage systems of the Sunda and Sahul shelves with sea level at −75m (After Vorkis 2000, and Field Museum, Chicago, 2000)	103
4.1	Human migration into Sahul with sea levels at −75m (Field Museum, Chicago, 2000)	118
5.1	Megalake phase of the Lake Eyre/Lake Frome system at 85 ky. (After Nanson *et al.*, 1998)	136
5.2	Distribution of late Quaternary megafauna quadrupeds (*Diprotodon, Zygomaturus* and *Palorchestes*)	156
5.3	Distribution of Late Quaternary megafauna macropods (*Procoptodon, Macropus, Sthenurus, Protemnodon* and *Propleopus*)	157
6.1	The Willandra Lakes system, western New South Wales	188

Tables

2.1	Asian landmass by country	page 46
2.2	Chinese, Indian and Asian populations per land area occupied (km^2)	47
2.3	Population increase and land availability during the Pleistocene	47
2.4	Estimates of Pleistocene world population growth ($\times 10^6$)	48
2.5	Estimates of Pleistocene population density (persons/km^2)	49
2.6	World land surface occupied ($km^2 \times 10^6$)	49
2.7	Population doubling times (Dt), where $Dt = 0.6931/r$	52
2.8	World population growth using various base populations ($\times 10^6$)	55
2.9	World population growth from 1 My using four base population sizes and a reproduction rate (r) of 0.000005	57
2.10	Pleistocene population growth in China, the Indian subcontinent and the rest of Asia (AsiaP) $\times 10^6$	60
2.11	Indian subcontinent population growth from a base population of 100 000 at 400 ky and a reproduction rate (r) of 0.000005	61
5.1	Prey weight ratios (5:1) for large crocodiles (after Webb and Manolis, 1989)	148
5.2	Prey weight ratios (5:1) for large lizards	148
5.3	Prey weight required by three top Upper Pleistocene scavenger/predators of the Lake Eyre basin	149
5.4	*Diprotodon* weight calculated using long bone circumferences of fossil remains from the Lake Eyre basin	150
5.5	*Diprotodon* weight calculated using fossil long bones from the Darling Downs, Queensland	150
5.6	Weight estimates for two major groups of herbivores	151
5.7	Annual prey weight required from each of four herbivore groups required to sustain the estimated Lake Eyre basin carnivore population.	151
5.8	Fluorine analysis of fossil bone	164–166

List of tables xiii

6.1	Cranial vault thickness (mm) at primary anatomical points on Willandra individuals (a) and other individuals and populations (b)	194
6.2	Percentage of diploeic bone in vault construction	197
6.3	Comparative cranial vault thickness ranges (mm) for various hominid populations and individuals	201
6.4	Supraorbital thickness (mm) and module for WLH Series compared with *Homo soloensis* (Ngandong) and *Homo erectus* (Zhoukoutien) and Neanderthals	205
6.5A	Willandra Series glenoid fossae dimensions (mm)	215
6.5B	Comparative glenoid fossae ranges (mm)	215
6.6A	Mandibular metrical data (mm)	216
6.6B	Comparative mandibular metrical ranges (mm) among three fossil and one modern Australian population	216

Preface

This book arose out of pure curiosity of where it might take me. I also wanted to write down a number of ideas that I have been thinking about for many years as well as some more recent ones. The bottom line of the ideas focuses on the origin and timing of the first people to enter Australia. The subject now, however, is infinitely more complex than it used to be because it seems the more we learn about the process the less we know. The subject leads into the origin of modern humans; what happened to more archaic humans in our region, the abilities of early humans like *Homo erectus* and how large the world's population was by the late Pleistocene. Indeed, how did humanity grow, where were the main growth centres and what were the characteristics and outcome of that growth? I have come to accept, however, that in this business for every question answered two more appear.

There are some accepted norms regarding Pleistocene palaeodemography that require challenge. So I have tried to challenge them and put forward some conclusions that to some may seem eccentric. The main object is to open all doors; put up the challenges, have a fresh look at the evidence before us and propose some what ifs. Perhaps the most interesting thing about archaeology is that is constantly surprises. Our ideas about the past are constantly challenged or turned on their heads. New discoveries alter paradigms and philosophy regarding our ancestors and their story. Just as we think we know what that story is, it changes sometimes 180 degrees.

The ideas presented here have emerged as an attempt to answer the many questions that surround Pleistocene palaeodemography and migration and how these contrived to place people in Australia. They also attempt to answer some fundamental question that always pop up during any discussion of this topic. Some of these include:

1. Who were the first people to arrive in Australia?
2. What did they look like?
3. Where did they come from?
4. Why did they come here and when did they arrive?
5. Was there more than one migration?
6. How many of them were there on that first landing?

7. Where did they land?
8. How long did it take them to get here?
9. Where did they go after they arrived?
10. What effect did they have on the Australian megafauna?

The list goes on but answers to all these lie in knowing something of world population growth and human migration patterns over the hundreds of thousands of years prior to the arrival of the first Australians.

Was there a natural curiosity among people that made them just explore, or were there more fundamental reasons such as population pressure or invasion of their lands by others? Perhaps a natural catastrophe made them move! To me all these are intriguing questions and, although some may have been asked before, answers continue to elude us. Some researchers regard them as too difficult to tackle and they should be avoided at all cost. It has been expressed to me that only the eccentric or foolhardy would tackle questions in prehistory that are almost impossible to answer. I do not believe it is eccentric or foolhardy to discuss mysteries of the past and try and apply our data, albeit very scrappy at times, in an effort to provide answers or avenues forward. This book does not hide from hard questions nor does it shirk from trying to provide answers, albeit that some might think some of the propositions I have made constitute a one-way ticket off the planet. The subject matter is difficult and will always be so, but that is no reason to ignore it. Questions arising are likewise difficult to provide answers to and this book does not provide definitive answers to them either and I doubt whether any single book ever will. Instead it takes a look at the broad perspective and tries to offer solutions using logical possibilities as well as the available data. My belief is that only by discussing the questions will we ever come close to solving them. I also believe that discussing seemingly unsolvable issues lacking data does not necessarily make one odd – just curious.

When I began writing, it seemed that I continually came back to one fundamental question and that was: How big was the world's population at the time the first people entered Australia? After all, why would people move here if someone were not behind them? Surely, nobody would want to be out there on their own? We know the world's population must have grown over the last million years to produce us and those that entered Australia. So, my curiosity moved along a chain of events back into the past to consider world population growth. That process inevitably required that I try and produce some idea of how many people lived on the planet at the beginning of the whole process – at least one million years ago. So I found myself beginning the process of trying to find out how many people may have been around at the time of the *Homo erectus* diaspora from Africa and what the growth of

Pleistocene populations might have looked like, particularly in southeast Asia. To do this I used some basic principles of palaeodemography and a detailed but parsimonious description of world population in the Lower Pleistocene. Without knowing something of these issues together with an assessment of glacial and interglacial environmental and sea level changes, later human migrations into Australia and the megafaunal extinctions that followed do not make much sense. I propose how these events might have taken place and their timing. I assess options available to people when they arrived in Australia and discuss early Australian demographics as well as colonisation tactics and options using results of my own research in Central Australia. Finally, I present a description of the best fossil evidence for the earliest Australian people also using my own recent research and from this I offer some thoughts on the origin of the first Australians and their culture.

The few attempts to answer these questions reflect, perhaps, a well-advised caution many have towards a 'theoretical' approach to prehistory and that is natural enough. Most attempts at modelling Australia's regional palaeodemography were written some time ago. They are limited in their scope and mainly concentrate on the progress of people moving into Australia under present environmental circumstances. Recent research I have carried out with others has begun to shed much light on the environmental conditions prevailing when people entered the continent, from the penultimate glaciation through to the last interglacial. It is hoped that this book brings together these new data in a synthesis that makes better sense of Australia's formative palaeodemography. No doubt the book will soon be superseded. I look forward to that because it will mean that fresh data will have been found that helps us on our way to understanding the first chapter in the human story of this continent. Although answers to many of the above questions still elude us, I hope that this book may make answering at least some a little easier, or make the reader think a little harder about them. Many may not agree with what I say, others may quote that now famous Australian riposte 'he's dreaming'. But 'dreaming' is something the first Australians believe tells them how they began; perhaps it's time to join them in this.

Introduction

Early in January of 1992 a thin, hungry and very weary man was found by station workers stumbling through open bush in one of the most isolated parts of the Australian continent. Conditions were unbearable with early afternoon temperatures around 47°C, humidity 85 per cent and monsoonal storms beginning to build. He carried no water and the nearest 'civilisation' was one of northwestern Australia's remotest cattle stations 25 km away. The badly dehydrated man was taken to the station and when sufficiently recovered he told his rescuers he was not alone. He had left companions to make their way in small groups across the baking harshness of the rocky Mitchell Plateau. He was walking across one of the harshest and least explored areas of Australia that stretches for thousands of square kilometres. Without knowing the cattle station was there he would have missed it altogether.

One by one the man's companions were found and when they had partially recovered from their exhaustion they related their story. They had been walking for days over the rugged, scrubby terrain with hardly any water and little food except some flour and a few lizards and snakes that they had caught on the way. They had no knowledge of bush foods, no weapons and no previous experience with the Australian bush or that type of environment. The unfamiliar landscape and conditions were doubly hard for them because they had entered this harsh wilderness at its worst, during the extreme mid-summer monsoon. But who were these people?

Using fairly sophisticated electronic satellite navigation equipment, the group had headed straight for Australia from China. Unfortunately the navigation system could not tell them that they were heading for the remotest piece of Australia's 36 700 km coastline and one of the most inhospitable parts of the continent. Their boat had grounded on mudflats covered with hectares of densely tangled mangrove infested with sandflies, mosquitoes and sharp spines, and patrolled by more than the occasional salt water crocodile. Rather than stay in this extremely uncomfortable environment and with little hope of refloating their worn out fishing vessel, the 60 or more crew and passengers began to trek inland in small groups in the hope of finding a town and rescue. When found, the first man had already trekked over 140 km in seven days wearing only thin sandals on his feet. Others including women with children had no footwear but had still walked many kilometres.

The 'boat people' phenomenon has swept many parts of the world during the last 25 years and it continues. Obviously, it is a source of escape for those devastated by prolonged and intractable wars, persecution, disease, death, poor living conditions and no future. Termination of hostilities often brings little peace, persecution for some, food shortages, refugee camps and many other hardships for people in their own country. To escape these, many have taken to the high seas or crossed borders into territory that is sometimes almost as hostile as the one they have left behind, but they have little choice and for many nothing will be worse than what they live in. So, they have everything to gain and nothing to lose by going.

Humans live by chances and the boat people calculate that their chances for survival would be better if they try to escape elsewhere, even if there is a chance of dying in the attempt. For many it is not just a chance for them to survive, it might just change their lives for the better, release them from economic circumstances that provide little and bring an early death. Some 'refugees' are trying to circumvent lengthy or impossible immigration procedures and regulations in order to start a new life. But to start out on a voyage is a start at a fresh life even if there are no guarantees of success. Contemporary ocean-going voyages are not without danger. Modern-day pirates, cyclones, overloading and the unseaworthy, leaky and rotting hulks often provided for these escapes have probably seen more voyagers perish than succeed in their journey. So, in January 1992, 60 more people arrived in Australia and this group was only one of hundreds that have set out to reach Australia during the last 30 years.

From an anthropological perspective, there is nothing unusual about these arrivals, however. They represent a single drop from a bucket of human migration that has been dripping all over the planet over the last two million years. During the Upper Pleistocene some would have headed for Australia. Although none of our modern group perished, they were not able to read their surroundings as the Pleistocene migrants were. What they both shared, however, is that they had the will to survive, they used different tools to do it; the situation drew out the will for 'survival' in them. Their determination and stamina came from attempting to save themselves, a desire shared by all humans whether they are separated by distance or time.

Humans are great survivors. When pushed they can achieve remarkable feats with very little, and in essence that is what this book is about. It is concerned with a phenomenon that binds modern people to their distant ancestors through the twin problems of why and how we move from one place to another. Human migration to Australia continues in much the same way as it has done for tens of thousands of years. Today it may be shorter because people know their destination, but their reasons for making the journey now may not be very different from those of the ancient mariners. Nor is the will and fortitude to carry it out very different, together with the strength to survive in the hope that a safe landfall will be made. Perhaps the pressure of those behind them made migrants of people in the distant past as it often does now.

There was an evening during the last interglacial or before when Sahul (Australia, New Guinea and Tasmania) was completely devoid of humans. Its beaches had never had an imprint of a human foot on them and the fauna had never been hunted. The only fires here were natural and the landscape had not heard the chatter or call of the human voice. By the next evening, however, the first camp fire lit up ancient faces and glittered in time with strange voices drifting across an estuary or along a very lonely and pristine beach. The first human footprints were being washed by the tide or sniffed by strange looking giant marsupials. Whenever this event took place it was the most significant and exciting event that Australia has ever known. It represents a time when humans came of age on open sea and they would now begin to explore the oceans.

1 *Going to Sunda: Lower Pleistocene transcontinental migration*

Introduction

By its very nature archaeological evidence is almost always deficient and rarely tells us what we would *really* like to know. Our inquisitiveness, however, drives us to seek new evidence, ask more questions and keep looking. After more than a century of looking, the fossil evidence continues to point to Africa as the development centre for humanity. While little evidence has been found to show how large early human populations were and how they subsequently grew, it is always taken for granted that they were small, actually, very small. Besides the equivocal definition of what is 'small', the premise that our formative populations were 'small' could be wrong. There may have been several population centres that arose from an earlier single centre, where the human line broke away from a common ape ancestor. As evidence emerges that hominins may have divided from anthropoids as far back as the Upper Miocene, it is increasingly difficult to believe that they lived in *extremely* small populations (Brunet et al., 2002). Research during the last decade suggests that early hominins were distributed over a wide area of Africa and were not necessarily confined to eastern parts as previously suspected. Perhaps more importantly, it shows that they could survive in savannah and close to semi-arid environments. They were not necessarily confined to rainforests or dense forest eco-niches (Vignaud et al., 2002). There may indeed have been a single geographical area where the separation of the human line from our common ancestor took place or it may have taken place in several areas. A parsimonious view is that the latter is less likely than the former. Whatever the case, it is likely that sooner or later groups budded off from founder populations and the ancestral line leading to humans began to grow as a distinct group separated from the apes. It is always going to be difficult to recognise this process because of the immense odds and sheer chance of finding the fossil remains necessary to piece the story together in any great detail. The budding-off process could have given rise to morphologically similar forms and this would have been underpinned by linkage in the form of intercommunication and territorial overlap which, in turn, facilitated gene flow. Of primary interest here is that population budding suggests population growth, but the fact that life may have been nasty, brutish and short for early hominins, the net outcome of such growth was minuscule. Nevertheless, it was with these early groups that a process began that would eventually lead to one of humanity's most frightening problems – overpopulation.

Competition or geographical barriers may have kept some early hominin groups apart for considerable periods of time. This is likely to have been one mechanism leading to such a complex branching tree of evolution from which the *Homo* line eventually staggered away. Genes would have been exchanged between most groups, as females were captured or one group overwhelmed another. Geographical separation of related lineages would have produced morphological divergence, even on a lesser rather than greater scale. We know virtually nothing about the reproductive biology or biogeography of proto-hominins or the later ubiquitous Australopithecines of southern, eastern and north central Africa. Naturally, we know even less about the fundamental demographic parameters of life expectancy, natality, rates of fertility and mortality, birth intervals, and the age and sex structure of Australopithecine groups or later populations. Using reproductive and demographic profiles of higher primates and anthropoid apes to reconstruct these parameters for hominins may mislead rather than inform us. To do this is similar to rigidly extrapolating demographic information gathered from modern hunter-gatherer people to hunter/scavengers living 1 My ago and less.

Because of these constraints it is difficult to understand the population dynamics of our earliest direct ancestors, including their demography, their migratory habits and, consequently, how their populations might have grown and spread. This situation is frustrating because the reproductive and demographic dynamics of these populations led directly to the later spread of erectine or erectine-like groups in the early Lower Pleistocene out of Africa. At this stage I am not too concerned with taxonomic propriety. It is immaterial for the arguments in this book whether *Homo ergaster*, *Homo erectus* or even *Homo antecessor* was the migratory species. I am only concerned with a hard-working 'erectine' as the protagonist, and I will refer to these migrants as 'erectines' throughout. The ability to know the size and distribution of these populations and their migration patterns during the Lower and Middle Pleistocene is vital to understanding the development of modern human populations, how they emerged and the eventual occupation of the planet by us.

Our knowledge of world populations in the Lower Pleistocene is as meagre as our knowledge of those that came before them and a resort to guesswork is all that seems possible. There is a desperate need, however, to understand humanity's palaeodemographics in order to develop a fuller understanding of the important evolutionary processes taking place during this phase of human development. It would also help us understand better how erectines successfully crossed the globe. Even basic questions seem beyond us, such as how large were these populations when they started out and how fast did they grow? Answers could provide a better understanding of the speed and direction of migrations, when they began and how fast the world was populated. It might also help throw more light on when and where we as modern humans evolved.

Therefore, it is my aim to tackle a number of questions relating to world population growth over the last 2 My. Naturally, it takes a largely theoretical approach because it is the only one open to us. In addition, I use as much knowledge of the biology of these early people as we know at this point. The model is, however, conservative and based on the most plausible parameters used widely in palaeodemography. This first chapter begins the task by examining how people may have arrived in and crossed Asia. A general discussion describing evidence for hominin distribution is presented, followed by a description of some possible palaeodemographic parameters based on what is known about the biology of erectines.

Leaving Africa

When the first hominins left the African continent, probably crossing the Sinai Peninsula, it sparked off a process of world migration that began in the Lower Pleistocene and has never really ended. Although hominins must have moved within continental Africa before this, leaving the continent marked the first transcontinental step for humans. Many steps later they had spread from the African savannahs through a wide variety of environments stretching into Europe, crossing the Middle East and into Asia. Almost 2 My later, migrating peoples stared out across the icy waves of the Southern Ocean, the Tasman Sea, the Beagle Channel and the North and South Atlantic Oceans. Our fascination with exploration continued until great Pacific voyages took people to almost all the tiny atolls scattered across that vast ocean. Effectively, our explorative journeys ended with the occupation of New Zealand a little over 1 ky ago. Exploration of the world by Europeans during the next 800 years or so was essentially re-explorations. With the exception of Antarctica, humans had fully explored the world's continents long before any galley, dhow, junk, longship, caravel or galleon had left its respective home moorings. Trans-world migration began with *Homo erectus* and ended with *Homo sapiens*, but when viewed in this way it seems that whatever we like to call these people it was always *us*, equipped with our inquisitiveness, cunning, adaptiveness, toughness, innate ability, athleticism, need to explore and our doggedness of purpose. We continue to explore and probably always will because of our natural curiosity, a special quality that readily flags up who we are.

The large collection of fossil human remains from China and Indonesia is mute testimony to the success of erectine migrations. The antiquity of the first erectines to leave Africa is not in dispute. It had certainly happened by the Lower Pleistocene, but what we lack is both detail of these events and the exact timing. Even though originally disputed, the almost 2 My date for the Perning (Modjokerto) *Homo erectus* seems to be holding the record for the oldest evidence for hominins outside Africa (Swisher *et al.*, 1994; Antón, 2003). Two million years is a controversial date for erectines to be living in Java, because to accept their presence in this far corner of the Old World at such an early time would almost predate the species in its African home (KNM-ER3228, 1481 ~ 1.9 My). It is not unlikely that some might even argue that they could have evolved outside that continent (see below). Other early dates for *Homo erectus* outside Africa include those from Dmanisi, in the Republic of Georgia. Recent discoveries there include two crania and a mandible dated to 1.7 My (Tchernov, 1987, Shipman, 1992, Gabunia *et al.*, 1999). Reassessment of the dating of the Sangiran fossils from Java has now produced a minimum date for them of 1.5 My (Larick *et al.*, 2001). The oldest firm date for fossil hominins in China is that for the Gongwangling (Lantian) calvarium (1.15 My) (An *et al.*, 1990). The upper incisors from Yuanmou in southwestern China still present problems for interpretation and it continues to be unclear whether they are morphologically closer to an *Australopithecine* or are erectine. Nevertheless, they are still believed to be Lower Pleistocene in age, and could be as old as 1.7 My (Wu and Poirier, 1995). But the recent discovery of stone artefacts in sediments in the Nihewan Basin, radiometrically dated at between 1.77 My and 1.95 My and those at Renzidong that may be as old as 2.25 My, probably puts the earliest hominin occupation of China beyond that of Java (Zhu *et al.*, 2001). Indeed, if these dates are found to be correct, then the whole debate about the origin of *Homo erectus*, or a similar hominin, will be open for a radical re-think, together with their direction and rate of spread across the world. Perhaps an *Australopithecine* origin for the Yuanmou incisors may then not be as fanciful as has been assumed in the past.

For now, however, we will stick to the original story and go back to Africa's front door. Ubeidya (1.4 My) is an artefact site on the banks of the Jordan River 1400 km southwest of Dmanisi (Tchernov, 1987). Although dated later than Dmanisi, it has been suggested that the site may eventually reveal dates predating the Georgian evidence. It is immaterial really because it is logical to assume that if erectines came out of Africa then those areas geographically closer to the dark continent should eventually produce the earliest dates for their exodus. It is also logical because Ubeidya is on a route that would have taken them north through the Sinai region. The Dmanisi and Ubeidya evidence is interesting because that is not the direction to go if you want to get to Indonesia or China, assuming you knew they were there. The data point strongly to an erectine radiation out of Africa, not only in a north-easterly direction but probably in all directions. This suggests a vital point for human palaeodemography at this time: if erectine populations were extremely small and they radiated in all directions, these hominins would have been exceptionally thin on the ground. That would not have been safe considering they were prey to wide range of large carnivores at that time. So, were there enough erectines to go round? In other words, how many erectines were there?

The earliest evidence for humans in Europe still has not broken through the 1 My barrier. Sites at Soleilhac, Vallonet Cave and Isernia la Pineta are the closest (Roebroeks *et al.*, 1992). Discoveries of erectine crania at the Grand Dolina locality, Atapuerca, in Spain and Ceprano in Italy show erectines were living in southern parts of Europe around 800 ky (Bermudez de Castro *et al.*, 1997; Arsuaga *et al.*, 1997; Manzi *et al.*, 2001; Antón, 2003). This evidence not only brings them into Europe at a much earlier date than previously thought, they are living at the far western end of the continent, close to the Atlantic coast, a position that suggests others were living between them and the Levant. Even if they had crossed at a land bridge between North Africa and Gibraltar, it means that they had spread far across North Africa to reach that crossing point, but the evidence from that region is not particularly old at around 400 ky for Thomas' Quarry and Sidi Abderrahman fossils in Morocco (Antón, 2003). The European dates are much younger than those from the Far East and at least 900 ky years after Dmanisi, but a successful entry and occupation of Europe may have been harder to accomplish in the face of the glacial conditions that existed there around 1 My ago. But why not an entry before those conditions prevailed? The Georgian evidence shows us that erectines were moving in the right directions. Nihewan tools have been found above 40° north, so cold may not have played as big a part in the dispersal patterns of these hominins as previously thought. Also, we must not forget that at the same time as people were living at Grand Dolina, their relatives were island hopping through the Indonesian archipelago using water craft (Morwood *et al.*, 1999). It is debatable when humans first began to adapt to very cold conditions, but there is the possibility of them exploring as far north as the Lena River (Diring-Ur'akh) almost at the same time around 700 ky ago. While far to the east they had reached deep into Japan (Kamitakamori), then a peninsula jutting east from China, again around 700 ky ago (Sekiya, 2000). A later date of 500 ky also puts them in southern Mongolia (Tsagaan Agui), attesting to their ubiquitous spread across the Old World and obviously evidence for erectines trying their hand in not too benign regions as far north as 60° (Derevianko *et al.*, 2000). If confirmed, the Lena evidence also puts them within 600 km of the Arctic Circle and 200 km from the southern limit of the present permafrost line.

The first human occupation of northern Europe probably did not take place until the mid or even late Middle Pleistocene because of cyclic glaciations; evidence from Boxgrove in England shows that there were successful hunting bands living there around 450 ky. They probably arrived in Britain via the land bridge from Europe during low sea levels. But in those times, enormous glaciers lay across northern Europe preventing entry into the northern half of the continent and northern Britain for long periods. Glaciers would also destroy any evidence of occupation in those areas, particularly open sites, which was probably very sparse in any case. Moreover, the colder sub-glacial environments as well as steppe-like expanses were probably not popular with erectines and why should they be when the empty world before you is your oyster and you can stay in warmer regions? Thus, the European landscape may have remained an unfriendly venue for occupation until cultural and/or biological adaptation to such environments had evolved. I reiterate, however, that erectines were living in northern parts of China long before 800 ky. As with many other areas, the European evidence is fraught with site interpretation and dating problems associated with the relative age and timing of the appearance and disappearance of stone tool and faunal assemblages. Evidence for erectines in northwest Africa is somewhat later than that in northern Europe, probably reflecting Middle Pleistocene migrations into the region due to population expansion into other areas in the south and east (see Chapter 2).

At present, the sparse evidence for the first patterns of erectine migration and exploration outside Africa shows a radiation confined to the northeast and then the east. The reason for choosing one direction over another will never be known. It may have been purely random, but it is interesting to see that they went in several directions and, with a few exceptions, that they were not daunted by different landscapes. Although present evidence helps define the direction of erectine movement, how long it took them to reach China and Indonesia and what routes they took to reach those destinations are elementary questions but seem impossible to answer. Nevertheless, those questions beg others, including: were Indonesia and China reached concurrently or at different times; were those migrations isolated events, involving single bands, or was there a chain of bands each keeping in touch with the next? Moreover, how many individuals were involved? Even with an obvious lack of empirical data, what are the most likely answers to these questions?

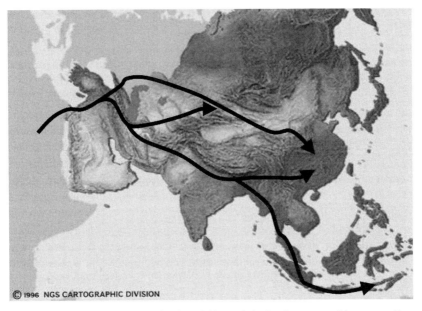

Map 1.1. Erectine migration from Africa to Asia showing two possible routes to East Asia and a third into central Russia.

Mechanisms and barriers to Lower Pleistocene hominin migration

A walk from Africa to China requires passage through most of the environments that our planet has to offer (Map 1.1). Today, these include desert landscapes, sub-alpine meadows, vast savannas, open forests, tundra and dense equatorial rainforests. No doubt ice was always avoided. The late Tertiary trend of planetary drying continued in the Pleistocene with a commensurate build up of ice at the poles. An accompanying drying in many parts of the world particularly affected mid and lower latitudes. As deserts expanded, forests and wet tropics contracted (Williams *et al.*, 1993). Many regions began to experience cooler winters and drier summers, although some fluctuations crept into this trend after 2 My ago. At least 20 major glaciations were experienced after that time, with fluctuating sea levels that alternately flooded and exposed vast areas of continental shelf. Some environments altered little, while others underwent enormous changes that redistributed and probably radically altered the composition of both floral and faunal populations. For example, a 20 ky exposure of the Sunda shelf between Indonesia and Thailand was probably enough time to produce a vastly expanded rainforest that blanketed the exposed continental shelf. Such changes presented an opportunity for species to expand, opening up new opportunities to adapt, evolve and proliferate according to the nutritional resources available. Moreover the emergence of large areas of rainforest would have enhanced the chances of the emergence of new species. With the commensurate increase in the number of eco-niches available at these times, there must have been population explosions among some species. Doubling the amount of rainforest would allow enormous potential for growth and spread of rainforest species. After all, if rainforests doubled in size, half would then not remain empty. These growth phenomena were probably more common in the equatorial, subequatorial and tropical regions than higher latitudes where, in many areas, glaciers replaced habitable land making vast areas uninhabitable. Another effect of lowering seas was that isolated islands became part of the exposed shelf, opening up new opportunities for gene flow, mixing of gene pools and reducing the founder effect that island communities experience more than those on continents. It is also worth remembering that for perhaps 80 per cent of the last 2 My, sea levels were low, probably below −65m. This allowed for genetic expansion and exchange throughout the terrestrial biosphere to a greater rather than a lesser extent. Overall, more land became available at these times, but as one part of the earth appeared from out of the ocean another disappeared under ice. Biomass derived from exposed continental shelves was significant, but in terms of Quaternary time scales the constant raising and lowering of sea levels caused vast ecosystems to be there one minute and gone the next. Well, it was not quite that fast, but such changes over the long time scales being discussed here changed optimum hominin migration routes as well as their degree of difficulty. Changes in the distribution of animal populations also would have affected routes chosen. Other elements affecting a chosen route were environmental barriers such as snow and ice, glaciers, deserts, dense rainforest, high mountain chains, the open sea and wide, fast-flowing rivers. All these were major obstacles to an exploring erectine and they must have dictated the routes taken at different times.

Erectine migrations are the first firm evidence for the unique human ability to explore, survive and occupy new and very diverse habitats and landscapes. They also demonstrate erectine capabilities, including that their almost modern physio-skeletal appearance provided them with the ability for endurance running (Bramble and Leiberman, 2004). This was important for tracking and pursuit of game, particularly small game, as well as for moving large distances to scavenger targets. The erectine expansions suggest that they were capable of occupying or at least traversing rainforests, deserts, tundra and steppe, although permanent occupation was reserved for those environments they felt comfortable in at the particular levels of cultural and biological adaptation they were capable of. Therefore, avoiding difficult terrain and marginal ecosystems would have led migrants only to those niches where life was more than tolerable and food freely available. The first bands could be choosy because the world presented a human-empty horizon for them and they could pick where to explore next and go down the safest avenue open to them. Equipped with an omnivorous diet, erectines could also prey on the animals they found living in different environments. This would not have been the case if they had a strict vegetarian diet, which would require specialised knowledge of plants in a variety of terrains. Their dietary reliance on meat and marrow was a central underpinning of their ability to make the migratory journeys they did, but it also fed into the development of the big brain that enhanced their abilities to optimise their migrational success.

While *Homo erectus* is not famous for technological achievements during the Lower Pleistocene, improvements in stone tool manufacture did occur particularly in the Middle Pleistocene onwards. The appearance of small scrapers, spheroid pounders and a growing trend for better chopping tools in the Acheulian tradition indicates an increasing efficiency in hunting and food processing, as well as improvements in skills and the ability to develop new ones. With these skills, the migrants of 1 My ago had an edge to better exploit certain environments and procure food than those living 500 ky earlier. Nevertheless, some groups must have perished in some regions where their simple technology was no match for the harshness and vagaries of more marginal environments and the animals that inhabited them. A sudden change in weather conditions, a chance meeting with new animals with different behaviour patterns, as well as the animals being unfamiliar with humans, and several erectines or the whole band could have perished. From time to time, packs of carnivorous animals, such as the big cats, giant bears and wolf and hyena packs, may have finished the exploration ambitions not only of individuals but possibly whole bands. Entering areas of abundant predators by tracking common prey would have brought the two together, thus requiring vigilance on the part of the travellers and adequate shelter and weapons for protection. The constant danger would have been an added incentive, however, to improve technology and/or hunting skills and techniques and to keep sufficient numbers of people around you for protection and as lookouts. Unfamiliarity with new species in some regions, as well as the animals' unfamiliarity with the humans entering their territory, must have presented advantages and disadvantages. The latter may have included a naivety that belied the skill of the human hunter, but among carnivorous animals this may have been the other way round, making for some interesting interactions between the two.

But what form did Lower Pleistocene migrations take? Modern human migration takes three basic forms. The first is a comparatively quick movement from A to B, involving hundreds or perhaps thousands of individuals in a continuous stream or rapid series of pulses. The second is a steady but continuous trickle, involving fewer people at any one time, but over the long term it may move as many people as the first example. The third form is a very slow version of the second, involving an almost imperceptible trickle. It is intermittent and forms a gradual 'background' migration that has probably taken place in many parts of the world for thousands of years. The number of people involved at any given time is so small that it goes just about unnoticed. There are, however, many permutations of these three basic forms and many other factors to consider. Migration, as a recent phenomenon, usually takes the form of the first example. Often, there is an accompanying image of 'illegal immigration', involving thousands or tens of thousands of people that, in terms of the time scales discussed here, move instantaneously from one place to another. To many who inhabit the countries targeted by these groups today, such events pose a threat that some might suggest constitutes a national 'break-and-enter'. It is unlikely that the latter tribal attitude began with *Homo sapiens*.

The time taken would have been an important factor in the overall success of some migrations. Very slow migration gave erectines enough time to become familiar with the landscape, terrain, animals and weather as well as to allow adaptive changes to take place and to remain in contact with others coming along behind. A choice of route may then have followed, with some assessment of the advantages of one way over another. Lingering too long in an area of climatic extremes would have been disastrous for the unprepared and particularly for the small group. Rapid movement avoided this danger, but it also reduced the opportunity to become familiar with a particular region and raised the distinct possibility of losing contact with other groups and bands. Moreover, it could place people in areas where extreme seasonal variability might suddenly deplete resources, leaving them stranded.

There is, however, a fourth type of migration but it no longer has an equivalent today: this is 'accidental migration'.

Erectines and their 'accidental migration'

It is difficult to conceive of anything like mass movement taking place during the Lower Pleistocene, nor a pattern fanning out across the landscape. Indeed, only the pattern of migration seems worth considering, but even this has to be modified. Except for natural barriers, Lower Pleistocene migration probably consisted of random or almost random movements that involved very small numbers of people in 'single lines'. We can only assume that there were no purposeful migrations with target destinations, because erectines did not conceive of them or, indeed, even have a basic idea of where they were heading. This, then, I have termed 'accidental migration'.

Seasonal hunting and herd trailing in the Lower Pleistocene could be examples of 'accidental migration'. They facilitated a natural rate of movement and exploration that sooner or later took people to new regions, as well as occasionally back to those visited previously. At some point, people may have stayed where they ended up, perhaps a few at a time, eventually to move on again. These processes took them into new territory that, on occasion, required slow cultural and physiological adaptation for them to remain. Herd trailing kept people within environments favoured by the trailed animal species. Almost imperceptible alterations in animal behaviour and migratory patterns, due to gradual and subtle changes in environmental and/or seasonal weather conditions, would, over time, have introduced variation. For example, herd animals may have increasingly moved north as ice sheets retreated during interglacials, only to be pushed south again when glacial conditions returned. Gradual accumulation of these sorts of changes over tens of thousands of years brought about migration that gradually shifted people as well as animals to and fro and ignited cultural and biological changes commensurate with better living conditions in those regions.

A theoretical approach to assessing the distance that people could have travelled in 1 ky is demonstrated in Table 1.1. It begins by assuming a certain parcel of land was occupied each generation. Parcels of various sizes are used together with the formula $\rho = \sigma\sqrt{2r}$ (after Hassan, 1981) to produce a range of distances travelled in 1 ky (D).

Table 1.1 demonstrates that forward movement could have been comparatively fast, even if small amounts of territory are occupied. It is likely that the trek would not follow straight lines, but would be convoluted so actual forward movement may not be as great as is implied by these figures. We can be sure that a bee line for a particular destination was not a conscious strategy over the long distance. It is also parsimonious to introduce 'extinction' into the picture, so some groups after achieving a degree of forward movement then disappeared. Apart from the accidental method of migration described above, there may have been episodes of deliberate migration instigated by curiosity or natural disaster, but these were probably few and restricted to a particular region.

The worst thing about this topic is that it continually begs further questions and suppositions that require more models. For example: can we be sure that erectines did not target a destination or, to put it another way, move deliberately and continually towards the east for example? Or were the

Table 1.1. *Distance travelled per generation by migrating* Homo erectus *bands*

1	2	σ	ρ	D
625	39.1	6.3	0.08	80
900	56.3	7.5	0.10	100
1225	76.6	8.8	0.11	110
1600	100.0	10.0	0.13	130
2025	126.6	11.3	0.14	140
2500	156.3	12.5	0.16	160
3025	189.1	13.8	0.18	180
3600	225.0	15.0	0.19	190
10 000	625.0	25.0	0.32	320
22 500	1406.3	37.5	0.47	470
40 000	2500.0	50.0	0.63	630

Notes:
1 – Area (km^2) occupied per generation (16 years).
2 – Area occupied per generational year km^2, e.g. 625/16 = 34.7km^2.
σ – Square root of column 2.
ρ – ρ = σ$\sqrt{2r}$ (*r* is reproduction rate, 0.00008) kilometres travelled per year.
D – Distance travelled in 1000 years.

farthest corners of eastern Asia reached only after a series of tortuous, stop–start journeys? By design or accident, migration would have moved in a convoluted path through many parts of the Middle East and Asian continent, probably avoiding deserts and arid regions, and possibly doubling the distance of a more direct journey. It is important that we know whether migrations were deliberate or random because each of these has different implications for the time taken to move from Africa to eastern Asia. A purely random pattern could take a very long time, but a deliberate forward movement might have taken only a few generations, perhaps even less than one.

It is most parsimonious to say that people would have taken the easiest possible route, while maintaining a particular direction for group reasons. Nevertheless, there must have been times when more difficult paths were either unavoidable or attempted because of ignorance of the conditions, because of curiosity or because animals were being trailed. The two basic routes taken were overland or coastal, or perhaps neither was particularly favoured over the other. A generalist hunting–scavenging–gathering economy would not necessarily confine people to the coast. The real situation might have been one where the best route was that which merely provided for the group, irrespective of where it took them. This is where the Dmanisi finds are interesting, because they show that erectines had at some point moved northeast from Africa and were heading into central Asia. They were not necessarily taking the most straightforward route to east Asia, through Iran and India. Obviously, we cannot expect them to take such routes because they did not know where they were going. They probably chose a hundred routes that took them away from Africa. The trouble is we have no indisputable evidence that they took any of them. Land bridges, for example, between East Africa and the southern Arabian peninsula and across the Straits of Hormuz could have provided a short cut for those heading east, but until evidence of a hominin presence is found there we have no way of knowing that they chose it. The Hathnora calvarium from the Narmada Valley in western India, shows that by 400–500 ky hominins were living in that area (Lumley de and Sonakia, 1985; Sonakia, 1985a, b). People had obviously pushed into the southern parts of India, but this is a rather late fossil in our story and has been assigned to an archaic *sapien* rather than an erectine. Nevertheless, it could have been part of a group slowly moving southward along the west coast, but it could also have been a descendent of a long-established Indian resident group. There are many possible origins for this individual, but it is comparatively late in the migration story and with people having passed to the east over 1 My before, so the latter explanation is most favoured. Let us now look at some quantifiable methods of describing the migrations that occurred in the Lower Pleistocene.

Today it is a difficult concept for us to believe that people moved about a world devoid of strict geo-political borders and barriers and that the first migrants had no other humans in front of them. The only regulation to travel then was to find enough to eat and drink. So, did a group move on continuously or did it do so only when a partially settled or stable population grew large enough to bud-off? The latter implies, however, that the world filled with people as the migration pushed inexorably onward in bow-wave fashion, leaving behind a landscape covered with people and moving ahead one generation at a time. Moreover, the process was complete by at least 1.9 My ago, with erectines entering Java and China in sufficient numbers for their fossil remains to be found. So, if we rely on the budding mechanism to supply migrants, then we have to accept that a fairly solid mat of people covered the landscape all the way from Africa to China and Indonesia by the end of the Lower Pleistocene. This clearly cannot be the case: a full Asia nearly 2 My ago cannot be seriously contemplated.

Lower Pleistocene migration processes

We still lack evidence for the exact timing of the Lower and Middle Pleistocene glacial/interglacial cycles. Nevertheless, this does not affect the arguments outlined here because they are not concerned with a particular time, only that these cycles began around 2.4 My. During these events the availability of land for exploration increased and a number of land bridges vital in the exploration process existed. The lower ambient temperatures of an ice age caused the uptake of oceanic water into massive glaciers, hundreds of kilometres wide and in some cases kilometres thick. As well as covering northern Europe, they extended across much of northern Eurasia, parts of Siberia and eastern Asia, and prevented exploration of northern regions and the use of the land covered by them. But these were fortuitous events because they did two things.

Firstly, areas measuring many millions of square kilometres became almost devoid of animal life and ice-free corridors funnelled herds along more southerly, narrowly defined and more predictable paths of migration. In the northern hemisphere glacial formation caused an almost complete shift southward of some faunal and floral populations. Secondly, the take up of oceanic water into glaciers effectively transferred vast areas of territory from one part of the globe to another: from higher to lower latitudes. People would have taken full advantage of lowered sea levels exposing continental shelves and adding almost 25 per cent more land to coastal areas and over 50 per cent more land to island southeast Asia. The result was the formation of the Sunda subcontinent that extended the Asian mainland southward by at least 1000 km, effectively transferring its southern shores from Singapore to the south coast of Java.

Because of their African origins, *Homo erectus* groups would have been biologically adapted to warm or tropical environments (see below). Therefore, if we assume that the earliest inhabitants were best suited to the tropics, they would not have been particularly well equipped to survive the rigours of cold northerly latitudes, particularly during glaciations. In these circumstances, they are more likely to have favoured the warmer regions of tropical southeast Asia, especially Indonesia. In order to follow the Sun, a southerly migration route across central Asia, the Indian subcontinent and down the Malaysian Peninsula is most likely to have been taken. During warmer interglacials, some may have been drawn north by ameliorating climates, but also with some pushing as continental shelves disappeared under rising seas and land bridges to southeast Asia were severed. This may have been how hominins were first drawn towards China.

Later, with continued migration from the west, new hominin bands adapted to and used the newly opened areas because others already occupied lower latitudes. The presence of abundant herds of reindeer, musk ox, bison, mammoth, horse and deer in the cooler regions would have been a further attraction; all they needed was a cultural and/or biological adaptation that helped them survive in cooler climates. Exploitation of these regions provided a selective force for cultural changes that included more complex tool kits, the use of clothing and fire and more sophisticated hunting behaviour suited to the open plains of the north. The latter was only a variation of something they almost assuredly could do already. The opening up of vast tracts of land would have encouraged more northerly exploration also, because of the northward movement of extant populations away from the shrinking coastlines. But how did hominins first move across Asia and how long did it take for them to cross such a vast continent?

Originally, there must have been a very small trickle of migrants from Africa into Asia probably following a 'single line' of sorts, but broken and sporadic in places with an irregular composition. I mentioned a 'single line' earlier without proper explanation. This concept posits a line consisting of small or medium-sized bands, keeping in contact with one another because isolation was an invitation to group extinction. There is security in having neighbours, hunting is easier and more successful and the group has a better chance of replacing lost individuals as well as increasing in size. Contact between bands of erectines in the line could have been constant, intermittent or non-existent, but each of these implies a certain population size namely large, medium or very small, respectively. Such a line would also be a conveyor for gene exchange.

Mating rules, if any existed, would have been difficult to observe in the isolated group situation. Concepts like female-exchange, patrilineal and matrilineal descent are most unlikely to have existed at this time, but let us not underestimate too much. It has been suggested that they may have had a loose social structure that included 'grandmothering' as well as the assistance of both male and female helpers (Antón, 2003). Very basic mating rules might have existed at some level, but, even if we assume the presence of a form of exogamous system of female exchange, group size would have been important. Isolation would require the group to be large enough to sustain an endogamous system, or perhaps it was small but in regular contact with others. The former case seems most likely, because the image of a multitude of bands stretched right across the Middle East and Asia in the Lower Pleistocene does suggest a substantial population to support the 'single chain' idea and that does not equate with our usual notion of human population size at that time. But this may have been the case to some extent because of the dangers of isolation that must also have been very real for erectines. If groups kept in contact with one another, even loosely, by the time they reached an area north of the Bay of Bengal, the line would stretch 5500 km back to the Sinai region. If each band area diameter was 150 km, and each band consisted of 30 people, this would equal 37 bands (to the nearest whole number).

Table 1.2. *Erectine migration from Africa to Asia, theoretical band size and spacing*

Suez to Bay of Bengal (Jalpaiguri)
5500 km ÷ 150 km dia. (471.2 km^2) = 37 bands of 30 individuals
1. 30 × 37 = 1110 hominins
2. 7 × 37 = 259 hominins
3. 12 × 37 = 444 hominins

Bay of Bengal to Beijing (Nihewan Basin)
3500 km ÷ 150 km = 24 bands
1. 30 × 24 = 720 hominins
2. 7 × 24 = 168 hominins
3. 12 × 24 = 288 hominins

Bay of Bengal to Java (Modjokerto)
5000 km ÷ 150 km = 33 bands
1. 30 × 33 = 990 hominins
2. 7 × 33 = 231 hominins
3. 12 × 33 = 396 hominins

		One chain	Five chains	Ten chains
Totals	1. Present study	2820	14 100	28 200
	2. Birdsell (1972)	679	3 395	6 790
	3. Hassan (1981)	1164	5 820	11 640

Therefore, a population of 1110 people would stretch from Sinai to the Bay of Bengal. With each band area being 471.2 km^2, the total area occupied would be 17 434 km^2 or roughly 17 km^2 per person. In the worst case, the widest separation of adjacent bands would be 300 km, close enough to meet for female exchange if they knew each other was there. Taking the chain beyond the Bay of Bengal to Lantian would require 61 bands and the population would rise to 1920 individuals. A similar extension to Modjokerto in Java (at low sea level), a distance of 5000 km, adds another 33 bands or 990 persons, which equals 2820 individuals (Table 1.2). This small number of people would stretch back over 10 000 km to Sinai.

The 15.7 km^2 per person used in this study is less than estimates made by other workers, which include 66.7 km^2 per person by Birdsell and 40 km^2 by Hassan. These latter figures also amount to fewer people, 679 and 1164 people living along the chain, respectively. These numbers seem extraordinarily low, but perhaps all those in Table 1.2 are. The model is also static. It assumes everybody stays in their respective band territories, no new people arrive, no one moves in or out and every band is composed of the same number of people. Moreover, no further movement out of Africa occurs once the chain is assembled and nobody moves on once the band ahead is established. It also suggests that regions either side of the narrow one-link chain remain unoccupied. Nevertheless, it does demonstrate the theoretical possibility of a precursor migratory population, assembled as a long string of interacting bands stretching from Africa across Asia and comprising very few individuals. What we should not do is fall into the old view of seeing the populations in Java and China as isolated from each other and the rest of humanity; they were not and I shall return to this in later chapters. One serious problem with theoretical approaches to this subject is that they tend to ignore the human aspects of the process, and indeed the nature of humans themselves. A point to be made here is that assembling static models, such as the one above, enables one to see them for what they are: un-natural (Antón et al., 2003). While they are useful to form some sort of picture, it could be said we cannot know what the real nature of erectines was, but the fact that they were very close to us should enable us to make some educated guesses about what their migratory processes actually were by delving into our own actions and reactions to the sorts of issues they faced.

Gatherings of two or more bands must have occurred from time to time. So too, individual bands must have known isolation on occasion and perished for one reason or another. Normally, new arrivals would replace them, but the model above does not allow for that. Moreover, territoriality may not have been a strong concept if a lifestyle of moving on was normal. In reality, such a process was probably responsible for constantly, if slowly, moving people around, introducing them to new niches and environments, facilitating gene flow and populating most of southern Asia by the end of the Lower Pleistocene. In other words, the palaeodemography of erectines must have been fluid, dynamic and ever changing, and the people themselves must have been very vital, robust and cunning to be as successful as they were in setting up the viable ongoing populations attested to by the fossil records. The fluidity of human demography, particularly involving a successful species like *Homo erectus*, essentially requires the movement of new groups along the chain and the spread of groups out from the chain and into new territory. In other words population growth, albeit small, must have taken place in order to cover the ground.

We also need to keep *our* feet on the ground here by briefly assessing a simple implication of these migrations, because this is going to be important for the whole tenet of this book. For example, if we have people stretching, even in a single line, from Africa to eastern Asia, where are the people coming from? Africa! Yes, but that implies population growth occurring in Africa to supply the migrants. If their origin is not Africa, then the 'chain population' is growing outside Africa in order to fill band areas and press on with its exploration. Then again, if there is no population growth in Africa, those leaving the continent effectively reduce its population size and so it could not produce more erectines for export. We will stop going around in these circles and move on for the moment.

Erectine band size: from Africa to Asia

The theoretical single 'chain' of bands above stretches across Asia from the edge of Africa, roughly where the Suez Canal lies today, to Dhaky, in Bangladesh, and then divides to form two prongs. One pushes northeast into China, as far as the Nihewan Basin, and the other goes southeast through the Malaysian Peninsula to Sumatra and then to central and eastern Java. The distances from Suez to Zhoukoutien and Suez to the Solo River are 9000 km and 10 500 km, respectively. I propose that hunters in the Lower and Middle Pleistocene would not need much more band area to survive than they do now, in other words around 16–17 km^2 per person. It is worth remembering that the protein biomass of those days was probably far greater than it is now, with more animal species around, many of which disappeared only during the last glaciation.

If two bands are placed each side of the single chain, making five chains, our population range is 5820–14 550, depending on the occupation area per person. The two extremes provide further flexibility to the model, with the lower figure allowing for a much wider spread of population, while at the same time keeping numbers to a minimum. On the other hand, if 14 550 is used, fewer chains of the same size can exist because of the finite population size envisaged for the Lower Pleistocene. Or is it finite? It is unlikely that vast tracts of land were ignored and, while whole regions may have remained empty, narrow chains of population were not a reality. So we have three choices: one, there were fewer people per square kilometre; two, there were more; or, three, estimates and perceptions of the size of Lower Pleistocene erectine populations outside Africa are far too low. If the first choice is correct, then they were spread extremely thinly across the landscape, suggesting dangerously low population numbers from which a viable population would be difficult to sustain. Such a population is also vulnerable even to minor disasters. Assuming extremely limited hominin numbers, the second choice automatically confines people to certain areas, limiting their distribution, leaving them isolated in vast areas of empty space and puts extreme pressure on their biological future. It does so because if more people existed across the landscape then the overall population size must have been far greater than we have so far suspected. The third choice really leads on from the second, but it is worth exploring further and I do this in Chapter 2. Before that, however, there are several issues worth pursuing in order to fill out our knowledge of the people whose demographics we are with.

Brain size increase in the Lower and Middle Pleistocene

The thousands of years it probably took for erectine people to reach the extremes of the Asian continent provided them with enough time to accumulate and develop many adaptive qualities that cannot be assessed or observed in the sparse fossil record. Their experience, for example, added to their tool kit, enhancing extant skills and producing new ones that enabled them to cope in many different environments. There is little doubt that for far too long we have underestimated erectine people in terms of their skills and capabilities. Some sophisticated qualities arose in these people between 0.5 My and 1.9 My, either in a generalist sense or occasionally there may have been regional or even local variations. The fact that we find these people living above the 40° latitude as well as in jungles just south of the equator is testament to this. A progressive increase in intelligence and the ability to communicate and interpret the world around them came from a gradual development in neurological complexity and organisation through increased cranial capacity. Increased brain size was probably the most prominent change that took place during this time, underpinning the ability to successfully move into many regions and adapt to their surroundings. Earlier dietary changes towards an omnivorous diet, if not a largely carnivorous one, allowed erectines to move with animal herds and not to have to rely on the availability of vegetable foods alone.

A study of 26 erectine crania from Asia, Africa and Europe dating from 0.2 My to 1.8 My found no statistically significant increase in hominin brain size during that time (Rightmire, 1990: 195–196). Alternatively, others have argued that cranial capacity increased significantly, particularly in an Asian and early African subsample (Leigh, 1992). The earlier work proposed that average increase ranged from 918 cm^3 to 1099 cm^3, a difference of 181 cm^3. Problems with analysis of this sort of sample include:

- the enormous temporal breadth of the sample;
- large temporal and spatial gaps in the fossil record;
- a limited fossil assemblage;
- a few individuals scattered over long periods of time, and
- few accurately dated specimens.

Figure 1.1 shows the linear increase in hominin cranial capacity among the 54 *Homo erectus* and early *Homo sapien* fossils listed in Table 1.3. A marked feature is the wide range of variation in cranial capacity over the last 1.5 My,

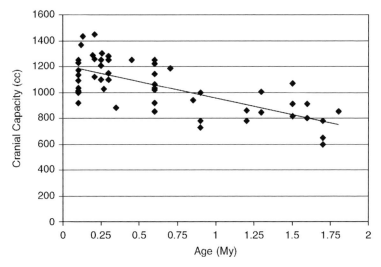

Figure 1.1. Evolution of hominid cranial capacity.

reflecting similar results to those of Leigh (1992). It shows also an obvious increase in cranial capacity over time, as well as a wide range of variation (400 cm^3) that emerges around 600 ky ago. The latter is a trait contiguous with what we know about the variation observed among modern human crania.

A single regression line plotted for the sample as a whole shows a slope of the same order as previously reported (Rightmire, 1990). The 1000 cm^3 variation among present-day human populations is obviously a reflection of a trend that goes back to the Lower Pleistocene. The widest range appears in the mid/late Middle and early Upper Pleistocene and it tends to widen over time. There is also a comparatively wide range of size variation among the earlier group dated between 0.5 My and 1.0 My that broadens out around 0.5 My. If this wide range of brain size is consistent, then we expect to find individuals dated to between 1.2 My and 1.6 My that will fall below the line and others between 0.7 My and 1.2 My that will lie above the line.

The palaeobiology and adaptive qualities of *Homo erectus*

Heat and cold

The East Asian erectine populations not only testify to the vast geographical spread of early humans, but they also provide good, if indirect, evidence for quite separate physiological and even early cultural adaptations to two quite

Table 1.3. *Hominin cranial capacity by chronological age*

Specimen	Date1	Date2	(cm³)	Country
KNM-ER3733	1.8 My	1.8 My	850	Kenya
KNM-WT15000	1.51–1.56 My	1.5 My	909	Kenya
KNM-ER3883	1.5–1.65 My	1.6 My	800	Kenya
D2280	1.7 My	1.7 My	780	Georgia
D2282	1.7 My	1.7 My	650	Georgia
D2700	1.7 My	1.7 My	600	Georgia
O.H. 9	1.5 My	1.5 My	1067	Tanzania
O.H. 12	0.78–1.2 My	900 ky	727	Tanzania
Daka (Bou-VP-2/66)	0.78–1.0 My	900 ky	995	Ethiopia
Buia (UA 31)	0.78–1.0 My	900 ky	775	Eritrea
Trinil 2	800–900 ky	850 ky	940	Java
Gongwangling	~1.2 My	1.2 My	780	China
Ceprano	700 ky	700 ky	1185	Italy
Sangiran 2	1.5 My	1.5 My	813	Java
Sangiran 4	>1.6 My	1.6 My	908	Java
Sangiran 10	1.2 My	1.2 My	855	Java
Sangiran 12	620 ky	620 ky	1059	Java
Sangiran 17*	1.3 My	1.3 My	1004	Java
Sangiran IX	1.1–1.4 My	1.3 My	845	Java
Zhoukoutien 2	580–620 ky	600 ky	1030	China
Zhoukoutien 3	580–620 ky	600 ky	915	China
Zhoukoutien 5	580–620 ky	600 ky	1140	China
Zhoukoutien 6	580–620 ky	600 ky	850	China
Zhoukoutien 10	580–620 ky	600 ky	1225	China
Zhoukoutien 11	580–620 ky	600 ky	1015	China
Zhoukoutien 12	580–620 ky	600 ky	1030	China
Bodo	600 ky	600 ky	1250	Ethiopia
Saldanha	400–500 ky	450 ky	1250	South Africa
Sambungmachan 1	50–150 ky	100 ky	1035	Java
Sambungmachan 3	50–150 ky	100 ky	917	Java
Sambungmachan 4	50–150 ky	100 ky	1006	Java
Ngawi	50–150 ky	100 ky	1000	Java
Sale	300–400 ky	350 ky	880@	Morocco
Kabwe	200–400 ky	300 ky	1280	Zambia
Ndutu	200–400 ky	300 ky	1100	Tanzania
La Sima de los Huesos	300 ky	300 ky	1250^	Spain
Arago 21	200–400 ky	300 ky	1150	France
Hexian	270 ky	270 ky	1025	China
Florisbad	260 ky	260 ky	1300	South Africa
Petrolona	200–300 ky	250 ky	1210	Greece
Steinheim	200–300 ky	250 ky	1100	Germany
Swanscombe	250 ky	250 ky	1250	England
Ehringsdorf	190–245 ky	215 ky	1450	Germany
Jinniushan	210 ky	210 ky	1260	China
Dali	190–220 ky	205 ky	1120	China
Narmada	200 ky	200 ky	1290#	India
Omo 2	130 ky	130 ky	1430	Ethiopia

Table 1.3. (cont.)

Specimen	Date1	Date2	(cm³)	Country
Leatoli (LH)18	125 ky	125 ky	1367	Tanzania
Ngandong I	50–150 ky	100 ky	1172	Java
Ngandong V	50–150 ky	100 ky	1251	Java
Ngandong VI	50–150 ky	100 ky	1013	Java
Ngandong IX	50–150 ky	100 ky	1135	Java
Ngandong X	50–150 ky	100 ky	1231	Java
Ngandong XI	50–150 ky	100 ky	1090	Java

Notes:
Date1 – Most likely age range. Almost every date has some problem and none may be correct.
Date2 – Average date used for this table.
* – The morphology of this specimen may indicate a younger age, perhaps between 400–500 ky.
@ – Sale has an unusual (pathological?) occipital morphology possibly affecting cranial capacity (Hublin, 1985).
^ – An average cm³ derived from crania 4 and 5.
# – An average of an estimated range 1155–1421 cm³ (Jurmain *et al.*, 1990).

different environments at a very early time. Hominins living in equatorial and subequatorial Indonesia endured high annual ambient temperatures and humidity, even during glacial events. Certain features derived from their African origins were probably adaptive for living in tropical climates and these continued to be selected for in southeast Asia. An efficient sweat mechanism that conserved body fluids, salts and electrolytes, a wide distribution of sweat glands and a slender body shape could have been some of these.

The cranial morphology of Indonesian erectines does not suggest slender body proportions, although their post-cranial skeleton is little different from that of modern people (Day, 1971, 1984; Delson, 1985). Although some minor differences are present, the overall functional abilities of these people, in terms of posture and gait, were the same as ours. A recent review of estimates of erectine height shows a range of 147–185 cm with a mean of 161 cm and a standard deviation of 13 cm (Antón, 2003). The highest of these estimates was found on the remarkably preserved and almost complete skeleton of the 'Turkana boy' (KNM-WT 15000) near Lake Turkana (Walker and Leakey, 1993). This 12-year-old boy was 160 cm tall and weighed 68 kg, well within an African adult erectine stature range of 158–180 cm (Walker and Leakey, 1993). With a growth spurt, the 1.5 My old individual may have attained an adult stature exceeding 180 cm, 10 cm above the average for *Homo erectus* and well within the top 17 per cent for modern human populations (Martin, 1928). Further examination of the skeleton has now forced a revision of its height at death from 159 cm to 147 cm, thus reducing substantially its projected height in adulthood (Antón, 2003). A crural index regression analysis of KNM-WT 15000 shows it was adapted to a mean annual temperature of about 30°C (Walker and Leakey, 1993). The body shape of the Turkana boy was adaptively fashioned to withstand high annual temperatures in an arid environment that changed little over 1.5 My. A combination of sweating, adequate water intake and ingestion of salt must have been additional physiological features of this population.

The ability to find salt through natural sources as well as its ingestion by eating meat supports the view that erectines were active hunters and scavengers. It has also been suggested that they had a physiology that enabled them to perform endurance running, perhaps as a method of running-down prey (Bramble and Leiberman, 2004). All these adaptations are to be expected among hominins successfully traversing half the globe. Southeast Asian erectines were shorter than those living in Africa. The small Javan sample of four indicates a range of 158–163 cm, while at higher latitudes the Chinese erectines were even shorter at 150 cm as were the Dmanisi erectines. Once again, the fossil record has us at its mercy because these figures are based upon very small samples amounting to four, two and one individuals from Javan, Chinese and Dmanisi, respectively. It is trite to say it, but the more evidence we obtain of these people, the easier it will be to estimate their biological qualities and, thus, their abilities to explore and live in different environments and I am sure that evidence will be found.

Another adaptive mechanism is skin colour. Besides having a positive regional correlation with the amount of UV radiation received, the evolution of skin colouration is probably closely associated with the loss of body hair and the increase in sweat glands among early groups of *Homo* (Jablonski and Chaplin, 2000). The onset, evolution and degree of melanisation among early humans is something that the fossil record will never help us with unless a *Homo erectus* is found preserved in a bog or ice. Dense melanin pigment would have helped screen out ultra-violet radiation from a tropical sun, thus preventing damage to dermal layers. A degree of melanisation was probably a feature of the early hominin populations that first left Africa with a later adaptive differential colouration emerging among occupants of high and low latitudes. A lighter skin would certainly have been maladaptive to tropical living, thus the long-term success of hominin populations in Indonesia must have partly relied on the emergence of a darker skin colour.

Indonesian erectines could have lived within or on the fringes of jungles as well as along relatively open littoral and riverine environments. Over a long period of time, they probably developed behaviours to cope with dense rainforests and high humidity. One strategy may have been to camp close to plentiful fresh water and food supplies. Whether the hazards of living out in the open tropical rainforest were as prevalent to them as they seem to us is hard to say, but it seems likely they were, only they became more used to being surrounded by them. Snake bites, being crushed by very large pythons, attacks from predatory animals, such as large tigers and salt water crocodiles, which were more common then, must have been constant hazards. Malaria may also have existed among primate populations that provided a suitable pool of susceptible animals to maintain the disease. It was not until hominin populations reached a certain size that endemic malaria became a problem for them, but occasional infections would not be out of the question. It would have been at that population threshold that selection for Hb variants, such as sickle cell, HbA and Thalassemia, may have begun (Webb, 1995, see Chapter 8).

Erectines living far from the Equator experienced a quite different environment and thus required a different set of adaptive qualities from their tropical cousins. This may have been when for the first time regional adaptation began to differentiate the same hominin type. Chinese hominins did not experience humid jungles, instead they coped with an open terrain of sub-alpine steppe, pine forest and more than the occasional subzero winter of cold rain and snow. In this environment, there were distinct advantages to having certain morphological as well as behavioural and physiological adaptations. These included a larger, rounder body form and thicker layers of subcutaneous fat, both important for reducing heat loss and protecting against the cold. They were short (150 cm), although the robusticity observed across erectine populations was retained, so a short rounder body may have certainly existed by the time erectines inhabited the Zhoukoutien cave site. The Dmanisi evidence also shows erectines entering high latitudes (above 40°) a million years before the Zhoukoutien people lived there. Body hair may also have been retained longer in cool climates as an insulator limiting melanisation, a clearly adaptive trend that enhanced Vitamin D_3 production in latitudes with reduced winter sunlight. Moreover, the possibility has been raised that melanin may predispose the skin to the effect of cold (Damon, 1977). High metabolic rates and/or vasoconstriction of peripheral arterial networks to conserve body heat would also have been useful to supplement changes in gross anatomical features. The relationship between the cold and small facial sinuses may have an expression in the diminutive size of these features among erectine fossils from Zhoukoutien compared with those from Indonesia (Steegman, 1977; Wolpoff, 1980). Narrower nasal passages for warming air was probably another adaptation to cold stress. The adaptive qualities of the epicanthic fold are in question, but the overall differences that may have begun to emerge between erectines living in high and low latitudes could have begun to emerge by the late Middle and early Upper Pleistocene.

Cold conditions may have been particularly extreme during extended glacial periods such as the long Mindel event (300–420 ky) that took up much of the Middle Pleistocene. A long-held behavioural adaptation to such conditions was to occupy caves and deep rock shelters that provided protection from bitter weather conditions as well as from hungry animals. It is difficult to believe that erectines thought any less of such conditions than we do today. With their narrow entrances, caves are also easy to defend against other groups and cold animals. A natural adjunct to cave dwelling in high latitudes would have been the use of fire. Fire use is certainly evident in later stages of occupation at the Zhoukoutien cave complex, around 250 ky. Interpretation of earlier evidence, such as ash layers up to five metres deep, remains unresolved (Jurmain *et al.*, 1990). The use of fire, however, may go back to the Lower Pleistocene, even if this was the transportation of embers from naturally occurring ignitions. Evidence for erectines living above 40° latitude by 1.7 My suggests adaptation to cooler or cold climates had taken place by this time. Perhaps what we are seeing is that the erectine penetration of higher latitudes did not take very long. This conclusion is further underpinned by the evidence from Nihewan, which places hominins above the 40° latitude before 2 ky. Therefore, their apparent first entry into Europe at 800 ky may now point more to the fact that we have not found their remains than their inability to cope with the environment there before that time. The Chinese evidence is a definite signal that cool to cold climates may not have been the barrier to erectines that they are thought to have been. It seems most likely that skins, fur and fire may have also been used by them to keep warm and underpin other biological and cultural adaptations for living in a cold climate.

In the light of the additional environmental stresses encountered by those living in higher latitudes, the pattern of feeding strategies in China was probably quite different from those used by Indonesian groups, even at comparable time periods. The harsher climate of the north must have been a powerful generator of selective pressures that drew more heavily on the cultural and biologically adaptive wits of human groups than the soporific tropics. Hunting or scavenging, for example, in cold, wet or snowy conditions is much more hazardous, less productive and generally more difficult than similar activities in the warm tropics. It is also worth considering that more wit may be required to hunt in open, snowy and rocky landscapes than jungles, because of the difficulties encountered from animals that can see and smell humans over long distances. The result of this is that the chances of going hungry are greater. Higher latitude populations needed to range widely, perhaps following animal herds and plotting ambushes in open terrain and rocky gullies. Faunal remains of bison, bear, deer, boar, leopard and horse have been found in the Zhoukoutien lower cave complex, suggesting that most of these animals were hunted by the erectine inhabitants, but, again, this interpretation has been questioned. An alternative theory proposes that the occupants were scavengers, pinching food from denning animals. You have to be quick, cunning and slightly cheeky to steal meat from carnivores, and in itself the practice suggests an ability to finely tune the planning and successful execution of the heist, let alone possess enough intestinal fortitude to carry it out. We do know, however, that considerable hunting skills were not unknown among erectines. Even more fortitude would have been needed to trap large carnivores, such as bear and leopard, and the hunting of horse must have required a great deal of group cooperation and organisation, possibly with pre-arranged traps and barriers. Fire may have been used to frighten and induce a stampede and adequate weaponry was required for both protection from angry animals and for the kill. In such circumstances, hunting in a moderately sized group would have been more common and the ambush used as a favoured tactic.

Hunting is strongly implied in the butchered remains of elephant as well as other animals found at Torralba/Ambrona, Spain, in the far west of the erectus world. The site is roughly contemporaneous with that of Zhoukoutien, and, while no human remains have been recovered, its inhabitants used hand axes, cleavers and scrapers, after the Acheulean tradition. Nevertheless, the evidence is testament to the cooperative hunting strategies of European erectines. Does this mean also parallel development of these skills at either end of the world? Evidence from Boxgrove in England lends weight to this conclusion. Here late *Homo erectus* people were hunting a variety of large game, such as rhinos and bears, in an organised fashion using sophisticated tools and weapons around 500 ky. Why should similar strategies not have emerged in the Far East at about the same time. Acheulean-like hand axes have been found in the Bose basin of south China, dated to around 800 ky (Yamei *et al*., 2000), a discovery dispelling previous suggestions that Chinese erectines were less capable than their western cousins. Moreover, it could be argued that Acheulean technology was being introduced (or traded?) from the west to the Far East by erectines carrying it from Africa, where it had been used almost a million years before. The more advanced flaking techniques of the Mode 2 Acheulean hand axes seen in China suggest that these advancements were being made contemporaneously in Africa and western Eurasia, as well as East Asia (Yamei *et al*., 2000). The discovery of an Acheulean hand axes quarry at Isampur, India, dated to between 400 ky and 600 ky, could indicate that, if the introduction of this technology drifted from west to east it would have arrived in India before 600 ky to be in China 200 ky earlier. That does not seem to be the case. Instead the Indian evidence probably supports the idea that hand axe technology was developing in different places at roughly the same time.

A slowly emerging picture of erectines is one of a very capable hominin, adapting, developing and planning a new technology to fit requirements. In other words there was a definite momentum in cultural and biological evolution. Moreover, if cultural development and behavioural complexity were keeping pace in widely separated erectine groups across the world, it meant that people were keeping in touch and *ipso facto* gene exchange was taking place across vast distances, even if this was slow and intermittent. Of course, these processes took some time to gain momentum, with very slow change over a considerable period of time, with intermittent speeding up of complexity (cultural and biological).

The possibility of a wide dispersal of hunting acumen at least half a million years ago is interesting from the point of view of language development. Verbal communication would have been a distinct advantage for cooperative hunting with vocal coordination ensuring a more successful outcome. From Spain to China, hunting skills may not have been so finely honed if they were not accompanied by some sort of verbalisation of plan, signal and intent. Attempts to describe and plan were probably the building blocks upon which the rudiments of the earliest human spoken language are likely to have begun. Ultimately, linguistic building blocks must have been based on the most fundamental of human urges, to eat and survive, as well as the need to communicate and stay in touch with others. After all we are not loners, we are gregarious by nature and need the company of others. I do not believe for one minute that Middle Pleistocene hominins grunted or just stood there and stared at each other. Language must have been present, perhaps unlike modern complex languages, but nevertheless a lexicon of sounds that accounted for a great deal of their activities and how these could be accomplished. Indeed, erectine cooperation and teaching of the young without detailed communication, even at a basic level, is not practically possible.

Having said all this, it was probably the lack of sophisticated biological as well as cultural adaptations to very high latitudes that prevented erectines from crossing the frozen wastes of north eastern Siberia and the Bering Bridge and prevented them from entering the Americas' during times of low sea level. The much more complex technological capabilities and biological adaptiveness of later sapien bands overcame the difficulties of coping with the extreme cold of these regions. Let us not forget also that the world's population had grown much larger by sapien times and people had to go somewhere and this is something I take up in later chapters. Moreover, the demographic and population pressures that existed for these populations probably added the necessary drive to undertake the journey. But earlier erectine groups did not posses the sophisticated technology or cultural and biological development required to undertake this journey. It could have been only the lack of adaptation to Arctic conditions that prevented them entering North America. It was certainly not just the cold, because those living in Zhoukoutien's lower cave complex must have experienced the odd frost-bitten foot and finger now and again, but it did not prevent them living there very successfully over a very long time. The curiosity of continuing exploration should, however, have pushed them towards the Americas, but obviously that was not the case. By the late Middle Pleistocene and during interglacial times, Chinese erectines may have been equipped to venture into areas to the north and northeast of China but no further; they had hit a barrier. Elsewhere and hundreds of thousands of years earlier they had, however, overcome another barrier in order to continue their expansion as far as they could.

Homo erectus beyond Java

The early hominin occupation of Indonesia provides a fascinating glimpse of the seemingly purposeful nature of erectine migrations. When for the first time hominins reached this enormous archipelago 2 My ago sea levels were low enough to join it to the rest of southeast Asia. It is most unlikely that erectines built watercraft allowing them to cross to Indonesia. If this had been the case, they would have probably continued their journey, eventually entering Australia and New Guinea before 1 My. The fact that they lived in Indonesia before 1.5 My clearly shows that they were not easily deterred by distance and were determined to continue their migrations as long as there was dry land to walk on. They continued onwards until they had reached the limits of their exploratory capabilities, Java's east coast. Once on the move, therefore, erectines were not to be stopped while there was land ahead. Obstacles to migration included lakes and wide rivers, but the visibility of an opposite bank or shore may have induced people to move on until they found a crossing where something like the age-old but unstable floating log may have been used, if the waters were not torrential and/or full of crocodiles. The open sea was a different matter: once reached, erectines turned left or right and continued along coastal routes. Or that is what was supposed.

Until recently, the most easterly fossil evidence for erectines is the well-known groups of fossils found in central west Java at Sangiran, Sambungmacan, Trinil, Perning (Modjokerto) and Ngandong. If not for the permanent water gap between Australia and this southeast corner of Asia, there can be little doubt that erectines would have reached here soon after their arrival in

Indonesia. To continue east from Java required crossing the ocean even at times of low sea level. Fossil discoveries during the last 110 years in Java have prompted researchers to assume that erectines did not have the capabilities to cross oceanic water gaps, even small ones. In the 1950s, however, archaeological investigation on the island of Flores, 500 km east of Java, reported finding pebble tools, retouched flakes and a small bifacial hand axe (Maringer and Verhoeven, 1970a and b). Some of these implements were in association with the bones of the extinct pygmy elephant Stegodon (*Stegodon trigonocephalus florensis*). After affinities were drawn between this tool kit and other Pleistocene assemblages, such as the Pacitanian, it was suggested that perhaps the most recent erectine people from Ngandong (Solo) had, indeed, managed to cross the permanent divide from Sumbawa to Flores (Bellwood, 1985). The timing of the Stegodon extinctions is still not known, although these animals do not occur in terminal Pleistocene deposits on Timor, suggesting a demise timed with other regional megafauna during the last glaciation.

In 1993 crudely fashioned chopping tools were also found on Flores (G. van den Bergh *et al.*, 1993). In contrast to previous discoveries, however, these implements were found in gravels and associated tuff deposits dated to the end of the Lower Pleistocene (~750 ky). They too are in association with extinct fauna, including Stegodon, a very large tortoise (*Geochelone* sp.) and a giant varanid lizard, somewhat larger than the modern Komodo dragon (*Varanus komodoensis*). The discoveries produced firm dates for the flakes and small chopping tools of between 800 ky and 1 My (Morwood *et al.*, 1998). The only person around at that time was *Homo erectus*, so unbelievably these hominins made it to Flores. As a result, this evidence expanded their known territory, highlighted their ability to cross water gaps of at least 20 km and brought them closer to Australia. Moreover, it requires a re-evaluation of the cultural and cognitive abilities of these hominins living well over 750 ky ago. They were certainly clever enough to carry out island-hopping because Flores has always been separated from other islands and Java by permanent water gaps, even when seas were at their lowest during the height of glacial maxima. The archaeological finds on Flores indicate a lot more than just erectines got there; it shows also that more than one person made the crossing, there was a viable population and it remained there for some time. To find stone tools from that time is rare enough, particularly when none that can be dated has been found in Java. The Flores find is, therefore, very compelling evidence for a viable local population.

At times of low sea level, it is comparatively easy to move eastwards along the Lesser Sunda chain, as long as you can manage the odd long distance swim. To reach Flores from Java today taking the shortest sea route requires at least eight ocean crossings using small inter-island islets. They include:

Java to Bali (2 km), Bali to Penida (14 km), Penida to Lombok (22 km), Lombok to Sumbawa (14 km), Sumbawa to Kelapa (4 km), Kelapa to Komodo (15 km), Komodo to Rintja (5 km), and Rintja to Flores (0.3 km). These distances are quite short, with 22 km the widest and none looses sight of land, but there are a number of them to negotiate. When glacial sea levels fell below -100 m the number of the water gaps to reach Flores from Java were reduced from eight to three: Bali and Penida (6 km), Penida and Lombok (18 km) and between Kelapa and Komodo (5 km). There is little difference between a crossing of 18 km and one of 22 km, so glacial or interglacial timing makes no difference: if you can cross 18 km you can cross one of 22 km. The presence of erectines on Flores suggests that the minimum oceanic distance they were capable of crossing was 18–22 km. With that ability erectines could have moved further east up the archipelago. A major difference between glacial and interglacial sea levels for this journey, however, is that all short crossings east of Alor during high sea levels disappear at -100 m. Therefore, once on Flores, *Homo erectus* could move on another 600 km to the eastern shores of Alor without getting his/her feet wet. With a demonstrated need to explore, therefore, it seems unlikely that they did *not* make it at least that far. The bathometric topography of continental shelves around the north coast of Timor, eastern Alor and Kambing Island are such that even with reduced sea levels of -100 m there is generally little shoreline retreat that might reduce the relative distance between them. However, the Ombai Strait between Alor and Timor may have narrowed to less than 20 km and is certainly the most likely avenue of the arrival of the Stegodon on Timor (Quammen, 1996). Therefore, *Homo erectus'* ability to undertake such journeys could have put them on Timor.

Erectines probably figured out the shortest distance across a water gap as long as they could see their goal. Observations of fauna (perhaps Stegodon?) swimming between islands might have provided an assessment of how safe such an exercise was, even if at times these were miscalculated. Our knowledge of Quaternary sea levels is improving, but there is little accurate information concerning the timing and extent of transgressions in the Lower and Middle Pleistocene. Therefore, what were the implications for inter-island movement during present sea levels and at -100 m at this time?

Two ways in which *erectines* might have crossed the 18 km of sea to reach Flores are either swimming or watercraft. At least half the crossings could be swum, currents, sharks and the odd crocodile permitting. Swimming long distances in equatorial waters is not recommended. They are not only hazardous in terms of the nasties but loss of stamina under a tropical sun and battling against fierce currents for hours do not make for success. Even if adverse sea currents and other hazards are ignored (and they cannot be), it would have taken the best of swimmers a minimum of nine or ten hours to make the Penida–Lombok leg. It seems highly unlikely that swimming was the chosen method used by erectines to cross to Flores, not to forget that there is a need for several of them (male and female) to make successful crossings for a viable population to emerge. Drifting on islands of vegetation or pieces of timber is somewhat less dangerous than swimming, but the voyager is entirely at the whim of the strong current passing between the Indonesian islands and the whirlpools that they cause: even modern fishermen in their modern craft avoid these areas. Logs are also unstable, non-steerable and tend to lie low in the water, particularly when saturated. Dangling legs also make fine lures for sharks, making even a walk up the beach difficult and something that would have cut erectine exploratory aspirations very short indeed.

The only safe crossing would be on a craft consisting of a platform of some kind. Food and water were not needed for the crossing, even if it took 36 hours. These crossings imply, however, that, while the basic principles of navigation were not required to reach a destination you can see, craft construction was understood and had been used before. Currents flowing between islands also demanded a steerable craft, because of their tendency to drag the craft out to open ocean. If a raft or boat of this sort was manufactured and used, when did these capabilities develop and is it possible that earlier crossings were made? Again, the presence of descriptive language as well as group cooperation and organisation are other implications of making this journey safely. Nevertheless, for those possessing limited manufacturing skills and an equally limited tool kit, a 22 km ocean voyage that has to overcome strong cross-currents between islands may have looked to erectines like a trip to the Moon for us.

Because water craft must have been used to undertake inter-island crossings, there was little to prevent erectines moving further east from Flores to the end of the Sunda archipelago, especially with the next five crossings all less than 12 km. At Alor they faced only a 16–20 km hop across to Timor. Their ability to cross 18 km of water must have stood them in good stead to make a successful crossing to achieve that. The next step east is to Liran (14 km) and then 3 km to Wetar. At times of high sea level, Timor is 560 km from Australia. That would have been a problem, or would it? During glacial maxima falls of -145 m produced many islets across the Ashmore reef area, suitable stop-off points 80–90 km from Timor. The exposed Malita Shelf on Sahul was 190 km.

In this chapter I have tried to bring a more human picture to the persona of erectines by assessing their technological, behavioural and linguistic skills and, thus, their general intelligence. The job is difficult, but I have tried to tackle it by viewing them through their achievements such as the fact that by 800 ky they had occupied Equatorial environments for possibly 1 My. If nothing else, we know now they were great survivors. In that time, they must have developed a close familiarity with their surroundings and the materials available to them, as well as developing suitable survival techniques. After a considerable history of occupation of one area or another, it is likely that they were not all the same. Indeed they must have diverged in terms of their adaptations to certain latitudes and environmental conditions, so that Chinese erectines were different from those living in equatorial areas and so on. So, one group may not have been able to live in another region without some considerable acclimatisation, adopting some adaptive strategies different from those they had been familiar with and a swift climb up a steep learning curve in order to know what they were doing.

If erectines on Flores were descendants of those living on Java, and we can only assume they were, then they were well adapted to and familiar with their surroundings so that associated adaptive behaviour was well in place by the end of the Lower Pleistocene. Millennia of littoral hunting and scavenging must have taught them how to read the sea, its currents and tides and assess its dangers. Many unsuccessful oceanic ventures may have occurred but gradually, probably through dogged trial and error, and not without the loss of life greater skills were developed, including the ability to negotiate narrow water gaps. It may have taken 200 ky or more for the foundation ideas to develop further. Our society is in a hurry, *Homo erectus* was not; there was plenty of time. Nothing was nor could be rushed. We should also consider the heightened senses these hominins no doubt possessed. Very acute hearing, sight, smell and taste, all far more sensitive than ours today, and used in a more acute manner than we use ours. They were the only means of spotting danger, and tell them what was happening around them: their lives depended totally on these. These senses also told them where they were heading. They were their greatest weapons and defence mechanisms, without using them to the full they died.

The Flores evidence suggests small ocean gaps were perhaps not the complete barriers to hominin movement in the Lower and Middle Pleistocene as has been generally believed. It seems more likely that Lower Pleistocene hominin migrations along the Lesser Sunda chain would have had greater success during low sea level. If crossings were made regardless of sea level, however, then movement could occur at any time, which does not rule out the spread of people to Timor and other islands requiring journeys of only 20–30 km. It may not be fanciful to suggest that other islands in Sunda separated by this distance, such as the Sulu Archipelago, the Palawan island group and even part of the western Philippines, may also have been visited. There was plenty of time for it to occur and the will was obviously there. One other point needs making. We have seen the movement of erectines across the world and now island hopping along the Indonesian archipelago. Are we to accept that they stopped there? The evidence very strongly says not. Why should they have stopped at Flores or Timor (if they got there)? Given enough time for further development of water craft to move across larger water gaps, let us be generous and say half a million years, they could quite easily have entered Australia during the Mindel ice age 350–400 ky ago. It has always been much harder to contemplate movement from Sunda to Sahul during the early Middle Pleistocene than movement of people to the Americas from Asia because the latter was at least joined up. The evidence is now compelling that the reverse may have been the case, because sea crossings were not only attempted, they were successful, not by one but probably by many. Obviously, human oceanic migration is not a *Homo sapiens* invention, it goes back much further than us. Perhaps this is the case with other capabilities that, till now, we have believed only we possess.

Time is important in building a picture of erectine demography. Any discussion of archaeology mentions it in many contexts, but what do we mean by 'time'? Climatic change takes 'time'; the movement of people takes 'time'; but each is different. So too are the various lengths of 'time' needed for economic, social and biological adaptation. The 'time' that humans took to actually move from the Middle East to eastern China is different from the 'time' that people living much later took to move from Sunda to Sahul; that was instantaneous compared with the former. 'Time' also means different things to different disciplines. An historian specialising in the Middle Ages would regard 100 years as a very large slice of time in which many things happened. To a modern historian a decade is equally as large. Time is a particularly special, complex and convoluted concept to the physicist and theoretical mathematician, where a second may be forever.

The archaeologist, anthropologist, biological anthropologist and palaeoanthropologist also use the concept of time for particular purposes and probably in a more peculiar and, perhaps, more 'dangerous' way than it is used in other professions because few ever consider the time scales we use so freely and liberally when discussing human evolution and the things that can happen during a given time slice. Consider, for example, the time frames of 'Lower' and 'Middle Pleistocene'. These cover not one hundred years but hundreds of thousands of years. Even errors in dating at these times can be 50 ky or 150 ky long. Exactitude is not paramount, because we are at the mercy of dating techniques that are not always as specific as we would wish. Geological stratigraphic dating naturally covers very large slices of time, and exact evolutionary processes and cultural changes are hard to pinpoint using it. This situation is exacerbated when discussing time 1 My ago and compounds even more the farther we go back, so that an error of ± 200 ky years is quite acceptable when discussing the dating of East African sediments. Such errors are really telling us that if we are within 400 ky of the time at which a particular event occurred, we are doing well and we should pat ourselves on the back – self deception on a grand scale I would suggest.

The concept of time and its use by the palaeoanthropologist needs to be reassessed particularly when discussing early Pleistocene palaeodemography. If, for example, we sprinkle the Asian fossil evidence within a two-dimensional framework of time and space, it becomes extremely sparse. If a few uncertain dates are introduced, our ability to put together accurate trends in morphological variation let alone predict the course of hominin demography, cultural development and behaviour over such vast periods of time and distance becomes very gossamer. This introduces a particularly interesting error into predicting early hominin behaviour, population size and migration patterns. Unless there was little alteration in human behaviour over hundreds of thousands of years, it is almost impossible for anyone to predict the sorts of changes that might have taken place over such a vast period of time as the Lower and Middle Pleistocene, let alone understand the maze of possible variations in these changes. With the introduction of regional differences in the speed of such events, the whole thing becomes infinitely more complicated and unpredictable. Is it possible, for example, to extrapolate the adaptive qualities and behaviour of people living at Zhoukoutien around 600 ky back to those who entered China 1–5 My before? Probably not, but should we throw the fossil out with the overburden? In the end what else do we have; what else can we do? In the discussion in this chapter, I am as guilty as any of swinging back and forward over vast amounts of time in comparing erectines from one geographical region to another and so on; perhaps comparing oranges with apples. The thing is they are all still fruit.

Where to now?

The trouble with the story of erectine migration is that, as so often happens, archaeological philosophy and visibility combine to confound the story. Archaeologists and palaeoanthropologists are often victims of their data and are sometimes forced to put together a story which, because of the constant introduction of fresh evidence, is often out of date almost before it is published (particularly in book form). At times, the researcher feels that what is about to be published will be superseded in a few months – if not before. Just when we think we have *the* answer or *an* answer, we find we do not after all; someone else does. At the other end of the spectrum, evidence can take years to gather or accumulate and in some cases no fresh data are added for decades. In the case of humanity's earliest migrations, we are sure that we do not know nearly enough about this topic to say anything very enlightening. For example, we do not know whether erectines moved into northern parts of the Middle East, central Asia, Siberia or even America, and we may never know. Therefore, we must assume that they did not. Is that good enough? Everybody knows that the absence of evidence is not evidence for absence. The inter-island migration of the erectines from Java to Flores would suggest it is not, nor do the discoveries of stone tools in northern China dated to over 2 My, or Acheulian hand axes in the Bose basin, southern China or the presence of a completely new, insular dwarfed, hominin (*Homo floresiensis*), a descendent of *Homo erectus* on the island of Flores living only 14 ky ago (Brown *et al.*, 2004; Morwood *et al.*, 2004).

Anthropology is, however, based on cumulative knowledge that continues to build even as I write these words. Tomorrow we may find the clues or the archaeological evidence to take this story a few steps or a very long way forward. Unfortunately, at this time there are many more questions concerning this area of study, which we must cast into the 'too hard basket' because of our complete lack of information, but answers to which may turn up next week. But we cannot leave it there. There is always room for speculation, so I will continue.

Why erectines left Africa and explored the Old World is unknown and, perhaps, unknowable. If their reasons were the same as those that inspired recent human exploration such as curiosity, and I like to think that they were, then that draws us together as hominins. In that case we could be seeing in these first explorers a hint of the obscurantism, the fickle, the whim that makes us humans what we are. The basic instinct that fires the human drive to push into an unexplored region cannot be studied objectively or quantified mathematically, at least not by those dealing objectively with the past. When faced with fundamental questions like those posed above, the anthropologist usually gives up on them, because they are too hard, or tries to find a more tangible answer among aspects of seminal human behaviour. In this chapter I have put forward some ideas concerning the way in which hominins moved out of Africa and travelled across the vast distances of Asia and offered some possibilities concerning their biological, technological and cultural abilities. The most important factor in any discussion of exploration and migration, however, is population size. Erectines accomplished what they did because of their numbers. A single band of explorers did not push across the world setting up communities from Georgia to Nihewan and Flores. Population increase was the driving force. The success and viability of the expanding populations of erectines suggests that population growth was taking place. But this really is one of the 'hard ones'. What was the size of world population in the Pleistocene, how fast did it grow and what were the consequences of population growth during that time? The next chapter tackles these questions.

2 *Pleistocene population growth*

The problem and the approach

Most people avoid the subject of Pleistocene world population growth. It has to be said that there are many obstacles to such study and few willing to venture into the distant and extremely hazy regions of early human palaeo-demography. Because of innate difficulties in establishing solid parameters for such a study, it is easy to fall into the wide grey area between fact and supposition or be reduced, as some might say, to mere 'arm waving'. Nevertheless, to venture in is not always a bad thing, if for no other reason than to get the discussion going or hopefully strike a few mental matches. I believe that answers to who we are and where we came from as modern humans are fundamentally bound up with how and where humanity grew during the Pleistocene and the size and rate of that growth. The processes and strength of flow and exchange of genes across the world is the key to the evolution of modern humans and their regional variation. To facilitate gene flow, however, you need populations, and the bigger they are the greater the flow within and from them. Knowing something about long-term world population growth is, therefore, critical to evolutionary processes as well as understanding human biological and cultural evolution, early hominin adaptation and the spread of humanity across the planet: all ingredients of the previous chapter. Moreover, it helps us better understand the growth and distribution of contemporary populations, particularly those that dominate world demography today, such as China and India. To view the world's present problems of overpopulation and rapid population growth as a very recent process may be short sighted as well as placing too much emphasis on the economic and cultural events of the early/mid Holocene as the fountainhead of these problems. Rather, the processes of Pleistocene demographic evolution must explain better modern patterns of population distribution and composition and size, particularly in Asia.

Besides the obvious lack of empirical data concerning early hominin populations, not knowing where to start is the usual reason for not tackling human population growth during the last 2 My and is reason enough for many to avoid the subject like the plague. To begin such a study, however, we need to know something of the fundamental parameters that enable us to reconstruct population size and growth rates. These include the reproductive biology, mortality rates and reproductive potential of early people. It is likely that we will never know exactly what these were, but, unless we intend *never* to discuss the subject, a 'best guess' carried out in a parsimonious fashion, using as much information as we can gather, may be enough to enable us to build a basic population growth model for the last 1 My at least. So, in the following study I apply the meagre information we have and form a conservative model of world population growth during that time period. To understand present population growth, the possibilities for the evolution, growth and spread of regional founder populations are explored, together with the possibilities of how and where Pleistocene populations grew, as well as the dynamics behind that growth. Pleistocene human reproduction parameters and their consequences for population growth rates are re-evaluated and the results are applied to a founding hominin population of finite size. From this, overall estimates of world population growth are compiled, using very conservative parameters. Subset estimates are determined for certain parts of Asia at various times throughout the Pleistocene that hopefully shed light on the size of Lower Pleistocene erectine migrations and set the foundation for an examination of population growth over the last 1 My in our most populous countries, India and China. In later chapters, these data are used to help explain oceanic and land-based migrations of *Homo sapiens* during the Upper Pleistocene and the 'problem' of the origin of modern humans in the Australasian region.

Pleistocene/Holocene population trends

What is most puzzling about prehistoric world population growth is why it suddenly exploded in the early/mid Holocene in the way it is always depicted. Was it really caused by socio-cultural and economic change, as is so often suggested? Was it as sudden or as explosive as is depicted? How big was the world's population when the explosion began? It is possible that the Holocene population 'explosion' is more a product of our misunderstanding of long-term world palaeodemographic trends than a spectacular early/mid Holocene phenomenon. It seems more likely that the Holocene population explosion was not an explosion but just the next step in a steady progression of population build up, stretching far back into the Pleistocene. Indeed, it is more likely to have been a continuum of those events that laid the foundations both for the explosion and humanity's main migrations into higher latitudes and eventually into the Americas and Australia. The reason may have been

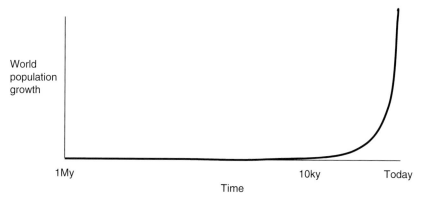

Figure 2.1. Standard world population growth trend over the last 1 My.

because of a far larger world population and a higher rate of population growth during the Middle–Upper Pleistocene than has been previously considered and which began at the end of the Lower Pleistocene.

Humanity's population growth curve is usually drawn as a J-shape, with major exponential growth taking place over the last 4 ky or so (Figure 2.1). The Earth's population has doubled in the last 40 years and is likely to continue doing so, with decreasing doubling times, for the foreseeable future. If unchecked, the problems arising from this will be catastrophic. Some commentators suggest that they may even be insurmountable without a drastic change to birth rates. For erectines no such problems existed, because the doubling time of the earth's population 1 My ago was of orders of magnitude longer than the present 25–30 years. In contrast, a persistent view of Pleistocene human population growth always begins with an extraordinarily small number of people and almost zero population growth that continued for at least 1 My throughout the Pleistocene. When viewed as a graph the usual published trend has four repetitive features. First, an initial world population is set at around 125 000 about 1 My ago. Second, the growth trend line barely sits above the X-axis, essentially without rising for 1 My until the Holocene. Third, a virtual zero-population growth trend continues until somewhere in the early/mid Holocene when a world population of up to 10 million people suddenly appears, depending on which writer one reads. Fourth, the growth curve then quite suddenly turns skywards at the beginning of the Holocene and soars at such a rate that by 2 My world population has increased from 20 to 40 times its size in only 8 ky. This pattern has been consistent in the literature over the last 40 years, receiving little or no adjustment or reassessment during that time. Today's world population has risen by at least one order of magnitude from that of 2 My. My question is: With such a small population at the beginning of the Holocene, how did such rapid growth take place in only 10 ky and what sparked such an increase in the first place?

The concept of 'the crowded planet' is now familiar to everyone. Familiar too are the two basic opposing arguments that explain the Holocene population explosion. The best known is that of Malthus (1798) who suggested that economic intensification through animal and plant domestication was initiated in order to supplement dwindling food supplies for an already burgeoning population. An opposing argument proposes that population increase was a *product* of domestication which increased through technological advancement that boosted food supply, which, in turn, encouraged population growth and so on. There are variations of these two arguments, but few if any give a good explanation of how such an extraordinary escalation in human numbers could have suddenly occurred in the mid/late Holocene.

I want to say at the outset that the idea of the sudden acquisition of skills to raise crops and domesticate animals, prompted by the vast increase in humanity, is not convincing. Neither are the suggestions that the dietary changes accompanying domestication enhanced human fertility to such an extent that enormous exponential population growth was the result. While both these factors may have contributed to some degree, the nub of both arguments rests with the idea that people voluntarily boarded a treadmill of continuous development for the sake of it. They then became entrapped, firstly, by an addiction to the benefits of crop and animal domestication, secondly, by the inability to reject it and, thirdly, by the compulsion to procreate fast enough to negate the extra food being produced and continue the upward spiral of technological change and innovation. Did humans really choose to launch themselves on such an insane rat race when they had the world at their bare feet as hunter-gatherers? Moreover, it is well accepted that economic intensification brought with it a significant downside: problems for human health. These included poor nutrition, higher frequencies of infectious disease, anaemia and parasitism, as well as a number of new 'crowd' diseases, many originating from a range of zoonotic types crossing to humans from their close association with the animals they were domesticating. All these were enough to reduce average human stature by up to 15 centimetres as well as life expectancy, the latter not conducive to producing high rates of population growth. Indeed, it might be expected that this pathological revolution had a negative impact on the burgeoning but still comparatively small population, but it had little if any on the growth curve proposed above. An accompanying increase in arthritic conditions that now included higher frequencies of osteoarthritis in elbows and shoulders in both men and women, because of changes to work regimes, longer working days and harder physical work was another added hazard of the agrarian revolution.

It is logical that the adoption or acquisition of the domestication treadmill was prompted by a growing need for food for an increasing population. Animal/crop domestication took place at different times, in different ways and in different places throughout the world. Most importantly, the need to intensify food production was prompted by a steadily growing population, particularly in certain areas. I contend, however, that world population growth was not even throughout the world; it did not occur overnight nor did it begin in such a dramatic fashion with the famous mid-Holocene exponential 'blast off'.

If the figures normally used in discussions of Holocene world population are anywhere near correct, then they amount to nothing less than a population explosion that is hard to justify within the parameters of the population sizes usually envisaged for those times. Why the breeding capabilities of humans suddenly responded in such a way to the appearance of agriculture and its accompanying raft of physical and health stresses in comparatively few population centres is not made clear. There can be little dispute that during the Holocene some populations did experience a great population increase. It is my contention, however, that in those regions where massive growth did take place, it happened far too quickly if the minuscule pre-Holocene population estimates are correct. It is proposed, therefore, that such growth does not make much sense if only five million people (an average of the stated pre-agricultural world population estimates) inhabited the planet 10 ky ago.

There is fairly reliable data for world population size in the late Holocene, but records are limited to a few regions. The discrepancy between these figures and those based on loose estimates of the population at the beginning of the Holocene requires further examination. I now want to re-examine the possibility of long-term origins of the enormous end-Holocene world population, which I will suggest began far back in the Pleistocene.

A view with room: reassessment of Pleistocene hominin population size

Only the very brave make estimates of the size of world population before 1 My ago. Numbers quoted for that time range between 125 000 and 800 000 people. If there were five million people living at the beginning of the Holocene, world population doubles from just under three to five times, depending on the initial population size chosen. With such small numbers, how were people distributed across the planet during that time?

Even if the upper figure of 800 000 is considered as a minimum world population 1 My ago, it is the equivalent of eight large football stadiums of people sprinkled over Africa through the Middle East to China, south to Flores and probably into Europe. With such meagre dispersal it is difficult to believe that people remained in contact with one another or even found a partner. If humanity at that time had spread itself so thinly, it is difficult to believe that it could survive. Such small numbers suggest small and widely dispersed groups, a situation that would put them in a vulnerable position and in considerable danger of extinction. Indeed, it is unlikely that such a sparse world population could thrive and maintain itself or mount any migratory expeditions that would have succeeded. Moreover, it seems unlikely that erectines would have spread out so that they exposed themselves to such dangers. But they did; moreover, the population actually increased and evolved and remained viable right across the Old World. The fossil record is and will continue to be testimony to that. In the previous chapter I suggested that the best way of surviving, maintaining the population and exploring was not to be isolated, but for this to occur everyone needs to be a neighbour of everyone else.

Without some level of steady population increase, it is logical to suggest that migration would have created human-empty areas as people left 'home' and migrated across vast distances. So there must have been people left 'at home'. Further replacement must have taken place to provide enough people to move out and keep in touch with those ahead. Therefore, the rate of natural increase at 'home' could have been a regulator of migration speed. If this was not the case with people moving off without regard for continuing links with others, then the result would be isolation of extremely small gene pools certainly open to and probably followed by extinction, preventing the successful exploration of the vast areas we know were explored. So, population growth at or close to zero from 1 My to 2 My ago is not possible, given the spread of erectines during that time. Therefore, 125 000 people as a world population 1 My ago would not be enough to either keep in touch across the Old World or fuel migration. The subsequent loss of hominins that did migrate in these circumstances would have probably sent world population numbers into reverse, let alone provided backup migratory bands to continue the process.

Land areas in Asia

For the purposes of this exercise 'Asia' is broadly defined as the continent east of Iran and south of Afghanistan and the Himalayas, excluding Sri Lanka, southern Siberia and Japan (Table 2.1).

Table 2.1. *Asian landmass by country*

Country	Area (km^2)
Bangladesh	142 766
Borneo	538 720
Brunei	5 765
Cambodia	181 036
China	9 560 985
India	3 287 606
Indonesia	1 919 270
Korea	220 839
Laos	236 799
Malaysia	330 671
Mongolia	1 664 075
Myanmar	678 034
Pakistan	803 944
Philippines	299 767
Thailand	513 519
Vietnam	334 333
Total area	20 718 000*

Note:
*Nearest 1000 km^2.

Table 2.1 lists present day land areas for countries comprising Asia as defined for this study. Table 2.2 divides Table 2.1 into three major groupings (China, the Indian subcontinent and the rest of Asia), so that regional population estimates can be made. The land area is divided by three estimates for past hunter-gatherer area-per-person (66.7 km^2, 40 km^2 and 15 km^2) and these produce an overall population size ranging from around 319 000 to nearly 1.5 people in Asia. If only 10 per cent of land area was used (2 070 000 km^2 to the nearest 1 000 km^2), then the Asian population would range from 31 000 to 138 000 people, depending on the size of area occupied per person.

By introducing an estimate range for the Asian population, it is possible to extrapolate those figures to the entire Old World and an estimate of world population is possible. The present land area of Europe, Africa and Asia combined is about 84 million km^2. If half of this area was not occupied or used around 1 My ago but dividing the remainder by the three land areas used above, a resultant world population ranges from 630 000 to 2 800 000 people at that time. With a land occupancy of only 10 per cent the population is reduced to a range of 63 000 to 280 000 people. Using only 20 per cent of a world population of 125 000 proposed by Deevey (1960) at 1 My, Table 2.3 shows the effects of population increase on land availability using the same

Table 2.2. *Chinese, Indian and Asian populations per land area occupied (km^2)*

	Area	P1	15 km^2	40 km^2	66.7 km^2
China	9 560 985	1 274 000 000	637 399	239 025	143 343
Indian SC*	4 234 316	1 277 000 000	282 288	105 858	63 483
Rest of Asia	6 922 828	579 527 000	461 522	173 070	103 791
Totals	20 718 129	3 130 527 000	1 381 209	517 953	310 617
10% land use	2 070 000	–	138 000	52 000	31 000

Notes:
* – Indian, Pakistan and Bangladesh.
P1 – Present population.

Table 2.3. *Population increase and land availability during the Pleistocene*

Asian pop. at 1 My	Area (km^2) per person	Area occupied (km^2)	Approximate % of total	Date BP
25 000	66.7	1 667 500	8.1%	
25 000	40.0	1 000 000	4.8%	1 Mya
25 000	15.0	375 000	1.8%	
50 000	66.7	3 335 000	16.2%	
50 000	40.0	2 000 000	9.6%	850 ky
50 000	15.0	750 000	3.6%	
100 000	66.7	6 670 000	32.4%	
100 000	40.0	4 000 000	19.2%	720 ky
100 000	15.0	1 500 000	7.2%	
200 000	66.7	13 340 000	64.8%	
200 000	40.0	8 000 000	38.4%	580 ky
200 000	15.0	3 000 000	14.4%	
400 000	66.7	26 680 000	*	
400 000	40.0	16 000 000	76.8%	465 ky
400 000	15.0	6 000 000	28.8%	
800 000	66.7	53 360 000	*	
800 000	40.0	32 000 000	*	350 ky
800 000	15.0	12 000 000	57.6%	
1 600 000	15.0	24 000 000	*	240 ky

three occupancy areas and that the smallest population living at the lowest population density (66.7 km^2 per person) would exceed the land area available (*) by the Middle Pleistocene. By the beginning of the Upper Pleistocene even the highest population density (15 km^2 per person) exceeds land area

Table 2.4. *Estimates of Pleistocene world population growth* ($\times 10^6$)

Author/Year	Lower Pleistocene*	Middle Pleistocene	Upper Pleistocene	Early Holocene
Young, 1979	0.100	–	–	–
Deevey, 1960	0.125	1.00	3.34	5.32
Birdsell, 1972	0.400	1.00	2.20	–
Coale, 1974	–	–	–	8.00
Pfeiffer, 1977	–	–	–	10.00[#]
Hassan, 1981	0.800	1.2 0	6.00	8–9.00
C-S et al., 1993	–	–	–	1–10.00[^]

Notes:
*Figures set at around 1 My.
[#]Figure set at 10 ky ref.
[^]Cavalli-Sforza et al. 1993.

available to continue at that density. This suggests that more land must have been occupied and limiting the above parameters is not practical.

What are the earliest reliable figures for the human population? The earliest censuses were taken around 2 ky ago in China, India and across the Roman Empire. Methods were not systematic or thorough, they did not afford complete coverage of a chosen area and are only a rough guide to population size. Poor communications encouraged neglect of remote regions and many thousands of people pursuing various nomadic lifestyles were ignored. The result was that these early censuses are gross underestimates of the actual size of their respective populations, particularly those in Asia. They do, however, represent absolute minimum numbers that always require upward revision (Borrie, 1970). Using a combination of estimate and census figures, authorities suggest that a total world population at the time of Christ could have been as low as 140 million (Deevey, 1960). Others claim 200–400 million might be closer to the real figure (Coale, 1974; Harrison et al., 1977; Revelle and Howells, 1977). Again, estimates fluctuate drastically depending which author one reads; these are shown in Tables 2.4, 2.5 and 2.6.

Archaeological evidence alone shows that by the late Pleistocene people lived everywhere from western Britain to Tierra del Fuego and from the edge of the Scandinavian ice sheets to Tasmania. If anything, the continuing trend of archaeological discovery is one of earlier and earlier pan-world human occupation with the exception of the remote Pacific islands and New Zealand. The world was, indeed, fully occupied before 10 ky ago and humans were as adapted to all the world's habitable environments as they are today. How would this have been possible with an extremely small world population at

Table 2.5. *Estimates of Pleistocene population density (persons/km^2)*

	Lower Pleistocene*	Middle Pleistocene	Upper Pleistocene	Early Holocene
Deevey, 1960	0.00425	0.012	0.040	0.04
Birdsell, 1972	0.015	0.032	0.039	–
Hassan, 1981	0.025	0.030	0.100	0.115

Note:
* – 1 My.

Table 2.6. *World land surface occupied (km^2 × 10^6)*

	Lower Pleistocene*	Middle Pleistocene	Upper Pleistocene	Early Holocene
Birdsell, 1972	27.0	38.3	57.5	–
Hassan, 1981	30.0	40.0	60.0	75.0

Note:
* – 1 My.

the end of the Pleistocene and before any population explosion took place? To answer this we have to go back again, to over 1 My before.

The reproductive parameters of *Homo erectus*

The main reason for so few attempts at estimating population 1 My ago is the lack of a rigorous method to make such a determination. Existing figures are based largely on educated guesswork and estimates of carrying capacity for different environmental circumstances (Hassan, 1981). Often added to these are the two basic and universally held demographic tenets for that time: low reproduction rates and high mortality.

There is little doubt that some time around 1.5 ky ago the planet supported 'comparatively few humans'. But what does that mean? Certainly, enough erectines were living in Africa even before this time not only to leave the continent but to provide successful founder groups that moved as far as eastern and southeast Asia. This occurred even in the face of what we assume to be extremely slow population growth and high mortality rates. Although the details of *Homo erectus* reproductive biology will always remain hazy, most aspects of their general biological make-up must have been very similar to ours. Post-cranial measurements of adult erectines suggest that they were of a similar size to us. Therefore, we can only assume they had similar physiological as well as adaptive capabilities. It has been suggested that *Homo erectus* is an unnecessary taxonomical separation of what is a *Homo sapiens* continuum going back at least 2 My (Wolpoff *et al*., 1994). This idea is an important point to consider for this discussion because it is precisely the perceived differences between erectines and ourselves that has, in the past, dictated a persistent view of them as a distant taxonomic relative rather than just an earlier version of us with a somewhat smaller brain. Indeed, two views now dominate ideas concerning the origin of modern humans and I discuss these in later chapters. In reality, the biological gulf between the roaming erectines of 1 My ago and modern people is probably far narrower than we might expect. Using a logical assumption that there is a fairly close bio-physiological relationship between modern people and erectines, we can add this into the range of other issues raised so far to begin the job of assembling erectine demography.

Firstly, at what age did erectine females become fertile? Recent work indicates that they were fully grown between 14 and 16 years, so females may have begun their active reproductive life at around or even before that age (Dean, 2001). If they reached an average life expectancy of 30 years and puberty at around 14 years, a reproductive span of 15 years was possible. Erectine females must have possessed a reproductive cycle similar to modern humans, suggesting they would have been fertile all the time. Then it is highly likely that pregnancy occurred much more regularly than it does, for example, among recent hunter-gatherers who deliberately space offspring for a variety of socio-cultural reasons involving highly developed customs, rituals, taboos and include expediency. Birth spacing in these groups often takes the form of an extended period of weaning, underpinned by a set of complex social relationships and belief systems that may also invoke post-partum sexual and nutritional taboos. There is, however, little to encourage extrapolation of such complex and socially controlled reproductive strategies back to Lower or Middle Pleistocene hominin bands.

Child spacing among erectines may have been possible through the natural contraceptive effects of breast-feeding, although the reliability of this process alone to prevent further pregnancy is not assured among modern humans and, indeed, is known to be considerably reduced after 30 months post-partum. Intermittent amenorrhoea or infrequent menstruation was much more likely to have prevented regular conception. Starvation or vigorous and/or excessive exercise may have been other reasons for low reproductive rates or at least reducing conception frequencies, working as natural checks on reproductive cycles. It is possible that females living 1 My ago may have had a menstrual cycle involving fewer annual episodes or one that was more seasonally based. I am not convinced we have to invoke such differences as the latter, however. There is no reason to superimpose these differences on erectines unless we think of them as wild animals or the systems themselves sound more primitive and, therefore, better suited to them. A menstrual cycle close to those of modern humans, albeit irregular at times, is the most likely that should be considered and is the one used for the arguments set out below.

Theoretically, it might have been possible for erectine females to produce one child per year. If the above-mentioned reproductive span (t) of 15 years is used with a three month spacing between birth and confinement, then each female could theoretically bear 15 children, which sounds unlikely but let us continue for a moment. Gross reproduction rates (Rg) are obtained by multiplying total fertility by a constant of 48.8 per cent or the percentage of females born into a human population. This produces an annual Rg of 7.3 females throughout a 15-year reproductive span. A net reproduction rate (Ro) can then be calculated by multiplying gross reproduction by female survivorship (Sv), or the chances of a female reaching a reproductive age.

Female *Homo erectus* survivorship has been put as low as 32 per cent, rising to between 30 and 50 per cent for later Palaeolithic populations (Nag, 1962; Acsadi and Nemeskeri, 1970). Even with high mortality rates, the vulnerability of females and a comparatively short life span, a net reproduction rate (Ro) of at least me was required to maintain population viability. Any number below me represents a population doomed eventually to extinction because of the non-replacement of females. *Homo erectus* must have achieved this to populate the world the way it did, as well as maintain its populations in the far corners of the Earth for a very long time. Therefore, using a Sv of 32 per cent in our calculation produces a resultant Ro of 2.3, which indicates a high replacement of females and, therefore, a viable population with strong growth potential. We can now calculate a rate of population growth using the standard formula: $r = \ln Ro/t$. The result is that r is 0.056 or an annual increase of 5.6 per cent, which produces a population doubling time (Dt) of 12.3 years using the formula $Dt = 0.6931/r$ (Hassan, 1981). However, *Homo erectus* did not double its population every 12.3 years; this is a figure yet to be achieved even by modern human populations. Appendix 1 presents a wide variety of calculations, using very small changes in both the net reproduction rate (Ro) and survivorship (Sv). The reason for including this was to demonstrate the wide variety of reproductive outcomes that emerge when minute variations of these parameters are made. I also included it to avoid the laborious task of working through the many combinations of variables that can alter reproductive profiles to produce the very long

Table 2.7. *Population doubling times (Dt) where Dt = 0.6931/r*

r	0.000005	0.000006	0.000007	0.000009	0.00002	0.00004	0.00005	0.00008
Dt (years)	138 620	115 517	99 015	77 011	34 655	17 328	13 862	8 664

population doubling times that probably existed in the early/mid Pleistocene. So let us examine population growth rates using very conservative demographic parameters.

Even small changes to such parameters drastically alter population growth rates and various combinations produce a diverse set of results. For example, if survivorship is 32 per cent and birth rate is 6.404 children in a ten year reproductive span, the net reproduction rate is 1.00005, which produces a population growth rate of 0.000005 or five persons per generation per million people. This is a very much smaller figure than the 56 000 persons per generation per million people used in the previous example. With a 0.000005 growth rate, the population would double approximately every 140 ky (actually 138 620 years). This is an extraordinary long time and represents one additional person, per generation, using 200 000 people for a founder population. Other changes also significantly affect the way a population grows as well as alter the shape of that growth. For example, if the chances of survival are raised only 3 per cent from 32 per cent to 35 per cent then, keeping the above factors constant, live births drop from 6.404 to 5.855 per female but produce a similar population doubling time. If the Sv is raised to 42 per cent, the birth rate drops to 4.879 but produces the same growth rate, and so on. The higher the chances of survival, the fewer live births needed to produce the same growth rate. Naturally, the growth rate does not alter significantly if the reproductive span is lengthened unless more children are born. Nevertheless, by doing this there is a resultant increase in the length of time in which a required number of live births could take place, thus adding greater spacing between births and adding more realism to the model. The original figure of 6.404 live births seems far too high for a ten-year reproductive span, although it might be a more reasonable expectation over 15 years. Table 2.7 shows how minuscule rises in the reproduction rate (r) can drastically alter population doubling times (Dt).

Fossil evidence suggests that female erectines may have had an average reproductive span of around 15 years. For example, the late Middle Pleistocene female hominin from the Narmada River valley in central India may have reached 30 years of age (Jurmain *et al.*, 1990). Analyses of the Zhoukoutien *Homo erectus* fossils show that 60 per cent of these reached maturity (>14 years) and 2.6 per cent may have achieved late, middle age

(50–60 years) (Weidenreich, 1943). Weidenreich (1952) also identified seven females among the Ngandong series from the Solo River, Java. These consist of three adults of advanced age, one young adult, two adolescents and one child. Positive sex identification of these is not conclusive and, if these are ignored, it suggests that almost 60 per cent of females reached adulthood. Not all researchers agree with Weidenreich's sexing, however. Santa Luca (1980) reduced the number of identified females to two, as well as leaving four individuals unsexed. Not having examined the original fossils, my identification relies on some of the casts and published photographs, but generally I agree with Santa Luca's results, although I believe there are three females in the group. The Ngandong group may be too young for the purposes here, but the results from other earlier Pleistocene remains suggest that female survivorship among Middle and Upper Pleistocene hominins could have been higher than is often believed. Therefore, a 15-year reproductive span may not be fanciful nor a higher female survivorship of 40–50 per cent; a more realistic figure than the 32 per cent proposed in the calculation above. Even if nasty and brutal, perhaps life in the Pleistocene was not as short as is often suggested!

Appendix 1 also shows that an average reproductive span of 15 years, a 40–50 per cent survival chance and between 4.098 and 5.123 live births will produce a population growth of around 0.000005 and a population doubling time of 140 ky. A small rise in survivorship to 52 per cent is required before less than four children must be born to each female to maintain this doubling time. Even to maintain zero population growth with a 50 per cent female survivorship, over four live births per female are required. Over a 15-year span this is equal to one birth every three years and nine months. Using these calculations any doubling time can be artificially created given a suitable set of reproductive conditions. What is demonstrated, however, is that even to maintain the population, a survivorship of less than 40 per cent requires a live birth rate per female that was probably unlikely for *Homo erectus*, as well as other Lower Palaeolithic peoples.

Having discussed how these figures work, we can now use the minimal growth rate of 0.000005 to show how various founding world populations at 1 My could produce a world population of between 256 million and 1638 million people at the time of the Viking voyages.

A model for world population growth in the Pleistocene

The growth curve of a biological population has three essential elements: increase, decrease and collapse. The human growth curve has never come close to a collapse, as some animal and plant populations undoubtedly have;

at least this has been the case for the last 5 ky or so. Even when the Black Death of the fourteenth century caused the loss of almost one third of Europe's population, human reproductive ability recovered admirably and very rapidly made up the numbers. With the complete lack of highly infectious 'crowd' diseases, such as bubonic plague, measles and smallpox, Pleistocene populations were not subject to pandemic impacts such as those experienced in the Middle Ages. But no matter how big natural disasters are, they have had little if any effect on recent human population growth. This situation may have been different in the Pleistocene, however, when small populations were affected by substantial human loss. Such disasters probably came from attacks by carnivores, tectonic and volcanic activity, and extreme weather conditions, such as droughts and floods, tropical storms, avalanches, Tsunami and severe, prolonged winter conditions; although, the sparser the population, the fewer people that were affected. It seems hard to imagine, however, that any of these made any lasting impact on the overall process of world population growth. Nevertheless, the loss of ten people out of 60 or 100 *is* a major disaster demographically speaking, particularly if this involves a disproportionate amount of females of reproductive age or young adults that make up the future generation. Warfare is another source of population decline, but with groups spread out these events must have been restricted to small affairs with the annihilation of a local band or group taking place only on rare occasions. Assuming, therefore, that mass disaster and warfare did not drastically affect overall population size, there seems little to prevent the world's population from steadily increasing during the Pleistocene.

Table 2.8 and Figure 2.2 present various world population growth rates in the Pleistocene and demonstrate how large populations emerge from very small progenitor populations using extremely long doubling times. Table 2.9 narrows the field by presenting four base populations, including three (1, 3 and 4) proposed by other researchers. The model uses the demographic parameters outlined above with a starting reproductive growth rate of 0.000005, a doubling time of around 140 ky and four very conservative world founder populations consisting of 125 000, 200 000 (this author), 400 000 and 800 000 at 1 My. It shows also a persistence of the same growth rate for 500 ky, at which time world population has reached 1.6 million. There is then an increase in the growth rate to 0.000006 due to the:

1. natural increase in world population which automatically increments growth rate,
2. exponential characteristic of growing populations,
3. reproductive capacity of an increasing population,
4. biological and cultural adaptation,

Table 2.8. *World population growth using various base populations* ($\times 10^6$)

Date	Base population size										
	0.125	0.15	0.175	0.20	0.225	0.25	0.275	0.3	0.325	0.35	0.375
1.0 My											
860 ky	0.25	0.3	0.35	0.4	0.45	0.5	0.55	0.6	0.65	0.7	0.75
720 ky	0.5	0.6	0.7	0.8	0.9	1.0	1.1	1.2	1.3	1.4	1.5
600 ky	1.0	1.2	1.4	1.6	1.8	2.0	2.2	2.4	2.6	2.8	3.0
470 ky	2.0	2.4	2.8	3.2	3.6	4.0	4.4	4.8	5.2	5.6	6.0
350 ky	4.0	4.8	5.6	6.4	7.2	8.0	8.8	9.6	10.4	11.2	12.0
240 ky	8.0	9.6	11.2	12.8	14.4	16.0	17.6	19.2	20.8	22.4	24.0
140 ky	16.0	19.2	22.4	25.6	28.8	32.0	35.2	38.4	41.6	44.8	48.0
62 ky	32.0	38.4	44.8	51.2	57.6	64.0	70.4	76.8	83.2	89.6	96.0
27 ky	64.0	76.8	89.6	102.4	115.2	128.0	140.8	153.6	166.4	179.2	192.0
10 ky	128.0	153.6	179.2	204.8	230.4	256.0	281.6	307.2	332.8	358.4	384.0
1 ky	256.0	307.2	358.4	409.6	460.8	512.0	563.2	614.4	665.6	716.8	768.0

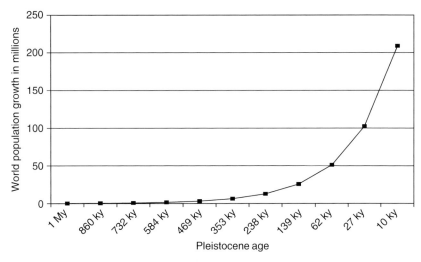

Figure 2.2. Proposed world population growth during the Pleistocene.

5. increased neural and thus intellectual abilities resulting in
6. increased technological ability;

all of which are enhanced by:

1. continued development of hunting capabilities and extractive behaviour, and
2. increasing success in exploration and occupation of new and more marginal environments;

The natural function of population increase to 1.6 million around 500 ky reduces the population doubling time to around 115 ky which coincides with the appearance of a wide range of variation in hominin brain size. From this point on, a more obvious and familiar exponential population growth curve begins to emerge. By the beginning of the Upper Pleistocene (150 ky), world population reaches a minimum of 16 million people and the population doubling time drops to between 80 ky and 90 ky. At the end of the Pleistocene, around 10 ky, world population is over 100 million and the growth rate is still only eight people per 100 000 per generation ($r = 0.00008$). It is interesting to note that without changing the chances of female survivorship, which sits between 40 and 50 per cent, live births have risen to between 4.103 and 5.129.

Table 2.9. *World population growth from 1 My using four base population sizes and a reproduction rate (r) of 0.000005*

Doubling time (Dt)*	Reproduction rate (r)	Date (ky)	$\times 10^6$			
			Base 1	Base 2	Base 3	Base 4
–	0.000005	1 000 000	0.125	0.2	0.4	0.8
138 620	0.000005	861 380	0.25	0.4	0.8	1.6
138 620	0.000005	722 760	0.5	0.8	1.6	3.2
138 620	0.000005	584 140	1.0	1.6	3.2	6.4
115 515	0.000006	468 625	2.0	3.2	6.4	12.8
115 515	0.000006	353 110	4.0	6.4	12.8	25.6
115 515	0.000006	237 595	8.0	12.8	25.6	51.2
99 015	0.000007	138 580	16.0	25.6	51.2	102.4
77 010	0.000009	61 570	32.0	51.2	102.4	204.8
34 655	0.00002	26 915	64.0	102.4	204.8	408.6
17 330	0.00004	9 585	128.0	204.8	409.6	816.2
8 665	0.00008	920	256.0	409.6	819.2	1638.4

Notes:
Base 1 – Deevey, 1960; Base 2 – Webb; Base 3 – Birdsell, 1972; Base 4 – Hassan, 1981.

During the Lower and Middle Pleistocene, developments in hunting capabilities included small, accumulative improvements in stone tool technology and its use, particularly Acheulian bi-face technology, the development of spears and increased use of fire. Biological adaptation continued and combined with behaviour and physiological changes that together allowed an increased exploitation of and movement into slightly more marginal environments found in rainforest areas, high country and higher latitudes. Increasing cranial capacity provided an additional complexity to the human cerebral cortex and a substantial increase in neural networks contributing to forward planning and refined exploitative strategies. This constituted a continuous development and honing of the behavioural and cultural elements of human society, leading to its ability to be successful in many environments, spread across the face of the planet. Moreover, it is likely that increases in neural complexity with increasing brain size enhanced language development which, in turn, underpinned improved tool manufacturing, hunting techniques and general survival chances. Perhaps the most important outcome of population growth would have been to greatly enhance the possibilities for gene exchange between populations, thus providing a greater impetus for increasing gene flow throughout the inhabited world.

Pleistocene population growth and the CIA (China, India and Africa) connection

On 22 September 1999 world population reached six billion and it is expected to reach nine billion by 2050, a frightening statistic. At the same time world population growth is beginning to slow for the first time in our history. Fertility is declining on most continents with an average birth rate of 2.7 children per female, down from five in the 1950s (United Nations, 1998). The birth rate in Africa has been reduced from 6.6 to 5.1 and in Asia from 5.1 to 2.6 children per female. The combined populations of China and the Indian Subcontinent make up nearly 70 per cent of the total Asian population with 2200 million people, which comprises more than 40 per cent of total world population. But why has the great rise in world population been focussed in these two regions?

One reason for writing this book has been to re-examine the widely held belief that our huge world population is almost totally a product of the last 5 ky. It is from this standpoint that other highly contentious issues concerning human migration and origins can be more easily understood. In order to throw light on them and the growth of the world's population, I need to introduce a proposition: that the enormous size of the Chinese and Indian Subcontinent populations is not a recent phenomenon, they have *always* played a prominent role in world population growth, particularly during the last million years and that the size of these populations today reflects their long history as 'population generators'.

The Indian and Chinese populations did not emerge overnight but grew over a considerable period of time and, therefore, they contributed more than their fair share to Asia's total gene pool as well as those of adjacent regions. The extent of those contributions cannot be underestimated if we consider how long they have been prominent in producing people. For example, there were enough people living in the Far East (China?) to populate the Americas with viable populations by the late Pleistocene and early Holocene. Such growth would have undoubtedly influenced patterns of regional human evolution and migration in Asia, as well as prompt the appearance of anatomically modern human characteristics in the region and direct human gene flow across a large part of the Old World.

Whether you are an 'Out of Africa' or 'Multiregional Continuity' hypothesiser, in comparison to China and India, Africa, as the home or 'cradle' of humanity, has been comparatively left behind when it comes to producing people. In 1995 Africa's total population was around 640 million, just over 70 per cent that of India and only 50 per cent of China. I hinted in the previous chapter that it is natural to expect that the continent that gave rise to humanity, produced enough hominins to travel to other parts of the planet and set up viable populations there, should have maintained a lead in population production. Indeed, recent hypotheses suggesting that Africa is the producer of waves of anatomically modern humans that swept out of the continent around 100 ky, enveloped the planet and replaced all pre-existing population groups, rely strongly on such a conclusion (Stringer and Andrews, 1988). Obviously, Africa produced the original migrants sometime before 2 My, but did it do it again 100 ky ago?

The 'Replacement', 'Out of Africa' or 'Eve' hypothesis (choose your own title) proposes that modern humans left Africa in such numbers that they were able to roam across the globe, eliminating and replacing all previous hominin groups (Stringer and Gamble, 1993). Such a theory has extremely important implications for previous and later world demographic patterns and population growth, no less than that of Africa, which, the theory would imply, was a *people generator* during the last half of the Upper Pleistocene between 50 ky and 150 ky ago. If this was so, what happened? Did it dry up?

Mitochondrial DNA studies, which form the basis for the 'Eve' hypothesis, do not seem to consider human legs, human behaviour and the human condition. Africa is not a people generator now and has not been so probably for a very long time. This is an important issue, but it has not been openly discussed in the literature. Important to this discussion is the continuous migration process implied in the 'Out of Africa' story that produced enough people to totally replace all external populations. Also there is the implication that once the annihilation was completed the enormous African gene flow implied in the idea, then for no apparent reason stopped. After producing the AMH 'armies', why then did they stop migrating out of Africa? If something stopped them so that more people could not leave, then why did Africa not fill up to the brim with those being produced internally? These issues are taken up again below, but before going on I want to go back to the original migration out of Africa.

The fossil record shows that by 1.5 My ago hominins had settled in the farthest corners of the Asian continent with the exception of higher latitudes. It is, therefore, difficult to believe that less than two jumbo jet loads of erectine people (679) were lightly sprinkled in a line across the vast plains and valleys of the Middle East and Asia, from Georgia to Java (see Chapter 1). The enormous difficulties faced by so few people spread so thinly across enormous tracts of land were discussed in the previous chapter. Moreover, it is difficult to believe that people would have organised themselves in the precise 'chains' described in that chapter, so it is clear that any attempt at a head count using the 'chain' method is unsatisfactory. What these chains do show, however, is the absolute minimum numbers required to produce

60   The First Boat People

Table 2.10. *Pleistocene population growth in China, the Indian subcontinent and the rest of Asia (AsiaP)* $\times 10^6$

Date BP	r	AsiaP*	China	Indian SC
1 My	0.000005	0.051	0.02	0.022
860 ky	0.000005	0.103	0.04	0.044
723 ky	0.000005	0.206	0.08	0.087
607 ky	0.000006	0.413	0.16	0.175
492 ky	0.000006	0.825	0.325	0.35
376 ky	0.000006	1.65	0.65	0.7
272 ky	0.000007	3.3	1.3	1.4
178 ky	0.000008	6.6	2.6	2.8
92 ky	0.00002	13.2	5.2	5.6
57 ky	0.00003	26.4	10.4	11.2
34 ky	0.00004	52.8	20.8	22.4
16 ky	0.00005	105.6	41.6	44.8
2.5 ky	–	211.2	83.2	89.6

Note:
*Thailand, Myanmar, Malaysian Peninsula, Indonesian archipelago, Vietnam, Cambodia, Laos and Korea.

the populations of hominins living in Asia 1 My ago. It seems, then, that 10 000 individuals is probably the smallest Asian population that can be contemplated at that time. Five times this figure would be a more realistic number, although even 50 000 people (half the crowd size at a Melbourne 'footy' final) spread over such a vast landscape as Asia, even if they were largely confined to narrow environmental settings, would be very transparent indeed. Nevertheless, let us suppose that there was a slow but constant trickle of people leaving Africa between 1 and 2 My ago. Let us also suppose that Africa had an extremely small intrinsic growth rate. By the time regional populations were thriving in Indonesia and China, the total number of hominins living in Asia could easily have been 50 000 or more.

Table 2.10 illustrates a likely Asian population growth pattern through the Pleistocene as well as dividing that population into individual growth profiles for the Indian subcontinent, China and the rest of Asia. It shows how a very small growth rate and base population of 50 000 could have produced an Asian population of just over 200 million at the beginning of the Christian era. A prominent feature of this growth rate is that it overtakes the rest of the world by the end of the Lower Pleistocene and marginally stays ahead till the end of the Pleistocene. Another feature is that, while growth begins with a quarter of the world's population, it constitutes over half at around 2000BP. A third point is that these calculations show even

Table 2.11. *Indian subcontinent population growth from a base population of 100 000 at 400 ky and a reproduction rate (r) of 0.000005*

Doubling time $(Dt)^{\#}$	Date*	$n(10^6)$	Reproduction rate (r)	Epoch
138 620	469 000	0.1	0.000005	
115 515	338 000	0.2	0.000006	Middle Pleistocene
99 015	239 000	0.4	0.000007	
86 640	152 000	0.8	0.000008	
77 010	75 000	1.6	0.000009	Upper Pleistocene
34 655	41 000	3.2	0.00002	
17 330	23 000	6.4	0.00004	
11 550	12 000	12.8	0.00006	
7 700	4 000	25.6	0.00009	Holocene
2 310	2 000	51.2	0.0003	

Notes:
$^{\#}$To the nearest five years.
*To nearest 1000 years.

when very long population doubling times are used, growth steadily increases so that large end-Pleistocene populations are not only possible, they are inevitable.

Table 2.11 presents an individual calculation for the growth of India's population from the Middle Pleistocene. The element that drives the increase is the growth rate itself and if those suggested here are close to the actual rates they were very subtle indeed. Compare, for example, the extremely small difference that occurs in gross reproductive rates *(Rg)* when the rate of population growth changes from 0.000005 to 0.000006 and then to 0.00005 for females with a 40 per cent chance of survival and a 15-year reproductive span. The respective changes are 2.5001875, 2.500225 and 2.501875 female children born. These are minute differences but enough, given certain population sizes, to kick growth along in such a manner that, over hundreds of thousands of years, large populations can be built; indeed, they cannot be avoided.

The above tables track one another in their broad characteristics with India achieving a slightly larger population by the late Holocene. The accuracy of the figures is not paramount, but the rate of growth and order of magnitude of their proportions is. It has been suggested that India supported 100 000 people around 12 ky, reaching 25 million around 2.5 ky (McEvedy and Jones 1978 – Atlas of World Population). It seems difficult to believe, however, that a population which probably took off 500 ky ago, took all that time to reach 100 000 (which amounts virtually to a zero population growth rate), would then, in the space of 9.5 ky, suddenly escalate to 25 million. Notwithstanding the above proposal that the population of the Indian subcontinent was over three times as large by 2.5 ky, a far more realistic view to achieve 25 million is to invoke a steady, slightly exponential, increase over a very long time as shown in Table 2.11. If these figures are anywhere near correct it would seem that an Indian population of 100 000 by 12 ky is an underestimate by a very long way and so too is the 25 million of 2.5 ky. The various population sizes proposed are certainly reasonable given the very small founder population, the equally small doubling time and the conservative parameters used in the model. I now want to show that the growth of these populations was the fundamental force influencing and shaping the migratory and evolutionary development of people living far from the centres of the two main Asian population generators: India and China.

Given their prominent place as population generators today, it is likely that the Indian subcontinent and China have been prolific sources of migrant populations and, therefore, gene exporters for a very long time. But these places were important for another intimately associated reason. Because of their ability to produce humans they have played a very significant part in Asian migration, particularly in the Upper Pleistocene, and provided Asia with many wandering bands of explorers. Africa must have also had a similar, although less productive role in the production of humans for its peripheral regions, particularly in the Lower Pleistocene, but the initial importance of its role decreased as the Pleistocene drew on. Europe has always been a net importer of genes from other regions until the last millennium when it too began to export them.

To explain the above statements requires a new assessment of the size of the Asian population during the Upper Pleistocene and particularly the time between 30 ky and 90 ky ago. I have already shown how it could have exceeded 10 million around 110 ky (Table 2.10). The steady exponential increase that followed saw a doubling of the population at 72 ky and again around 43 ky during a period which marked a very important time in the demographic and cultural development of modern humans in Asia. These times coincide also with major migration events, heralding the movement of people from Asia into adjacent regions no doubt as an alleviation of a population pressure welling out from growth centres. The process is likely to have been a natural movement of hunter-gatherers responding to what might be termed 'population pressure' within the terms of that lifestyle, although the pressure did not come from millions of people. The Chinese and Indian gene pools were unable to move into each other's territory due to their high genetic pressure gradients where the two pools met that maintained their fundamental population characteristics (Map 2.1). The resultant migrations were probably

Pleistocene population growth 63

Map 2.1. Population growth centers in China, India and Africa shown with Upper Pleistocene migrations.

not sudden or speedy and they did not involve thousands of people at a time. Instead they took the form of slow trickles in a subtle pattern of pulses, over extensive periods of time that alleviated population pressure pockets in some regions. The pulses were released into areas that could absorb or funnel people away from high pressure centres.

Genes and migration

Physical and genetic barriers normally restricted land-based migration in the Upper Pleistocene. Physical barriers included high mountain ranges, swamps, impenetrable rainforest areas, gorges, valleys and deserts. They are lateral restrictors: when any are encountered they present little choice of direction but to follow the easiest route available around or through them. Steep valleys, for example, restrict movement to along their floors, which is easier than trying to scale a mountain range on either side. The flow of people becomes concentrated along narrow and totally proscribed routes and direction of movement is determined to a large extent by the natural shape of the terrain. Without watercraft, large river systems and coastlines facing the open ocean too may also have acted as barriers to gene flow, although these are not lateral restrictors because people can spread laterally in one or both directions along the system until they find a place to cross. The genetic consequences that emerge from lateral restrictors are that:

- genes are transported relatively quickly along narrow paths because lateral spread is restricted and all energy is put into forward movement.
- confinement of a migrant group in this way naturally maintains a high frequency of its genetic compliment, and
- it sees relatively quick placement of genes in extended target areas through comparatively rapid movement of the migrant group rather like a genetic tunnel.

Without lateral restrictions, however, migratory gene pools can spread thinly across the landscape; this also increases chances of extinction. Flat and open country is more likely to entice people to spread out provided that food and water are available and the landscape is suitable for movement. Thus, the greater the distance covered in this terrain, the thinner the population becomes and the more vulnerable the migrant gene pool becomes, unless, of course, new people arrive. Without backup groups the greater the distance travelled from the original gene source, the greater the chances of losing people and, therefore, of the weakening of the genetic pool. There is also the threat from any other, larger gene pools that might be encountered that could then dominate the smaller pool. Without backup from the original pool, then, lateral expansion can predispose the migratory pool to the threat of extinction. Genetic loss occurs also when sub-populations divide and break away from the main group and this is more likely to take place among laterally expanding groups than laterally restricted ones. This is compounded if small groups become isolated and genetic loss occurs through group vulnerability to predatory animals as well as encounters with other populations, which brings us to the second major type of barrier: the genetic barrier.

A genetic barrier includes encounters with new populations, particularly those living in areas of high genetic pressure. A high-pressure genetic barrier is one formed by high concentrations of people sharing the same origin or gene pool. In this case the smaller pool can disappear by its incorporation into the larger one. These barriers are best thought of as similar to high and low pressure centres surrounded by gradients of decreasing intensity, similar to the meteorological isobaric patterns of high and low pressure. Areas of high genetic pressure are difficult to penetrate, they dominate smaller, weaker gene pools that try to infiltrate them and they occur in population growth centres. Genes flow from these centres outward to places of low genetic pressure. When relatively close to one another, two high pressure centres can cause a different sort of barrier, where their respective populations confront one another; this was mentioned earlier with reference to India and China. In this case migratory pressure from two directions can result in a narrow zone of mixing, where both gene pools merge but neither moves very far into the opposite zone. This is compounded if a natural barrier, such as a large river or mountain range, separates the two and it is then that the natural barrier is recognised as the accepted division between the two pools. Lateral genetic flow on either side of the natural barrier moves genes into areas of lower genetic pressure, resulting in a stark boundary between the two pools. Gene flow will continue to take place, however, in places of low genetic pressure, where there are sparse populations, and where the interdependence of people who need numbers to survive cooperate with each other. Areas of high genetic pressure must have been rare during the Lower and Middle Pleistocene, but became more prominent as the Upper Pleistocene drew on.

The position of genetic barrier gradients differs with the respective pressures of the neighbouring gene pools that varied over time. Unequal or low pressure between two different gene pools will relax gradients, allowing the movement of genes one way or another and these could move back and forward over thousands of years. Low-pressure areas or unoccupied regions encourage natural gene drainage and, thus, migrant settlement centres and potential places for the development of migrant gradient pressures. These can be high or low depending on the rate of migration and the influence of neighbouring gene pools. The migratory gene pool would effect a strong influence as it emerges from its genetic tunnel, but as it spreads out it weakens as it disperses across a broad landscape.

As barrier and migrant pools meet they can either avoid each other or mix. In the first case the migratory group maintains its founder frequency and continues its journey, obviously in a different direction and possibly dissuaded from further encounters with the barrier group. In the second case the outcome depends on the relative strengths of each gene pool. If the barrier pool is weaker, the migratory pool will dominate and the barrier pool will be freshly assorted and take on characteristics of the migrants. The migratory pool might be have a strong impact initially, however, but the greater the distance from its ancestral pool and/or the lack of backup genes will see it weaken over time. This situation will be compounded even if the barrier pool is weaker than the migratory pool at first encounter, but is able to maintain a fairly strong flow from either a source pool nearby or closely related neighbouring pools. Both groups will depend also on their respective distances from the centre of population pressure. In this case, the migrant stock will dominate for some time before its influence becomes increasingly weak as barrier gene frequencies increase through genes continuing to enter the barrier/migrant area from a high-pressure zone. Therefore, the frequency of the genetic mix depends also on conditions at the point of mixing and the geographical distance from both the barrier and migrant gene sources. Such encounters would always be difficult to interpret or observe in the skeletal or archaeological record, unless they involved very different morphological and/or cultural types.

Another variation is that a weak source flow for either group could allow the gradual filtering of genes in either direction as well as almost equal mixing to take place. Following contact, the awareness of others, the use of conflict, or even more subtle mechanisms, could reverse the gene flow along pre-existing pathways. Genes would move down routes used as a backup by either the original barrier population or migratory group, but be taken over by the group with the highest genetic pressure, creating a clinal effect. This flow would heavily depend on the strength of gene flow in the opposite direction. A strong flow from both migratory and barrier groups could also prevent the exchange of genes beyond the immediate mixing area. The reason for this would be the inability for genes of either group to penetrate very far into the high-pressure area of the other, thus resulting in the barrier described above. Clearly, the consequences of these events would have been modified even further over time by basic evolutionary mechanisms, such as random mutational events, isolation, genetic drift, founder effect and adaptation. Such a complex fabric of genetic change and exchange must have played a very important part in the production of differences in cranial morphology and types, the general biological as well as cultural evolution of human populations in Asia and the exchange of genes across the globe. Variation in outcomes from these encounters under different circumstances could produce a morphological variety that would be difficult to interpret, if only a few fossil specimens were all that remained. But in the final analysis all these processes depend on population sizes and the rate of regional population growth.

A high genetic pressure barrier: India versus China

To illustrate the above model, I want to return to India and China, the two major areas of high population pressure. Upper Pleistocene gene flow from both these areas would have naturally followed lines of least resistance, gravitating towards low-pressure regions where population densities were very low or non-existent. The direction of gene flow from these centres was similar, although India's is more restricted. The Himalaya Mountains restricted gene flow to India's north and it was later adaptation to high altitude and colder climates that allowed any sort of penetration into this remote and hostile region. Southern gene flow was impossible because of surrounding oceans that prevented flow till the trading of the last two or three millennia. Therefore, India's genes could only move west or east, then south (Map 2.2).

A similar barrier to India's northern gene flow existed in China. Until cultural and biological adaptation to living in high latitudes took place, movement of genes north into the cold wastes and steppes of eastern and northwestern Siberia were prevented till the late Upper Pleistocene. Eastern movement was not possible and there was a limit to northern flow along the coast. Only western and southwestern flows were possible. These genes freely drifted down into a less populated southeast Asia, through Vietnam, Cambodia, Laos, Thailand and south along the Malaysian Peninsula. Over time Chinese genes flowed west meeting Indian genes flowing east, meeting at the border of present-day Assam/Bangladesh with Myanmar (Burma) (Map 2.2). Thus, a build-up of high-pressure zones formed either side of the Ganges/ Bramaputra River, which today sharply demarcates two morphologically distinct peoples, Caucasians and Mongoloids. The Himalaya mountains barred the flow into and out of northern India. The passage or penetration of each by the other became increasingly more difficult or impossible as the pressure from both these areas steadily rose by migration of their respective growing populations. Where they met, little mixing took

68 The First Boat People

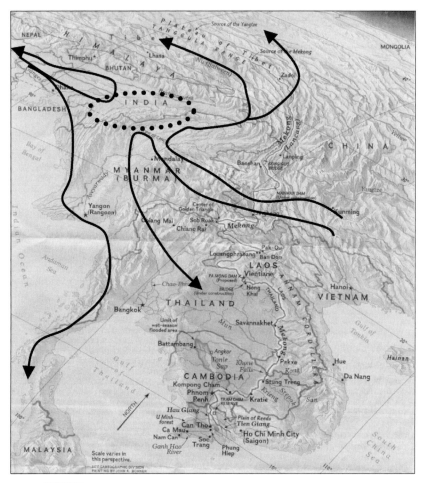

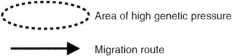 Area of high genetic pressure

———▶ Migration route

Map 2.2. A genetic pressure barrier on the Indian/Myanmar (Burma) border.

place. Consequently, India's genetic outlet has been largely to the west, through low-pressure avenues, whereas China's outlet took the empty genetic paths northeast, northwest and south. Northwestern expansions were separated from India by the Himalayas, which acted as a natural wall of separation from the Indian gene pools on its southern flank. The process eventually took Mongoloid genes into western, central and northern Asia and Siberia, perhaps bringing them into contact with outlying groups of Neanderthals. Indian genes penetrated far into the Middle East, Asia Minor, probably into eastern Europe and may have even reached North Africa.

Carrying capacity and population size

The model presented above shows how subtly and quite naturally large populations could have arisen in India and China and provided a far-flung genetic influence over the Middle East and Asia, respectively. It proposes also that their respective sizes were orders of magnitude greater than has ever been suggested before. Naturally, the next question is: how did such large populations feed themselves and where is the archaeological evidence for them to be found? I need to say from the outset, however, that, while they are natural enough, these questions are based on a set of criteria that have probably been used by too many for far too long.

One method of estimating prehistoric population size in the past has been to use the concept of carrying capacity. So, by estimating the carrying capacity of Asia, we can know how many people live there. To estimate the carrying capacity of Asia, we would need to know the floral and faunal circumstances of more than a dozen different environmental settings in great detail in both glacial and interglacial conditions. We still have no precise data concerning the exact timing, duration and consequences of glaciations before 150 ky ago. In particular, the Lower and Middle Pleistocene glacial events remain largely unstudied and we have no way of estimating the fluctuating carrying capacity of India, China or anywhere else during that time. There are also indications that no two glacial events were the same and indeed the trend during the last four was to a greater swing in the environmental pendulum that may indicate increasingly severe conditions certainly as indicated by data gathered from Antarctic ice cores (EPICA, 2004). The impact of these episodes may have brought about a variety of environmental changes that directly or indirectly affected the availability of different food resources in ways we cannot even guess at and changed the pattern and composition of populations dozens of times during that period.

Also, we would need to know the limits of the extractive potential of humans living in all these environments as well as predict how flexible their changing extractive strategies were. We have two other basic and very broad variables to consider. The first is the adaptive, behavioural, cultural and technological skills of humans that continuously changed over a very long period of time. Also, we need to consider consequent variation in human choice, emphasis on and demand for different foodstuffs, all of which fluctuate in response to population change and human societal development in general. In turn, adaptation to more marginal environments may have taken place to a greater extent than we might give the ancients credit for. Cold and semi-arid regions, for example, may not have been the barriers to occupation or use that we might expect. Moreover, the variable success brought about by the developing and ever-changing wit and cunning of the Pleistocene hunter-gatherer is, I suggest, beyond our capabilities to know or estimate to an extent needed to assess carrying capacity.

Another factor in the equation is that even if a particular carrying capacity can be known, it may not tell us anything about population size. Estimates of the carrying capacity of Chad and Somalia, as well as other largely traditional-based economies in existence today, would suggest population numbers that bear no resemblance whatsoever to those that actually occupy these countries. So, no amount of estimating Pleistocene carrying capacity will tell us how many people lived in a particular area or region the size of Asia at that time. I suggest we will never know how it fluctuated during glacial episodes of varying severity across 1 My! The impossibility of estimating carrying capacity is highlighted further when the enormous environmental and sea level changes that took place during glacial episodes are considered. For example, what was the carrying capacity of the exposed Sunda shelf during periods of low sea level? Those several million square kilometres of extra land could have supported a variety of environments, all very productive. They offered, no doubt, a great attraction to bands of humans and the vast majority of those humans lived on coasts, estuaries and along the rivers that stretched across the exposed shelf. To estimate the carrying capacity of this region is now an impossible task. Moreover, inundation of this shelf has probably destroyed all evidence of its human inhabitants. Consider further that, if all interglacial sea levels were as low as they were during the last interglacial, modern shorelines only existed for about 10–15 per cent of the time. In other words, if riverine, estuarine and coastal environments on the exposed continental shelf were favoured by people during the Middle and Upper Pleistocene, they would have had only 100–150 ky of opportunity to live along what we regard as the modern coastline during the last 1 My. This leaves 850 ky or more of living areas that are now totally submerged, most below 50 m of water.

Today, Australia's landmass is 25 per cent less than it has been for 95 per cent of the last 130 ky. Coastlines and river estuaries, even hundreds of kilometres of river course, all favoured areas of the hunter-gatherer, have disappeared under water. What our continents look like today is not 'normal'. So much land that must have been utilised by humans for the vast majority of history is no longer available for study or answering these sorts of questions.

Archaeological debris and population size

The quantity and distribution of cultural debris, debitage, living areas and stone flakes are often used as a loose, proportional indicator of the size of prehistoric hunter-gatherer populations. These are the only indicators of the 98 per cent of humanity's life on earth as the hunter-gatherer. Rather than nicely stratified and datable sequences, or convenient foundations of buildings with rooms, living areas and neat concentrations of life's rubbish, the majority of archaeological remains consist of smashed stone scattered widely in variable proportions and found on surface sites that cannot be dated or are of dubious age and awkward distribution. A few flakes, some long-abandoned fire places and scatters of broken stone are all that is normally left for interpretation about the life of the makers and how many of them there were. The delineation and definition of what is, more often than not, enormous but intermittent stone scatters often confounds rather than enlightens the broader interpretation of what has taken place across large tracts of time and space. The difficulties associated with interpretation of such scatters usually result in much archaeological 'data' being swept into the 'too hard basket' for want of definitions and interpretative methodology to unlock the secrets of stone scatters. The other problem is that of how discrete an archaeological site might be. Australia is a prime example of this. Rather than a series of archaeological sites, most of the continent is one big archaeological site. Java presents a different picture. Without the presence of fossil hominin remains, there would be no accurate indication of their presence in Indonesia except for a few undated stone tools. There is not one living area, rock shelter or cave site of Lower or Middle Pleistocene age to suggest that humans were there during those times. Even the Upper Pleistocene Ngandong remains are not associated with cultural objects or living sites. Therefore, the lack of archaeology is not evidence for the absence of humans.

The sheer size of the world's population has, over many years, trampled, disturbed, destroyed as well as unwittingly and deliberately buried untold layers of cultural garbage left by bygone generations. Moreover, natural processes disturbances and disasters have also contributed to this enormous loss of cultural debris, as well as the enormous environmental changes that have taken place over tens of thousand of years. Floods, glaciers as well as volcanic and tectonic activity have long since swept away or crushed tons of archaeological debris or scattered it in such random and diffuse ways as to completely destroy any hope of making sense of it. They have redistributed it so that it takes on the appearance of something completely different from its original form. Another factor is that in the cultural refuse tip stone survives against all else. In the end, however, its survival does not really enlighten us much when it comes to determining the size of the group, band or population that made and used it.

We know that there is an exponential loss of cultural items as we go back in time. This is compounded by the increasing simplicity of those items, the older they are. Hunter-gatherer archaeological debris is even more cryptic when it comes to using it to determine population size from ancient campsites. How many artefacts and campsites are now under many metres of water? The vast majority of the time humans have been out of Africa, they probably utilised coastal environments because they offered good camping and productive food sites. Consequently, we have lost inordinately more evidence of our ancestors than we will ever find, but even if stone debris were littered everywhere, what would it really tell us? For example, unless it can be accurately dated, a scatter of stone flakes always has the tag on it that says: 'Who made it . . . ten people over a thousand years or a thousand people over ten years?'.

One interpretation of the absence of evidence is that it was never there, but it is quite obvious that is not so. Naturally, our interpretations of the past are based on what we find not on what we do not. What we know about the past is constantly being upgraded and reinterpreted. Occasionally, our ideas or theories are overturned when fresh evidence, sometimes minimal in size but vital in essence, is discovered. We may have to wait for future archaeology to show the existence of populations of the size I have suggested. I believe, however, that we will, because it is difficult to accept that, unless Upper Pleistocene populations were much larger than has been generally believed, the growth in cultural complexity and the adaptation and migration of *Homo sapiens* into all parts of the Old and New World and Australia could not have taken place the way it did and when it did. More importantly, nor could humanity have grown so rapidly in such a short time during the late Holocene. Perhaps the archaeological evidence for such populations is staring us in the face but lacks recognition, that is another problem often encountered in archaeology.

3 From Sunda to Sahul: transequatorial migration in the Upper Pleistocene

Population centres and edges

The steady migration of people to all parts of the western world, north and southeast Asia, Australia and the New World between 30 ky and 90 ky was probably a product of increasing pressures from a slowly growing world population. Small incremental improvements in implement manufacturing, planning and strategy building, economic diversification and cultural modification helped these movements. People were now being soaked up in marginal regions, including high latitude and semi-arid environments, which enhanced the migration process, enabling them to enter regions not previously explored and the use of watercraft culminated in island hopping and the comparatively quick spread of people into island southeast Asia and Australia. Moreover, amelioration of environmental and weather conditions during interglacials opened up vast new areas for occupation and to explore, although, with the rise in sea level, continental shelf space was reduced. Migration was not an imperative, but a hunter-gatherer way of life required space, so migration and exploration went hand in hand. A small push here, a foray there and valleys, rivers, deserts and steppe country gradually became home to many. In empty territory, migrants entered in slow trickles not bow waves; they did not fan out but penetrated in rivulets. It is difficult, however, for people living in a world that has been populated for some time to move to an area where others do not already live. Compared with today, people living in the first half of the Upper Pleistocene were spread thinly and probably in patches across the landscape. Nevertheless, there was a scattered density differential that depended on the availability of food, landscape type, demographic history, cultural development as well as general environmental conditions. I cannot envisage rapid movement of groups across empty regions that had always been devoid of people. It would take time to learn about the subtle and not so subtle changes in landscape and differences in subsistence gathering as well as what vegetable foods were edible and what were deadly. Of course, this depended on how different one area was from another and that was governed by speed of movement. If groups moved quickly, change was comparatively rapid and that required increased landscape learning, but I can only envisage such rapid change occurring for those travelling by watercraft, particularly island hopping (see also Meltzer, 2002). Moving into empty regions also implies the lack of neighbours. That is a disadvantage, because people rely on each other for a variety of support functions, genetic exchange and just for good old 'more heads are better than one' when it comes to optimising survival chances. Occasionally, however, small groups must have enjoyed quite large regions to themselves, but when others arrived it may have become comparatively overcrowded because of limited or scattered resources that required a large area to feed the human group. As hard as it is to believe, there must also have been other times when there was limited space for others to move in, such as some valleys and marginal niches in semi-arid regions, thus causing conflict or the pushing of one group or another onwards.

Taking up the threads of the previous chapter, population growth centres could have emerged in central and northeastern China, which much later encouraged population aggregation that eventually inspired large-scale sedentism along the Yangtze and Huang (Yellow) Rivers and the age-old high protein diet became a high carbohydrate one. The processes of Upper Pleistocene migration from population centres cannot be envisaged in the same way as migration from today's highly industrialised centres or even those of the mid-Holocene. There would, for fundamental reasons, be a certain element of urgency in the later two situations that did not exist in the former. Upper Pleistocene migration did not consist of hordes of people moving along tracks or roads, carrying enormous bundles of personal belongings. Neither is it possible to imagine speedy movement covering tens or hundreds of kilometres in several days (with the possible exception of island hopping), or journeys that had specific destinations and goals. These migrations consisted neither of mass nor speed; it was an almost imperceptible gossamer thin shuffle through the landscape, often without a deliberate direction except, perhaps, to choose the easiest route. This almost subliminal way of alleviating 'population pressure' took the form of a constant re-positioning of people as they moved away from population centres. Nomadism or incipient herding particularly encouraged such movement by their very lifestyles, but migration does not necessarily mean rapid technological development or the introduction of innovative and more efficient extractive methods in order to feed people. People fed themselves by moving on and gathering experience. As they moved they learnt: new environments meant new knowledge that was piled on top of what you already knew. Over time knowledge was gained, stored and some was lost as circumstances changed – two of these circumstances were the weather and the environment. As people moved into new environments, so their adaptive strategies altered to some extent to cope with the changing conditions associated with the new region. This could bring about an intergenerational discard of knowledge that had once been useful but was no longer relevant to the new regions that required different extractive strategies and approaches. Allied to this was the requirement to adapt to climatic changes, which naturally accompanied the new environments into which people moved. The variability by which groups captured information on climatic and environmental conditions has been described by others (Gunn, 1994, Meltzer, 2003). The process is governed by the length, severity and frequency of climatic change, as well as the time over which the given conditions are experienced. This process of gaining knowledge has been described as 'capturing'. The mobile colonist would, for example, have a low frequency of capture or well-developed knowledge of weather and environmental conditions, particularly those conditions which occur over decades or centuries, because they are not around long enough to gather a deep understanding of conditions. But as mobility slows and people settle down, then the level of capture, or accumulation of knowledge regarding climate, mounts through generational experience of extremes.

To return to technological change that might accompany migration, it is likely that developments always took place to some extent, regardless of the movement of people from one region to another. From time to time, they would also be sufficient to hone extractive potential, bring about a fuller exploitation of food resources and thus underpin the travellers' ability to live in more marginal environments. It would also enable them to tolerate living in centres of slowly growing population. Technological innovation probably occurred more often in these centres because (1) there were more heads to think of them and (2) there were the commensurate needs of the larger group. Moreover, there was also an increased efficiency required for dealing with issues arising from everyday living in such growth centres.

Genetic pressure gradients would come into play, with ideas moving at variable rates from high to low pressure regions. In other words, a general flow of ideas from the middle of population centres towards edges where local ingenuity and innovation were absorbed when and where appropriate. The geographic position of population centres may have changed from time to time, with emphasis on growth fluctuation between certain regions. The overall result would be a gradual outward movement of people and their ideas from a centre to an edge, where, over many generations, people would experience increasing pressure from populous regions that might slowly push them into new territory. For example, those living to the north of China would feel the push of people from farther south as one or two extra bands arrived in the vicinity every 50 or so years. This might then result in the arrivals or the original group moving on a little further. It could also mean warfare, which left one or other group depleted and incorporated into the other.

During these sorts of discussions, it is easy to fall into arguments concerning the rate of expansion/migration at this time and many would urge the need for it. By trying to explain these gradual changes and the mechanisms behind them, the process naturally becomes speeded up in our dialogue, but it should be remembered that the process is very subtle and is taking place over very large slices of time. They may have been so subtle that it might have taken several thousand years just to put enough people into a new region so that they would be archaeologically visible today as a new population.

Population valves

I have proposed that world population growth during the Lower and Middle Pleistocene gradually built populations in eastern and southeastern Asia and in and around the Indian subcontinent. Thus, by the beginning of the Upper Pleistocene there was a need for population expansion to relieve economic pressure on the hunter-gatherer way of life. Bands spread into peripheral areas accompanied by cultural and biological adaptations that allowed them to occupy more marginal habitats. Cultural adaptations may have included technological change as well as social change in the form of ceremony and group/tribe ritual interaction and trading. Biological adaptation included a range of physiological and morphological changes, allowing people to live in very cold and very hot semi-arid and arid environments. The emergence of a true complex language would have underpinned these processes. Overall, alleviation of growing population pressure in major growth centres eventually gave way to a series of migratory events in the Upper Pleistocene that took place over a very long time (Figure 3.1). Five such movements are proposed, including three from China and another two from India.

The Chinese migrations moved essentially in three directions:

1. *northwest*, into central Siberia through the Altai;
2. *north* and *northeast*, into the Siberian Far East, the latter setting the scene for later entry to the Americas via the Bering Bridge; and
3. *south*, spreading through southeast Asia and eventually into Sunda (Australia).

The Indian migrations included:

1. *southeast*, down the Malaysian Peninsula towards island southeast Asia; and
2. *west*, into central Russia, the Middle East, eastern Europe and possibly as far as North Africa.

Transequatorial migration in the Upper Pleistocene

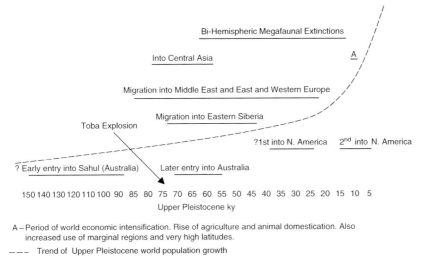

Figure 3.1. Human migration in the Upper Pleistocene.

All these would play a significant role in the later exploration of the entire globe as well as influencing evolutionary trends in areas distant from their migratory sources. These migration events acted as valves alleviating the two main centres of population growth in Asia. They took place over the next 100–150 ky stopping only when New Zealand was occupied 1 ky ago.

A population pressure gradient existed along migratory routes in a drawn out form of those described earlier. The longer the migrational route, the shallower the pressure gradient became between the migrant group and progenitor groups or centre. In many instances, the direct influence of genes in population growth centres may not have extended far. The reason for this was that, as migrants moved as far as they could, the back flow of pressure from occupied areas against further migration would have halted or severely limited gene flow along the lines of migration. A gradual clinal effect in population gene composition would have emerged along migration lines and in some cases this situation could have produced a founder effect in the migrant group. This would have been dictated by the initial gene frequencies of the particular group and, in turn, distance from the original source, environmental circumstances, drift, founder effect, local mutation rates and selection, together with any effects of other gene pools encountered. Meanwhile, world reproduction rates were increasing adding to the rate of gene exchange and flow in many regions (Figure 3.2). Eventually this resulted in

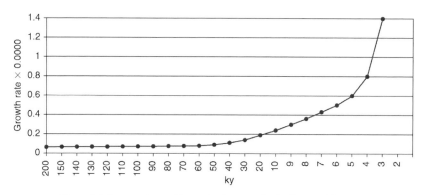

Figure 3.2. Increase in world reproduction rates in the Upper Pleistocene.

population increases in those centres that in turn gave impetus for economic intensification and eventually initiating full sedentism and the development of large towns and cities.

Gradual exploration of and adaptation to new regions would have been a natural outcome of migration in the Upper Pleistocene, none more so than northern expansions. These had little choice but to enter marginal environments, subject to long winters, low temperatures and a fluctuating seasonal food supply that dwindled precariously for at least half the year. The ability to overcome these difficulties honed the socio-cultural development of the population, but some regions did not undergo the extremes we might expect today. During the last interglacial, the central and eastern Siberian environment was warmer than at present, with an almost ice-free section stretching to the northern coast. Indeed, much of Siberia was ice-free even during the height of the last glaciation. While freezing winters might have been expected, seasonal fluctuations triggered time-honoured migrations of herd animals that wandered far into northern regions and hunters would have followed these. There was no need to be fully adapted to subarctic conditions as modern populations are, instead partial adaptation to cold conditions would suffice in a similar way that Neanderthal populations were able to cope with Europe's ice age conditions. While the known easterly expansion of these people is Teshik Tash (44 ky), in Tadjikistan we may yet find them further east. Moreover, Asia may have also produced populations similarly adapted, considering they had had people evolving in China throughout the Pleistocene. Indeed, while obvious differences do exist between Maba and Neanderthals, it has been suggested that the former is an east Asian equivalent of the latter. Full bio-cultural adaptation to high latitudes certainly existed later when, unable to go south due to continuing population growth there, people were forced to remain in and occupy even colder climates where they spread out and adapted or perished.

Gradual movement of peoples out from growth centres also allowed them to adapt economically as well as culturally to changing conditions. Movement into higher latitudes was almost certainly a slower process than crossing the Equator ever was. Enduring long-term sub-zero temperatures requires certain biological as well as cultural adaptations. Body shape, controlled use of fire, specialised hunting techniques, clothing, a diet heavily reliant on meat and fats, as well as a constitution that can cope with such a diet, are a few bio-cultural requirements for living in central and eastern Siberia. Moreover, it is interesting to speculate at this point how long it took for the epicanthic fold to develop among modern peoples who came to populate these cold regions. Did this go back to erectine times or was it a development associated with modern humans? Certainly, this trait is not easily explained when African bushmen and Inuit share it, unless either it was taken to Africa from Asia or it evolved through parallel evolution in two completely different environments.

Stone artefacts from Nihewan and Renzidong in northern China have been dated to at least 1.8 My, indicating an erectine presence as far as 40° latitude even during the Lower Pleistocene. It is no surprise then that archaeological evidence from the 400 ky old Diring site on the Lena River at 60° latitude shows a further spread of erectines into higher latitudes by the Middle Pleistocene. It is likely that these forays were opportunistic, carried out when environmental conditions became milder during interglacial periods. Although archaeologically singular events, they obviously did not consist of one person, and other people, perhaps two or more bands, were involved in these excursions. They may not have been permanent, but once again we see the adventurous spirit of erectines pushing into marginal territory. On present evidence, however, it took the arrival of modern humans to actually stay there and adapt to the conditions, but erectines took their journeys as far as they could with what they had. Indeed they had explored widely before modern humans then took the story a little further. The timing of permanent entry into cold climates is still being debated, but the oldest evidence for modern human habitation in southern Siberia is 50–70 ky in the Altai (Derevianko and Petrin, 1992; Derevianko and Shunkov, 1992; Derevianko and Zenin, 1992; Derev'anko et al., 1993). Recent evidence for human occupation of the European Russian Arctic has, however, shown that these extreme environments may have been used on a more or less regular basis. The presence of side scrapers, bifacial stone tools and a mammoth tusk bearing cut marks at Mammotovaya Kurya and Byzovaya, both well within the Arctic Circle and dated to 35–40 ky, suggest either that Neanderthals were already adapted to and exploiting these harsh environments or more modern people had moved into them by that time (Pavlov *et al.*, 2001).

Pleistocene migrations achieved three very important things. Firstly, they were primary vehicles for human gene transfer and genetic intermixing, keeping humanity in touch with itself over vast distances, as well as enriching biological variability and facilitating cultural spread and interchange. Migration corridors also allowed the dispersal of adaptive mutations back and forth between widely separated populations, albeit the process must have been extremely slow in the early and mid-Pleistocene. Movement of humans across the globe has been largely responsible for the development of a range of climatic adaptations that we all share. To some extent we all have the unique ability to live anywhere, although certain groups of us can endure cold, heat and humidity in differing amounts through inheriting different degrees of those adaptations and that has enabled us to successfully occupy most of the world's environments. Our explorative behaviour was the unwitting force driving Pleistocene migrations and added a drive to much of our cultural and technological development, which is exemplified by the movement of people into Australia. The ability to conduct trans-oceanic travel is surely one of the salient markers of our general intellectual development. With the discovery that Indonesia's *Homo erectus* groups had that ability at least 800 ky ago, albeit it embryonic compared with later voyages, it is becoming harder to pinpoint when that milestone was reached, how many milestones it took to be a modern human or how much each milestone overlaps with the previous one. It looks increasingly, however, that voyaging could have been something developed and shared by humans over a long period of time rather than a particular spark of genius that emerged only with the appearance of *Homo sapiens*.

Trying to understand regional Pleistocene migration is like trying to view all the facets of a diamond at once. Because it involves so many different elements, it is a truly multi-dimensional subject and a juggling act for those who try. Salient elements include environmental change, sea level fluctuation, biological adaptation, regional population growth patterns, socio-cultural change and behavioural modification over both the short and long term. Migration patterns are even more complex when dealing with the movement of people through island southeast Asia because different combinations of the above elements, coupled with the variety of possible migration routes through the region, could substantially change who would have eventually crossed into Australia.

By the Upper Pleistocene a steadily growing world population equipped with increasing technological skills spread its genes widely. It did this through its growing ability to travel greater distances successfully and to facilitate interaction between groups. The obvious import of faster migration would have been the relatively quick transfer of adaptive characteristics in a manner that would eventually reduce differences in skeletal morphology between regionally differing populations, while increasing genetic diversity. This was accomplished through the bringing together and mixing of previously semi-isolated groups. Nevertheless, time, environment and partial but variable degrees of geographical isolation maintained some regional morphological variability until that time.

Many have commented on the subject of the earliest migrations to Australia (Keesing, 1950; Lindsay, 1954; Birdsell, 1957, 1967, 1977; Bowdler, 1977; Jones, 1979; Horton, 1981; White and O'Connell, 1981; Theil, 1987; Andel, 1989; Butlin, 1989, 1993: Mulvaney and Kamminga, 1999). It has not been discussed, however, against its wider background of world population growth and human expansion into other regions. I have little doubt that people did not choose to come to Australia, a mixture of natural inquisitiveness and population pressure brought them here – we were just another destination or safe haven. Where they moved to was probably inconsequential to them at the time. On that level, the journey may not have been as significant or momentous as we might see it today, because it may not have been a deliberate crossing, just a natural end product of a continuous process that had been going on for some time. History is full of markers that suggest people reached as far as they could using the cultural baggage they possessed at the time. In view of this, there is room for fresh discussion of the journey that essentially addresses humanity's first large-scale sea venture and marks a salient cultural and behavioural landmark in the capabilities of *Homo sapiens* and, perhaps, earlier types of people. The journey is as significant as the invention of the hand axe, the ability to control fire, the development of spoken language, space travel and unravelling the human genome. So, who were these people? Where did they come from? How long did it take them to reach Sahul? What routes did they take, and how did they achieve such a significant journey?

Which way? Traversing Sundaland

Coastal routes

For this discussion, I define Sunda as all land and exposed continental shelf of southeast Asia below 15° north, excluding Wallacia, Australia, Tasmania and Melanesia (Figure 3.3). At times of low sea level, Sunda was little more than a series of moderately narrow land bridges linking the Malaysian Peninsula with

82   The First Boat People

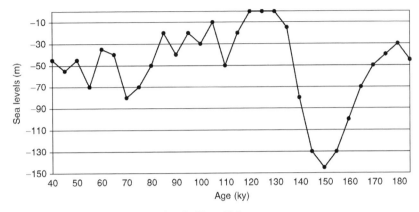

Figure 3.3. Sea levels during the Upper Pleistocene.

Sumatra, and Sumatra and Java with Borneo. During glacial peaks, when the sea dropped to around −150 m, it became a vast subcontinent that replaced the Gulf of Thailand and Java Seas and expanded into the South China and Banda Seas (Figure 3.3, Map 3.1). The amount of exposure and the shape of these prehistoric coastlines changed as sea levels fluctuated between extremes.

It would be only logical to assume that those who first journeyed to Australia had had a long association with the ocean and watercraft. Although they may have favoured a coastal lifestyle, they were probably not so coastally focused as to exclude them from living along rivers or inland, which east of Java is never very far from the coast. Watercraft made these nomadic 'littoralists' very mobile, able to move freely and they may not necessarily have lived in one place for very long. They would have been able to navigate coasts and rivers, moving in on others with variable success, while using shallow bays to shelter, regroup and perhaps settle for a while. Mobility meant comparatively rapid movement. It would not have required thousands of years to travel down the Sunda coastline, although the first law of migration at that time was probably to keep in touch with others – total isolation is dangerous.

Jungle is not conducive to migratory movement, so coastal travel may have been the favourite mode of movement, with forays into denser undergrowth only undertaken when required. This would also have been part of the behaviour of those exploring and living on river systems. It is also likely that they did not develop their basic coastal adaptation and navigation skills as they migrated; rather, these would have had their origins in their original homeland somewhere above latitude 16° north. Migratory experience, however, was cumulative and it enhanced their basic capabilities.

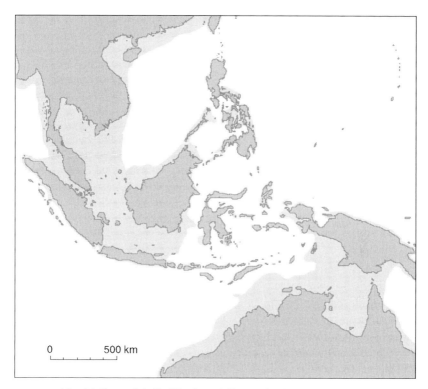

Map 3.1. Exposed shelf of Sunda at −145 m sea level.

The west coast of Sunda

When Sunda existed its east and west coastlines met on the east coast of Bali. The western coast of Sunda was the shortest. Beginning at the Irrawaddy delta, it included the Gulf of Martaban then went south, following the Burmese (Myanmar), Thai and Malaccan coasts as far as modern Kuala Lumpur, where, at sea levels below −40 m, it turned southwest, linking up with northern Sumatra. The length of the coast changed as sea levels fluctuated, so that at −40 m it was 6000 km long and at −130 m shortcuts emerged to reduce this to 5200 km. Distances have been estimated by measuring the coastline at these levels and adding 10 per cent for detailed coastal contour. The difference between them is not significant and reflects the relatively minor changes in west coast morphology at times of low sea level. Similarly, the south coast of Sumatra was largely unaffected by glacial sea level change. Here, the continental shelf is very narrow and quickly dives into the Indian Ocean. The major difference on the west coast was exposure of the Strait of

Malacca and the Gulf of Martaban, but this occurred only when the seas dropped below −100 m. At these times a short cut from the Thai coast across to Sumatra would have been possible, but it only existed between 142 ky and 160 ky and again during the last glacial maximum. Half way across the Malacca/Sumatra bridge was the outlet of the Malacca Strait River System (Voris, 2000) (Map 3.1).

Sunda's west coast was much straighter than the east coast, thus providing a more direct route south. It was essentially directed north–south in an L-shape, which curled out eastwards towards the Lesser Sunda archipelago. In the north, the Andaman Sea coast is permeated with many small islands, making a complex bathometric contour and a comparatively sheltered and pleasant environment for those travelling the area by small watercraft. The shape of the Malay coast further south would have drawn coastal navigators into an increasingly narrowing Gulf of Malacca. The coastal contour here naturally turns coastal migrations around to face northwest to continue their journey along the north Sumatran coast. An alternative was to follow the Malacca Straits River System southeast which would have taken people into the heart of southern Sunda. Much of the western Sumatran and all the southern Javan coasts face the Indian Ocean and are subjected to high energy seas as well as strong currents and rips. For those continuing along this coast such conditions would have made coastal cruising difficult and landfall hazardous. Another hazard comes from tectonic activity along this coastline, which can produce tsunami that can penetrate far inland.

The east coast of Sunda

Sunda's east coast is very different from the west coast. The major difference is that its length varied considerably with sea level change. At −150 m the coastal length was around 6200 km, but at −40 m the distance almost doubled to around 11 500 km, which would have made the journey twice as long at a given rate of travel. The shape of the east coast would have changed radically with quite small sea level adjustments and inordinately more land was added at times of low sea level than on the west. This convoluted coast would have provided a greater choice for semi-permanent settlement, thus tending to slow migrations moving along it and providing much more area for exploration and exploitation.

Any migrations out of China could either travel inland or down the Chinese and along the Vietnamese coasts, turn west, then northwest into the Gulf of Thailand. Turning west again, then south, they would follow the Thai and Malaysian peninsulas before turning east and crossing the South China Sea roughly at the Equator. At sea levels of −75 m, however, a short cut was possible across the Siam Plain (Gulf of Thailand). There, the Siam River System flowing from the north filled two large lakes and swamp areas before picking up tributaries of the Kampar and Johore rivers and emptying at the edge of the exposed Sunda shelf (Voris, 2000, Map 3.4). Continuing along this coast would take them northeast and, after crossing the 12 km wide Balabac Straits, on as far as latitude 12° at the north end of Palawan. At this point, it is questionable whether migrations would have continued to follow the coast or whether they moved across the Mindoro Strait and into the myriad of 7500 islands that make up the Philippines. Before reaching the Philippines, however, the North Sunda River System could have drawn people south into South Sunda to the area where those moving down the Malacca Strait River System would also have ended up. The complexity of migrational possibilities through this region could have swallowed up many migrations and equally as many people and cannot be dealt with here, but it does not stop there. Another large river complex, the East Sunda River system, could have carried these people out of southern Sunda's heart and placed them on the east coast close to Bali.

In Sunda's extreme northeast, coastal migrations would naturally arc around the Sulu Sea. On the eastern arm, the Sulu Archipelago is likely to have attracted those venturesome enough to make the 40–50 km wide crossing of the Sibutu Passage. From there a series of extremely short, island hopping exercises would have taken them to Mindanao via Jolo and the Basilian islands. This southern entry to the Philippines was shorter than the northern route across the Mindoro Strait and may have been preferred after exploration of the area. As with all these proposals, however, the decision to take one route over another by people who were completely unfamiliar with the regional geography is probably one that lies more in the mind of the writer than with the will of the ancient mariner. It is obvious that to know the 'shortest route' is to know that all other routes are longer, a doubtful possibility for first timers.

Continuing south brought people to the southwestern corner of the Celebes Sea and the first of several places that could have been stepping off points into Wallacia. Cape Mangkalihat provides a jutting approach to the Wallacian island of Sulawesi, but this crossing was never less than 100 km even during sea levels of −150 m. Cape Mangkalihat marks the beginning of the northerly migratory route proposed by Birdsell (1977). It passes through a series of islands stretching eastward towards New Guinea, but, with an average interglacial sea level of −40 m, it seems more likely that it required a 60 km crossing throughout the 60–180 ky period envisaged for Australia's first arrivals. A shorter distance offered a safer crossing, but, with the possibility of success over this distance, we must accept people could have negotiated the southern entry into the Philippines.

There are several important differences between the two coastal routes. Firstly, although the west coast is shorter, it is more dangerous. Secondly, it does not offer any stepping off points to Wallacia until the end of the voyage. Thirdly, west coast migrations around 75 ky would have been directly affected by the enormous Toba volcanic explosion in northern Sumatra that occurred around that time. I would suggest that anybody within 200–300 km northeast and northwest of the volcano would have succumbed to debris, ash and gases (see below).

Although twice as long, the east coast was by far the safest and most opportunistic way of reaching eastern Sunda. It would offer many more places for settlement and would have distributed people throughout the area. But it would have taken far longer to reach the Bali coast and may have naturally brought them to places where further eastward spread towards Sahul could begin without the need to move along the Indonesian island chain to continue on to Australia. Landing on the western end of Papua New Guinea at times of low sea level was in fact a landing on Australia as the two were joined. East coastal migrations could also have been sidetracked away from a generally eastward movement by other possible excursions; one of these was entry into the Philippines. If we assume that when migrations reached a suitable place they took the opportunity to cross into Wallacia and continue their journey towards Sahul, then equally they could have pushed their way into the Philippines. The Philippines would have acted as a sort of sponge, probably soaking up many people who had about 375 000 km^2 of uninhabited land surface to explore among the thousands of islands.

Previous chapters proposed the development of major population growth centres in India and China during the Pleistocene and outlined how these may have contributed to various migrations out of and within Asia. One of these was a southerly flow into Sunda and on to Sahul. The section above described how this might have taken place using two coastal migration routes, so it would have been possible that coastal migrations from India moved along the west coast and another from the China region would have naturally adhered to eastern routes, both remaining discrete migrations in terms of mixing till the respective gene pools moved far into southern Sunda or had even reached Sahul itself. The two routes kept the migrations separated, but it is worth considering that the adventurous human nature could have produced mixing at various stages in this process. The overall simplicity of the model, however, belies the natural complexity of the actual migratory processes, but I believe it serves to present the principles of migration across the region at that level.

Inland routes

Coastal travel would have brought people into the estuaries of the large river systems mentioned above. Moving up these would bring them into extensive catchment areas. When both inland and coastal migration is contemplated, the picture of the possibilities for genetic intermixing becomes very complex indeed and the pattern and direction of migration infinitely more difficult to predict and describe. The routes people might have taken are almost endless, therefore any discussion must accept that the general direction of travel was always southeast. Another factor worth considering is that migration routes in the Upper Pleistocene were probably determined by difficulty of access. It is only human nature that places offering fewer obstacles must have been favoured over those where the going was difficult or dangerous.

Changing sea levels controlled the availability of the landmass upon which people lived and moved. During low sea levels, a mix of grassy plains, rainforest and gallery forest were found along the river channels that spread across the exposed shelf. Increased landmass meant increased biomass as a vast array of animals expanded their range and proliferated. In turn, increases in animal numbers would have attracted and fed the new migrants. During glacial episodes, tropical climates in southeast Asia generally became drier, savannas increased and rainforests shrank. It is likely that these changes were less pronounced closer to the Equator, where little change occurred between glacial cycles. There was also an increase in the availability of fresh water through the newly exposed catchments and the extended river systems reaching out across exposed continental shelves. Low-lying areas on the shelves became swamps or large lakes; two examples of these probably existed in the forms of very large lakes in the areas of the Gulf of Thailand and western Java Sea (Butlin, 1989). The natural abundance of birds and animals in such niches would also attract humans. How these ecosystems were then exploited depended upon the technological abilities of the group and the numbers of people that focussed on them. These two factors constantly changed throughout the Upper Pleistocene and must have been the most important influences on the development of regional human populations during that time.

But how did sea level fluctuations affect human populations in the area? As continental shelves became exposed, people and wildlife would have moved on to them and spread out. In some areas sea level change must have been almost imperceptible, while in others perceptible change took place within a few years. In areas of long-term change, new populations would have settled down, accepting the area as the natural situation. In areas where rapid change took place, adaptation to the changing pattern would have been required, resulting in moving people on quite quickly. When continental shelves began to sink below rising seas at the end of each glaciation, people and animal populations were slowly squeezed out and on to higher ground. Naturally, those already living in the latter regions would have new neighbours a process producing higher density populations per square kilometre. A natural mechanism to relieve this situation would have been migration into unpopulated areas or into those with fewer people in them. The whole process acted like a pump, allowing population flow into empty areas at low sea levels, only to squeeze them out and into smaller areas at high sea levels. It is well not to forget, however, that living on continental shelves was the norm, because to some degree they were exposed for 80–90 per cent of the Quaternary. Inundation at today's levels only occurred for about 20 ky of the last 160 ky. When driven off shelves, there is either the possibility of increased conflict as populations come together and have to compete or there is a need to migrate away from bottlenecks. We can see that as seas rise and fall, so populations expand and contract. Add this to natural growth of world population and it is easy to envisage a pump that continually pushes out people at the time that seas reach modern levels. There are also other reasons to migrate, such as disasters.

Toba: a natural disaster

Above and in the previous chapter, I briefly mentioned types of natural disaster that might have overtaken Pleistocene populations. Major disasters probably played a decisive role in temporarily checking the growth of early human populations, particularly at the local group level and those living in the Lower and Middle Pleistocene. Larger Upper Pleistocene populations were more able to withstand and/or recoup numbers following such events. While attacks by carnivores may have been a constant hazard, they would not have had the effect on small populations that extreme climatic conditions and tectonic and volcanic activity could have wrought. These would have occasioned the loss of significant numbers of people, particularly at the local level, where the loss of hundreds of individuals may have had a significant effect on short-term population growth and migration. Sparse populations would have been both advantaged and disadvantaged in that fewer people would have succumbed, although small populations were, at the same time, extremely vulnerable to such events and liable to extinction. In Chapter 2, I described how the loss of females would have been of particular concern to these groups. Having introduced this theme, I want now to outline one natural disaster that must have caused some considerable impact across a large section of Sunda during what may have been a significant time of human movement through southeast Asia in the Upper Pleistocene. This event was the eruption of the Toba volcano.

Lake Toba is crater lake, measuring 100 km by 35 km, lying just above latitude 2° north in central, northern Sumatra (Map 3.2). It is a resurgent caldera that exploded around 75 ky producing the largest volcanic eruption recorded in the Quaternary (Ninkovich et al., 1978). Not even the massive size of the caldera, however, can convey the enormity of the explosion and consequent power of the eruption. To give some idea of the size of this event, it is worth documenting some of the associated statistics.

The force of the Toba explosion is estimated to have been at least 3000 times greater than the Mount St Helens eruption in 1980 and may have been 6000 times as great. It was somewhere between 20 and 40 times the magnitude of the Krakatau eruption in 1883 and 30 times that of the Santorini explosion that destroyed the island of Thera in the Aegean Sea around 4 ky (Rampino et al., 1988). The height of Toba's eruption column may have reached 80 km, at which time the magma discharge rate has been calculated at 10^6 cubic metres per second with a total discharge of about 2800 km^3 of pyroclastic rock compared with 13 km^3 for Santorini (Ninkovich et al., 1978). The eruption column lasted between 9 and 14 days, resulting in a rhyolitic ignimbrite tuff layer covering over 20 000 km^2 surrounding the caldera, although this is closer to 120 000 km^2 according to a map of the affected area (Ninkovich et al., 1978). The distribution of the ash layer takes the shape of a huge fan, stretching out northwest from Sumatra and northeast into the South China Sea (Map 3.2). It covers a minimum of 5×10^6 km^2 of the Indian Ocean between Sumatra and India, most of the Bay of Bengal, Sri Lanka and stretches as far as northern India (Francis, 1993). Toba ash has also been found in the Vostok ice core from the Antarctic. The thickness of the layer within the fan has been measured at 30 cm and 10 cm, 1000 km and 2500 km from source, respectively. No oceanic tephra from the Krakatau explosion has been found even 200 km from source. It is interesting to speculate that had the ash cloud been driven in the opposite direction, Australia would have had a tuff layer that might have played a significant role in dating early and, no doubt controversial, archaeological sites, particularly those in the north and northwest of the continent.

Toba's statistics are impressive in their own right. The possible implications of this spectacular and devastating eruption for humans living in the region, however, were immediate and lasting, at least in the short term. The immediate consequences are quite straightforward in that a sudden eruption of such magnitude would have provided little or no prior warning to those living within the 20 000–80 000 km^2 immediately surrounding the Toba volcano. Even if some pre-eruption disturbances did take place, it seems unlikely that

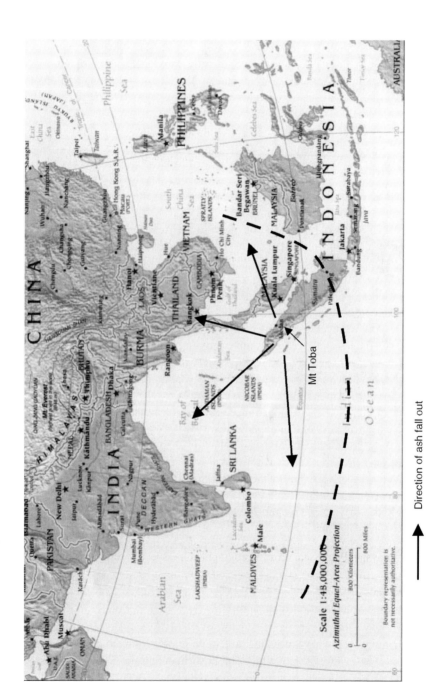

Map 3.2. Boundary of ash fan from the Mt Toba eruption.

these would have attracted anything more than cautious glances and a flurry of explanations possibly invoking the supernatural. The dynamics of the eruption strongly suggest that it was not a gradual event, neither did it consist of comparatively quiescent lava flows that could be avoided. A pyroclastic lava flow, identical to that from Toba, can:

> ... sweep over 100 kilometres from its source at more than 100 kilometres per hour, depositing a layer of pumice many metres thick over thousands of square kilometres. An area the size of Belgium could be obliterated in a few minutes.   *(Francis, 1993: 294).*

If this sort of aftermath followed an explosion big enough to create a caldera of 3500 km^2, the entire population in the immediate vicinity would have been obliterated. Other populations living throughout the region must have succumbed also. Moreover, suffocating ash, rock-fall, acid rain and gas clouds would have brought about a rather sudden and choking end not only for humans but also animals right across the Sunda peninsula from coast to coast, and food stocks would have been destroyed in the face of such an enormous disaster.

Stone tools of the Tampanian industry pre-dating Toba have been found in the Perak Valley of central Malaya underlying the ash layer (Collings, 1938; Walker and Sieveking, 1962). Although early interpretations suggested that these artefacts had a Middle Pleistocene origin, recent interpretations have proposed a closer association with the eruption (Harrisson, 1975). People were obviously living in the region before the eruption, but there is little to directly associate archaeology with the actual timing of the event. In light of the enormous amount of time hominids have lived in this region, it is worth noting that the 75 ky eruption was the last of four eruptions that occurred during the last 1.2 My each at 340–400 ky intervals (Francis 1993).

The area of Sunda affected by the Toba eruption is small compared with the total area covered by ash-fall. Of the five million square kilometres under ash-fall, only 350 000 km^2 of the modern land mass in this northwest corner of Sunda was involved. This includes the northern half of Sumatra and the western half of the Malaysian/Thai peninsula. As we have already seen, the size of the Sunda peninsula at 2–4° latitude varied enormously during the Upper Pleistocene. Around 75 ky sea levels rested at −70 m, joining Sumatra and the Malaysian Peninsula. From a geographical perspective, although more land was available, a given number of people were more closely confined at this time than they might have been during times of lower sea levels. Neither were they confined to the present coastal boundaries of northern Sumatra. The latter is significant because the tuff layer reached either coast, producing secondary explosions from pyroclastic flows entering the sea and cutting off northern Sumatra from the south. We have no data concerning the movement of poisonous gas clouds, except to suggest that they would have probably blown in the same direction as the ash-fall, but they would also have had a devastating impact on the people living in the region. The ash fan crosses the Thai coast at about 9° north and extends in a southeasterly direction as far as 1° below the Equator in central Sumatra. About 2700 km² of the west coast of Sunda was affected by ash-fall and this would have been severe because of its proximity to the source. The result would have effectively closed the region to movement, probably for some considerable time. The eastern coastal strip of Malaya would have been the only region on the Sunda Peninsula safe from the effects of the Toba eruption; the long, open U-shaped arc describing a boundary that any prospective migrations would keep east of. The shape of the ash-fall area would further encourage a southerly movement away from any lateral drift of volcanic debris and gas clouds.

The secondary effects of the event may have been as bad as the immediate effects. Historical accounts of the climatic and environmental effects of volcanic activity are well documented (Rampino et al., 1988). There is little need to describe these here, but the results of large volcanic explosions have imposed appreciable influence on the world's climate even in areas thousands of kilometres from the source. There can be little doubt that the largest explosion in the Quaternary was not only spectacular but must have had a very significant impact on world weather patterns for some time. It has been suggested, for example, that a 'volcanic winter' may have followed the eruption and it may have lasted six years (Rampino et al., 1988). Sulphur, in the form of aerosols, injected into the upper atmosphere, is the key to these climatic changes. Aerosols block the Sun's incoming radiation, thus cooling the lower atmosphere while increasing stratospheric temperatures through back reflection. Sulphur-rich magma causes the largest volume of atmospheric aerosols and the greatest amount of lower atmospheric cooling. The higher these are injected into the stratosphere and the longer the eruption continues, the greater the long-term effects. But even with comparatively sulphur-poor magmas, long eruptions with high columns may cause similar effects.

Although the Toba magmas were sulphur-poor, it has been estimated that between 1000 and 5000 million tonnes of sulphuric acid aerosols were injected into the stratosphere.

The atmospheric after-effects of a Toba sized explosive eruption might be comparable to some scenarios of nuclear winter, although the aerosols are expected to have a longer atmospheric residence time than would the nuclear winter smoke.

(Rampino et al., 1988: 90).

Besides atmospheric cooling, a high level of stratospheric dust pollution reduces the transmission of sunlight. Even if only 10 per cent of the estimated

20 000 megatonnes of fine ash exhaust from Toba entered the upper atmosphere, total darkness could have lasted weeks (Rampino *et al.*, 1988). At best, ash cloud pollution from such an event would have caused sunlight strength equivalent to an overcast day. At worst it would have been somewhat brighter than a full moon, dark enough to prevent photosynthesis.

It is difficult to predict what sort of effect a super-eruption like Toba had on the behaviour, migration and settlement patterns of people living in the region. Secondary effects would undoubtedly have lasted many weeks, if not months. The terrors of extended ash-fall and semi-darkened days must have induced a natural response to escape, given one knew which way to go. Mainland areas covered by pumice and ash would have been avoided because of the inhospitable environment and the complete lack of plants and animals, and it may have been decades before people ventured back into these. Migration routes would have been created away from the disaster area, thus avoiding the desolate landscape. Perhaps myth and superstition took over to explain the disaster and the ravaged terrain that resulted. These may have extended to avoidance of these particular places, leaving them uninhabited until almost all trace of the eruption had disappeared and the stories themselves long forgotten.

Clearly, this rare event must have caused local demographic changes with people avoiding and moving away from the area, but it could also cause severe disruption to populations living much further afield. It may or may not be a coincidence that the Toba eruption occurs at the end of a warmer interstadial and precedes a fresh glacial advance because there is 'excellent correlation between glacial phases and preceding episodes of volcanic activity' (Williams *et al.*, 1993: 26). Claims that this disaster produced a population bottleneck, however, must be greatly exaggerated. Certainly it did nothing to disturb a population of Homo floresiensis, living almost next door to the Toba explosion. It could, nevertheless, have had a far-reaching affect on populations living and moving along the Sunda Peninsula; its enormous scale could not have been ignored. This was not just the greatest volcanic explosion in the Quaternary, it was also the greatest volcanic explosion ever experienced by humans, and it happened during a critical stage in human movement to Australia.

Sea crossings and routes to Sahul

The mere mention of the first sea crossing to Australia can induce mass yawning among some colleagues who may argue also that yawning is better than arm waving, which, they would also argue, is a common trend among those interested in this topic. While we have little empirical evidence for the way in which the formative migrations to Australia took place, there is always room for regular review, the incorporation of any new evidence, meagre as it might be, and modification of the theoretical framework that describes these events. Moreover, there are enough logical possibilities and finite components in this story to justify their examination again.

The first landings here could be viewed as a further step in the accumulation of a set of intellectual, manipulative and technological skills that when combined allowed people to cross substantial distances of open ocean. The question remains: did Homo erectus use a steerable craft to cross to Flores, because you would certainly need one to reach Australia. So, the journey marks the end product of adaptive and cultural development that had been maturing and used on and around the southeast Asian continent during previous millennia. This argument might be supported particularly by those who feel that these oceanic explorers were Anatomically Modern Human groups, completing a long journey that had begun in Africa at least 50 ky before. This is a neat story, but was it the case?

The oldest evidence for humans in Australia stands at 60 ky for a habitation site at Malakunanja in Arnhem Land, northern Australia (Roberts *et al.*, 1990), Despite some blaming termites for stirring up the sediments from which the T/L dates were taken, Malakunanja is very convincing evidence for humans living 200 km inland from the coast at that time. The oldest dated human remains in Australia are those of the WLH3 skeleton from Lake Mungo in western New South Wales (Bowler *et al.*, 2003). At one time it was thought that the earliest date for humans in Australia was automatically the oldest for the use of water craft. The evidence from Flores put an end to that, but the earliest Australian date would still mark the first movement of people across very large water gaps even at low sea levels. That trip began an even more lonely, dangerous and overwhelming chapter in the process of human endeavour: trans-oceanic exploration. These people were, indeed, the first boat people moving into Australia, but whether they were fleeing for their lives as more recent arrivals have is another matter.

The venture in itself provides us with a measure of the capabilities of those involved as builders, navigators and explorers. The skills required are almost never considered in discussions of their contemporaries living in other parts of the world. Although not considered fully modern humans by some, were Neanderthals capable of navigation, either coastal or open ocean? There could have been contact between North African and southwest European populations, but we cannot infer this in the same way we can for those who sailed to Australia till a Moroccan Neanderthal turns up (Simmons and Smith, 1991). Ironically, of course, we have considerable skeletal and cultural evidence for

Neanderthal groups, whereas we have almost nothing of the sort for Australia's mysterious seafarers (Stringer and Gamble, 1993). Therefore, unlike Neanderthals, we can only guess what these early sailors may have looked like; how tall they were; whether they were big or small, muscular or slim. The tool kit they used to build their craft as well as their culture remains a mystery. So, is it possible that some of the old petroglyphic art, pecked into Australian rocks in southern and central parts of the continent, disowned by modern Aboriginal people, is a later expression of the cultural baggage of these first arrivals? Did the complex burial practices and ceremonies of some of the Lake Mungo people have their origins on Sundaland or were they developed in Sahul? Finally, were the very first arrivals pre-modern people? This question is the crux of arguments concerning the 'multi-regional continuity' model and Out of Africa hypotheses for anthropologists and archaeologists in this country.

Even if they had some raft-building abilities, it is unlikely that the erectines that made crossings to Flores were cognisant of the need for and advantages of food and water storage if they were to make ocean crossings. But am I underestimating their capabilities again? On the face of it, it seems unlikely that either by accident or intent they did make it to Australia, because inter-island sailing is a far cry from sailing to Australia. It is also difficult to believe that erectine people, and archaic *sapien* groups such as *Homo soloensis* did not reach the end of the Indonesian archipelago and even move on to Timor. The dwarfed appearance of *Homo floresiensis*, however, suggests strongly that this individual did not wander far from its home on Flores island. Certainly, a continued successful journey to Sahul required a safe, purpose built craft and enough of these would need to complete the journey so that even with 50 per cent losses a successful landing would eventually be accomplished. Middle Pleistocene hominids have always been denied the intelligence to manipulate their world and manufacture a material culture that has any sort of sophistication, however minimal. This idea has been overturned with the discovery of *Homo floresiensis* of course (Brown *et al.*, 2004 and Morwood *et al.*, 2004). I believe this error of judgement has severely hampered our ability to open up discussion of the movement of Indonesian hominids to Australia before the appearance of those with distinctive skills from other regions. Given enough time and attempts and an allowance of another 500 ky from the time of their first voyage to Flores to develop their sailing skills, it is in the realms of possibility that the erectine determination to explore and survive, attested to by their presence right across the Old World, could have put them on Australian shores before 300 ky.

The greater the sophistication of the craft (steering, sails, etc.), the more navigable it would have been and the longer it could stay at sea, reducing the need to take the 'shortest' route, whatever that was, as well as increasing the choice of routes open to them. Moreover, with sails and steering boards, there is little point in discussing prevailing winds and sea currents as a factor in cross-correlating a 'most likely' route to Australia with the season when it took place. The level of craft design did not need to be too sophisticated, however, for people to move just about anywhere. In turn, this makes the pattern of possible migrational routes very complex and not necessarily logical. It is possible that the odd castaway may have made it here, similar events have been recorded in recent times (Calaby, 1976). But these events cannot seriously be considered as providing enough people to establish a viable population. Logs are not a very safe means of travel across open sea. They are unstable, not suitable for a sail, cannot be steered properly, food and water cannot be stored and, when saturated, lie low in the water where submerged, dangling limbs make fine lures in the warm but shark infested and crocodile patrolled waters between Indonesia and Australia. Bamboo is the best material for such craft and there is plenty of that in Indonesia and Wallacia. Bamboo logs are hollow, comprising a series watertight natural floatation chambers. Strapped together as a flat platform, with several compartments used to carry water and dried food, they would allow for the transportation of several people. Those with the skills to live off the sea, as these people undoubtedly had, would not have starved while at sea. An adequate, steerable water craft allowed a crossing at any time, whatever the sea level. It is also reasonable to assume that the earlier the crossing, the less sophisticated the water craft. Small craft would have carried few people and were probably less successful.

While very low sea levels were an advantage to people coming to Australia, they may not have been essential in the migration equation. Even so, with the exception of 10–15 ky around 120 ky, sea levels were always lower than modern levels from about 180 ky to 8 ky, although no one would claim that people waited for them before setting off. All voyages to Sahul had to cross the permanent Wallacian and Weber Deeps that divide Sunda from Sahul. Such voyages were deliberate acts specifically aimed at oceanic venturing, where people could not see their destination and, once out at sea 20 km off shore, they could not see their departure point. It seems to us now that they almost knew Sahul was there and that was their destination. Some sailings may have been initiated and/or guided by smoke from massive, naturally occurring bushfires either on Australia, Papua New Guinea or on the distant continental shelves when exposed (Dortch and Muir, 1980). More realistically, however, the voyages, while they may have been deliberate, were probably random events.

The fact that the Sahul shelf jutted out towards Timor made the journey somewhat easier than it would be today. At sea levels below -60 m, the wide

Arafura Plain straddled the divide between Australia and New Guinea. In the middle of what is now the Gulf of Carpentaria stood the enormous fresh to brackish Lake Carpentaria, which varied from 28 000 to 165 000 km^2 (Torgerson *et al.*, 1983). Both these features occurred at 55–62 ky, 68–80 ky, briefly around 112 ky and from 138 ky to 168 ky. The shallow bathymetric contouring of the Arafura Sea would have caused rapid shape changes to the coastline with only small changes in sea level, while at times much of it disappeared. In turn, these changes affected landing placements, the ease and direction of movement across the plain, the direction of routes inland and, consequently, points of entry into Australia and the subsequent pattern of colonization that followed.

The concept of 'shortest distance' has often been invoked to show how quickly and easily Australia could have been reached and by what route. Moreover, it emphasises lowest sea levels and thus the best time for people to have arrived. Arguments state that migrations used the shortest route, between islands and to Australia, while taking advantage of low sea levels as optimum voyaging times to shorten the journey. The shallow submarine contours of the Australian continental shelf allowed the emergence of dry land, sometimes with minimal drop in level. It can be argued also that oceanic reductions apply more at either end of the journey, where, as sea levels dropped, the wider continental shelves of Sunda and Sahul moved out to meet the nearest islands. The distance between inter-Wallacian islands not perched on continental shelves was not reduced by much because of their steep submarine contours. Similarly, minimal reduction in the distance between the eastern and southern Irian Jaya coast and nearby islands during transgressive events was small enough not to affect the chances of successful crossings there. Lowered sea levels did produce shorter crossings in some places and there is little doubt that lowered ocean levels did optimise voyaging, particularly when seas dropped to -145 m around 150 ky. This marked the first optimum voyaging time in the Upper Pleistocene. I have stated before that it is unlikely that people knew what the shortest route was or that sea levels were at their optimum. The chances of randomly picking the shortest route to Sahul is extremely slender and the concept of the 'shortest distance' was probably never in the minds of the voyagers, if for no other reason than they were unaware of any sea level that was not prevalent during their lifetime. Moreover, if inter-island distances during low sea levels could be overcome, then the 10–30 km added to the journey on some legs during times of higher level would have been of little consequence. Having said all this, I want to discuss the shortest ocean crossings to Sahul during low sea levels because before 60 ky high sea levels only existed between 115 ky and 132 ky. Another factor in oceanic migration is point of departure. Adding this makes the issue infinitely more complex, and our ability to predict migration routes takes an inordinate leap towards the impossible. But it also takes on a more fascinating aspect and one that presents a unique challenge.

With a suitable craft, direct entry into Australia would have been possible from Timor. A direct crossing from Timor's south coast to the Sahul Rise, a line of limestone rocks off northwest Australia, was at least 160 km (Mulvaney and Kamminga, 1999). A group of continental islands (Sahul Banks) formed a line 110 km from Timor and were also exposed during low sea levels. Staging here would split the crossing into two legs of 110 km and 62 km. Except for catching the breath, the Sahul Banks would not have offered much in the way of food and water resources, however. The first leg is not impossible, it is only five times that achieved by *Homo erectus* 800 ky before. Crossings would have been more successful, however, if they took place further north, where the island-hopping distances from Aru, Ceram, Tanimbar and Halmahera islands to Sahul are much shorter. So these are the two best guess points of departure.

How many?

If optimum sea levels were important for migration, then the first people arrived around 150 ky. But even a successful crossing need not necessarily have signalled the beginning of Australia's human populations. The chances of individuals blown away from the Indonesian coast on some form of flotsam and arriving safely on the Australian coast are very slim. The logical extension of this argument is, of course, that *Homo erectus* might have reached Australia in a similar manner. A lot of accidental castaways could arrive over tens or hundreds of thousands of years, but it does not necessarily mean that a viable population grew from such occurrences. Viability is the key to understanding initial colonisation. It is not just arrival, but survival.

Certain conditions needed to exist before a viable population took a permanent foothold. One of these is a requirement for the right number and mix of individuals to begin the process. Even with a successful voyage, a lone sailor would eventually be doomed to die alone. It also seems unrealistic that a landing by one or two rafts would begin a viable population. Many dangers faced a very small founder group, least of which in Australia at that time was one creature that would have been very difficult to live with, *Megalania prisca*. This giant relative of the Komodo Dragon was the largest reptile since *T. rex* and if it hunted in packs, as the Komodo Dragon does, then small human groups would have been very vulnerable. A seven or eight metre *Megalania* had no enemies and is something that only two rafts of people are unlikely to have survived if attacked by packs of these animals. Many other hazards faced such a small group, but the best chance to develop a thriving population comes from a series of closely associated landings occurring at or around the same time.

The rules discussed in previous chapters for erectine migrations are now broken for the first time, because, as opposed to land-based migrations, those taking place across large expanses of sea do not necessarily keep in touch. For example, it is most unlikely that a steady stream of rafts made this journey. Therefore, migration would have occurred in stochastic bursts, where, firstly, the migrant group or groups were not backed up by others and, secondly, they are not guaranteed that they can keep in touch with each other on the open sea when sailing for days at a time. The level of danger from migration now reached a new high. There is little doubt that the underestimation of technological and behavioural skills of past human groups has on many occasions clouded our ability to understand and interpret prehistoric events. Moreover it has made us totally unprepared for discoveries like that of *Homo floresiensis*. Oceanic migration forces us to recognise the sophisticated skills required, and the earlier such voyaging took place, the farther back those skills were in place. It strongly suggests also that those first boat people had survival skills that cannot necessarily be measured by examining their tool kits (whatever they looked like). Raft building using bamboo, vegetable twine/rope and bamboo knives would leave no trace in the archaeological record. The challenge before these people was not craft building, it was survival, with enough numbers that could produce a viable population.

The numbers of people involved in these first migrations will never be known, but the question of how many people were needed to begin Sahul's population has drawn many answers. One has suggested that it was not necessary to invoke large flotillas of rafts landing on our northern shores to start the population: 'Perhaps even one young pregnant female was sufficient to do the job' (Calaby, 1976: 24). It might be theoretically possible for a pregnant woman on a raft to have begun a continental population, but it seems extraordinarily unlikely in my view when we consider the dangers of living on a large continent and facing predatory animals without any experience of humans. It was once thought prudent not to invoke anything more than one or two people moving about the globe during the early Upper Pleistocene, because, of course, the world's population was so extremely small. The whole tenet of this book is that this idea is no longer acceptable, that many people were around at that time and many of those were on the move.

The use of the success of contemporary small island populations to explain the possibilities of population success arising from a dozen or fewer people is not a useful paradigm for this case. Discussions around the early arrival and subsequent populating of Australia have referred to the demographic experience of islands such as Pitcairn, Furneaux and Tristan da Cunha, as well as from ethnohistorical reports of isolated Aboriginal people living in remote or uninhabited regions around Australia (White and O'Connell, 1982). The long-term genetic consequences of these events are hard to envisage, but certainly adaptively narrow. The desperate extrapolation of these rare and unrepresentative examples to explain the origin of Australia's population is to deny the deliberate behaviour and demographic flux of the Upper Pleistocene and the abilities of the people that lived at that time. However, I am not discounting the initial success of the small group that is later joined by others.

Watercraft size must have limited arrival numbers to two or three families spread across three or four rafts. Infrequent crossings would produce a very slow trickle of people, particularly if one family or a few individuals were involved. This would reduce survival chances and the viability of the founder group, particularly if some craft were lost at sea, became separated and blown into isolation or if individual members of the group perished. Alternatively, there may have been a continuous flow, but this would depend upon the size of the original or regional population; the urgency to move; the whim of those involved, numbers of backup migrants and the time that migration began. Obviously, more people were available to take part in Upper Pleistocene migrations than would have been possible in earlier voyages. The success of voyages depended also on craft construction, route chosen, weather and sea conditions, and all these would depend on timing. Before continuing we have to return to basics. Who was it that was making the journey and what was their story?

To set the scene for the first people we have to consider who they could have been: initial human migration into Australia points to four possibilities. One, they were descendants of Javan hominid groups; two, they were a mix of modern humans; three, they were associated in some way with *Homo floresiensis*, or four they were not any of these but *Homo sapiens* from elsewhere in southeast Asia. Complex assortment would be one of the major genetic consequences of the movement of people through island Indonesia. The coming together of two populations with quite different ancestries would make sorting out who the first voyagers were even more complicated. In this case, we have to consider the mixing of archaic and anatomically modern humans along the archipelago. Obviously, that would be a mix of genes that had little in common with one another, with a common ancestor stretching back some 2 My earlier.

Regional Upper Pleistocene migration was probably multi-directional and relatively slow. Longitudinal migration (migration over a long period of time) conveys and distributes genes and, through adaptation, produces new ones and founder effect can produce new variation. The direction of migration is not necessarily the direction of gene flow, which moves in many directions including back up migratory routes. Movement into large, uninhibited landscapes initially isolated people, perhaps in some cases for long periods of time with the likelihood of multiple founder effects. Some populations were then locked up with their own founder effect and genetic drift to establish themselves in their own particular fashion. This continually changing situation has to be viewed against the vast background of time. Time contributes substantially to the variability of genetic frequencies and the emergence and proliferation of types through the five basic evolutionary mechanisms of drift, mutation, founder effect, local and general selection and adaptation. These processes in turn produce biological and cultural fluctuation, particularly if two or more gene pools from widely separated geographical areas are brought together. This could lead to an increased morphological variability among the resultant population. To illustrate what I mean, it is worth considering the possibilities for population movement, gene transfer and mixing over 15 ky. The chance of substantial change in gene frequency over this period is very good, but that period may only lie between 70 ky and 85 ky years, for example. This is a comparatively small time frame within the Upper Pleistocene and a temporal slice often difficult to delineate exactly using contemporary dating techniques. Therefore, however much we may try, we cannot view gene transfer in anything but the coarsest manner over even a comparatively large time span. If the picture of migration is complex over the short term imagine the genetic consequences of this process over 30 or 50 generations! When these factors are seriously considered, the resultant complexity of the permutations of human variation should make us question our ability to construct hominid lineages and ancestral affiliations from examining a limited fossil record assemblage spread across vast distances, and periods of time, and often lacking precise dating. The recent discovery of a completely new human species, the first in 100 years, is testament to this (Morwood *et al.*, 2004).

An infinite complexity

To try and demonstrate the picture I envisage, I want to introduce a base migrational model that can be viewed in three ways: statically or simply, temporally, and spatially and environmentally. It involves six elements, including four populations, two genetic mixing points and a resultant migrant population that migrates away from Sunda to Sahul (Map 3.3). The two migrant populations, 1 and 2, derive from the large population centres in India (1) and China (2). They may or may not consist of people born in those

102   *The First Boat People*

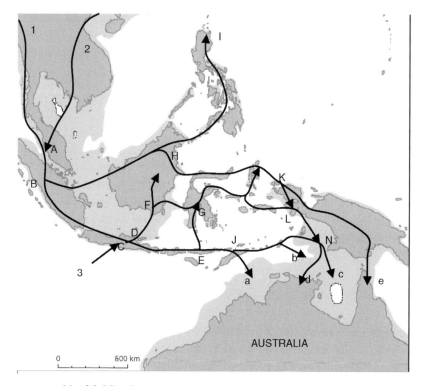

Map 3.3. Migration patterns and genetic mixing through island Indonesia.

areas but represent migrant groups with direct genetic affiliation to them. Mixing point (A) is an arbitrary point that could be placed anywhere across that part of the Sunda neck and indicates a possible first mixing of populations 1 and 2. At low sea levels they may not even have met each other, instead following the east and west coasts of Sunda, respectively. In that case the unmixed genetic complements of group 1 would end up in Java and that of group 2 in Borneo and the Philippines.

Point B indicates an unmixed Indian group moving down Burma, Thailand and the west coast of Malacca, crossing to Sumatra and continuing on a west coast route to Java. C is where new migrants meet up with indigenous Javan groups represented by *Homo soloensis* (3). The latter had probably explored north and northwest before any groups from outside the region ever appeared, pushing into peripheral areas of southeastern Sunda. Continuing on, a genetically mixed population moves eastwards through D, E, splitting at E and F. This could also represent a variable genetic relationship between the new and

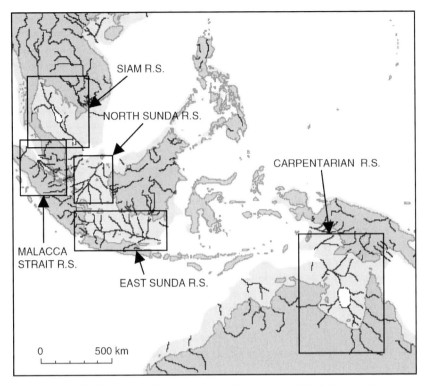

Map 3.4. Major palaeodrainage systems of the Sunda and Sahul shelves with sea level at −75 m.

older indigenous people. The resultant migrations leaving Sunda for Sahul are shown as populations 'a' perhaps following in the footsteps of the indigenous Javans.

Group 2 represents two possible migration routes taken by travellers from the south China area. One is the coastal route, while the other moves inland, then follows the Siam River System into the Sahul neck (Map 3.4). The coastal movement of the second group also takes it into the neck, but it arrives there by following the very large North Sunda River System. This system allows it to go through southeastern Sunda by linking up with the East Sunda River System (A–D). Following this system brings them to the east coast of Sunda, close to Bali. Another possibility for those determined to stick to the coast is to move up the Borneo coast and on to the Philippines (A–I). A splinter group is shown crossing Borneo and who then island hop through Sulawesi, Halmahera and on to the Irian bird's head (A–H–K). This route could be joined by others following the east Borneo coast (D–H) and those already on Sulawesi (G–H). The popularity of island hopping by people who knew what they were doing could be a rather quick and less hazardous migration tool and could have brought people to Sahul in innumerable ways; three are shown here. The first (H–K) was in the far north, while a second comes through an island chain taking in Sula and Ceram (F–L) but the third, through Java Flores and Timor (C–E–J), is favoured. I assume here that any migratory group moving this way would have encountered *Homo floresiensis* – what an interesting encounter to have been able to observe. The reason for favouring this avenue is that it directly stretches out towards Sahul with a 16 km crossing to Timor and then a 90–100 km crossing to the Sahul shelf (A), so it is probably the safest. Not that the migrants knew that, but it has a greater chance of putting people on Sahulian shores. We have then the possibility of people entering Sahul via Timor direct to the Kimberley shelf (a) or through other islands eventually landing on the Arafuran palaeocoast (b) later to be drawn by the enormous fresh water outflows of the Lake Carpentarian system into various parts of Australia (C and E).

Many changes to migratory gene pools were possible during their wanderings through eastern and northern Sunda. These could have altered morphology, which in turn depended on gene frequencies, mixing, founder effect and isolation and the possible addition of indigenous Javan genes. These then produced the morphology for those entering Australia, where no further genetic admixture was possible.

Map 3.3 shows only a fraction of the migratory routes available. In reality patterns of migration must have been far more complex, particularly if they extended over tens of thousands of years. While this model is probably some way from reality, I believe it is useful for demonstrating the principles of genetic change and exchange that could have taken place throughout the region during periods of migration. One component that could be added to make the model more realistic is 'time'. Populations take time to move, and, the earlier they appeared, the more likely movement was slow because of the fewer people involved. Speed and direction of gene migration depends on adjacent population pressure gradients, as explained in the previous chapter. Pressure would be relieved intermittently through people moving away until a further build-up took place. Over time this would have caused an irregular, migratory pulsing of variable speeds. Moreover, populations 1 and 2 may have entered Sunda at the same or at different times, each composed of groups of different sizes, moving at different speeds in different directions. There would also have been times when migration ceased altogether. The dominance of genetic expression of one group over another could fluctuate under these conditions. Initially, the founding groups 1 or 2 could appear as a single event, a series of pulses or a continuous stream, although the latter is unlikely.

They may have travelled as a broad front, a narrow path, or a random sprinkle of isolated units moving intermittently. Each of these would produce a different genetic outcome when proceeding through the routes outlined in Map 3.3. Moreover, each of these would have a different outcome on indigenous gene pools. Naturally, the consequence of all this cannot be discussed here, but, if 1 or 2 was composed of a few groups moving slowly and irregularly along narrow avenues over a long period of time, then the impact on indigenous groups would have been less than that of larger incoming populations. The impact is reduced further if the indigenous population, living in central and southern portions of Sunda, was composed of moderate to low numbers of individuals. A constant trickle of migrants, however, would inexorably overcome and dominate any indigenous populations in terms of genetic expression. This 'swamping' would depend also on the differential rates of population growth of populations 1 and 2 and further gene flow to them from source areas. Higher local growth rates among the migrants would compound the dominating process over indigenous populations and this would be exacerbated if population growth among the latter was low.

The spatial and environmental components in this model depend on transgressive events. Sea level change made certain routes more or less possible and, to some extent, determined who ended up where. The impact of sea level change on the Sunda shelf shape largely dictated the migratory patterns of populations 1 and 2. Changes in available land area can also be classed as another temporal factor. The variable-sized exposure of Sunda's continental shelf for almost 90 per cent of the Upper Pleistocene provided over one million square kilometres for people to explore. Depending upon the amount of time involved, local adaptations, random mutations, genetic drift and founder effect would have acted on migrational bands entering peripheral areas. For example, the people moving towards Timor would have already been subject to founder effect at 'E' and are now headed for the effects of relative genetic isolation and drift as they move along the island chain. With further expansion through Indonesia a clinal genetic rippling might later bring fresh genes into the area along the Sumbawa/Timor/Tanibar island chain. The rippling would only occur, however, if people kept in contact with one another, which, in itself, requires local population growth of some degree. Population splitting and recombination during these migrations and the temporal gap between the groups could, over time, have produced morphological differences between those eventually setting off for Sahul (a and b) and others moving north.

The above discussion clearly shows that to understand the genetic and, therefore, the morphological consequences of the Sahulian migrations, the following factors have to be recognised. They include:

1. the large time scales involved,
2. the numbers of individuals involved in and entering the migration process,
3. the timing of migrations, and
4. the complexity of genetic change and exchange occurring along the migration routes,
5. the influence of certain biological and demographic parameters,
6. the differential frequencies of isolation and contact over time,
7. the various group/population sizes, and
8. the reproductive success and survivorship.

There were many migratory avenues open to people crossing to Sahul and the routes chosen would have depended on the stepping-off point and direction taken for the crossing. Speed and direction of travel probably varied, routes could have been convoluted or direct, random or deliberately chosen. Movement might have been a slow trickle, moving continuously but slowly in a steady stream with people taking hundreds or even thousands of years to move along coasts and go from island to island, or it could have been a series of intermittent pulses where movement was rapid taking only weeks or days to cover quite large distances. The last leg to Australia has to have been one event that lasted only a few days. People are either in Sunda or on Sahul; there is no in between or stopping off point. The latter suggests a sudden appearance of humans in Australia as well as an abrupt transfer of culture that would then exist on either side of the Wallacian gap. With further migrations that situation would probably remain until locally derived cultural change took place on either side.

It is difficult to determine what indigenous people from northern areas of southeast Asia would have looked like before 100 ky. The Mapa cranium from southern China dates to this time and is a contemporary of *Homo soloensis*, but, apart from prominent brow ridges, its build is far less robust, strongly suggesting contrasting morphological gene pools. Besides these two groups, there is little else to indicate what the people that landed on Sahul before 100 ky might have looked like. It could not be anatomically modern humans out of Africa because the earliest evidence for the African exodus is in the Middle East (Qafzeh 92 ky). Unfortunately, the lack of cultural items associated with the Mapa fossils prevents evaluation of the cultural level and capabilities of either population, but the 380 cm^3 cranial capacity of *Homo floresiensis* with its associated material culture must make us seriously question previously held beliefs regarding the capabilities of archaic groups. However, we can only deduce that their 'footprints' seem to have been very light on the landscape. This begs the question: if there are no stone artefacts, does that really mean there are no people? I suggest not.

Timing was always a more important consideration for entry into Australia than it was for entry into the Americas. That is because of who first entered Sahul. The earlier people arrived, perhaps before 100 ky, the more likely it was regional hominids from Indonesia or other parts of southeast Asia. It is unlikely that the proposed Anatomically Modern Humans out of Africa arrived in the region before about 70 ky on present evidence. The earlier the first crossings to Australia were the less likely they involved people from northern regions of southeast Asia, because population growth was probably not strong enough to push people to the periphery of southeast Sunda. This is not to say, however, that a gradual spread from that region could not have taken place before 100 ky. Again, this points to the early migrants to Sahul being *Homo soloensis*, because they were 'local', derived from hominids that had lived in the region for a very long time and, no doubt, also had migratory tendencies as the Flores expansion of erectines shows.

Following the first landings on Sahul, the form that later migrations took is important, because each new migration brought fresh genes that would have impacted on the small gene pool already there, assuming they met each other. New genes may have had the same or different origins to the first people. Initially, each landing would have had the effect of altering the extant gene pool to some degree. Thus, incoming groups provided a constant updating of Sahulian gene frequencies, affecting the way in which the skeletal morphology of future generations might form and vary in different areas.

Movement of people inland away from coasts would do two things. Firstly, it would deplete the pre-existing gene pool at the point of entry. The next influx of genes would then have a bigger influence on the depleted pool than might otherwise be expected. Secondly, those moving away would constitute a classic founder group with all the genetic effects that implies. Changes would also depend on the size and composition of the 'mother' and 'daughter' gene pools. I use the plural because there may well have been several or more of such pools, depending upon the geographical distribution of landings and how isolated from each other these were. For example, three landing points might have been on the Arafura Plain, stretching between Australia and Papua New Guinea, the Arnhem Land coast or the shelf extending out from the Kimberley region in the northwest. A number of rafts could leave the starting point over a short period of time, but they may not have ended up in the same place, reducing the strength of the flotilla. Because of the vast distances and physical barriers between areas, a group landing in one area would have been isolated from and probably unaware of another in a different region, particularly those on the Arafura coast and the Kimberley.

Viable populations were less likely if successful crossings were limited in size, few in number, lacked backup from further migration, were scattered or became isolated. The loss of even one or two members without replacement, particularly female losses, would compound these problems. 'Extinction' is also an evolutionary mechanism and it must have played a part in genetic assortment among early migrations. Some groups would have perished from time to time from storms, cyclones, accidents and encounters with dangerous animals. Some may never have reached Sahul's shores and even successful landings may not have survived to pass on their genes. The effects of infant mortality, infertility and low birth rates would add to these problems, leading to local or regional group extinction. Using these tenets, it is most unlikely that a single man and two women began Australia's population.

Several attempts at settlement may have been made before a viable population finally took a firm foot hold on the continent. Initial success, for example, may have been thwarted generations later when, even after years of settlement, an initial population, for a variety of reasons, finally succumbed. It would certainly be more likely to occur if initial migration was slow; large time gaps existed between arrivals and/or the migrations involved small numbers of people scattered along thousands of kilometres of coastline. A threat to viability could also have come from the demographic patterns and strategies people adopted when they got to Sahul. Exploratory groups moving inland away from the main founder band on the coast began to experience environments, such as savannah, with which they were totally unfamiliar. In this case two things become important: the composition and the destination of the exploratory group.

The budding process of exploring groups reduces numbers in the founder band. The bud group is also smaller than the founder, thereby putting both at risk. The situation remains until numbers in the founder group are built up again. Budding groups most likely contained younger people, because older people were less likely to move away. The destination of the exploratory group is important to these processes. It might stay comparatively close to and/or in contact with the founder band, in which case the split poses little threat. On the other hand, it could venture some considerable distance, drawn on, perhaps, by the vast, empty continent that lay before it. In this case its chances of survival are reduced and the likelihood of its extinction is greater without backup from others. The immediate effect on the founder population is that, without backup, continued exploration, through a budding off process, means a weakening of that population, so that it may not be able to sustain itself. The in-built weakness of minimal numbers in the exploratory group predisposes it to dangers also and reduces its ability to thrive. Small groups are then faced with special dangers, where the loss of one individual is a disaster, two or more and the group is doomed. The genetic consequences of this are an impoverishment of the founder gene pool, reducing its genetic influence and its ability to contribute more of its genes to descendent groups. An alternative is to wait till the founder group grows large enough to allow the budding process to take place without compromising its safety.

The earlier the first landings were, the less likely they involved large numbers of people, thus reducing their chances of viability, particularly if groups became separated. For the palaeoanthropologist, this process would produce a morphological discontinuity in the skeletal record that would be hard to explain, even if there was an excellent fossil record from the period. If a spatial component is added, separating these events by large distances, the result would be even more perplexing.

Generations and morphology

Sahul's palaeodemography and cladistics mean more if they are measured in 'generations', which also better signifies genetic assortment, loss, remixing and passing time. Between 70 ky and 100 ky, for example, there would have been 1875 such assortments, if 16 years is used for each generation. The extent of assortment and rate of change in an isolated group depends upon its size. In non-isolated people, both these are much greater for the population as a whole because of the greater variety of combinations available. A series of migrations would provide continuous input of new genes, helping to change the pool begun by the founding population. Small isolated founding gene pools were particularly vulnerable to change and particularly were susceptible to swamping by gene flow from larger populations. Upper Pleistocene gene flow into Sahul was governed by the numbers of people involved in each landing as well as the frequency and distribution of landings.

Migration frequency probably varied throughout the Upper Pleistocene, although 'substantial' numbers of landings were unlikely till the last 20 ky or so. A typical landing frequency variation around 70 ky might have been ten boats one year, 25 the next, none for eight years, then two. They may have all landed in one place or been scattered along hundreds of kilometres of coastline, if weather, local conditions or local people, if any, prevented landing. Conversely, the new arrivals may have been stronger than local groups, so replacing or driving them off. This is where genetic 'swamping' with continued isolation of the invading gene pool would take place. The swamping process would alter the genetic composition of the original population suddenly, particularly if the new group had different ancestral origins from the smaller group. In the latter case, an ideal fossil collection would show instantaneous morphological variation and change temporally, but a wide range of variation spatially.

It is likely that morphological change would speed up through founder effects among migratory groups compared with that in ancestral progenitor populations. This could take place during a series of divisions and recombinations in the migratory gene pools, brought about by the routes chosen as they moved through island chains. In this way migratory groups with common ancestry could have come into contact with each other many generations later, each appearing somewhat different from the other in one way or another. The rate of change between the two would depend upon the time they were isolated, whether they had mixed with other groups, and the difference in genetic composition of each group from the parent group so that in time the people landing on Sahul could have had morphological characteristics different from those they left behind. So, new arrivals could bring fresh genetic compliments, combinations and adaptations, which would themselves have undergone changes that were added to those already here. Additionally, the five evolutionary cornerstones would be constantly operating in different proportions over the next 3000–4000 generations. Further changes took place through more migrations into Australia, which produced an extremely complex biological pattern.

The above discussion is a little on the dry side and somewhat repetitive, but I do not apologise for this. The interwoven nature of the 'ingredients', the complexity of the feedback processes that go on in these events as well as the many possible permutations of the Migratory process itself (and migratory I have mentioned only a few) during the Upper Pleistocene makes this whole process an extremely difficult thing to come to terms with. I reiterate, it is the simplest of pictures of micro-evolutionary mechanisms that no doubt became infinitely more complex during the Holocene in Australia and Melanesia. The implications of such processes for those researching 'origin' and 'affiliation' questions in Australia, using traditional methods of skeletal biology, are manifold. Studying modern skeletal samples to answer questions concerning the affiliation and origin of Australia's first arrivals will, therefore, mean very little indeed, although they still produce tantalising if extremely scant visions of those past migrations.

The majority of Upper Pleistocene crossings to Sahul probably took northerly routes (into Irian Jaya and Papua New Guinea), although sporadic landfalls may have been made directly from places like Timor. Demographic factors would have imposed their own variations and restrictions on numbers, composition and direction of movement down into Australia. To set up a viable population in Sahul required many successful voyages, requiring a trickle of people beginning the journey and a suitable size population to maintain it. Not all landings were in the same place and some attempts must have foundered at sea. The risk of this increases with simplicity of craft, length of journey and the difficulties of the chosen route, oceanic conditions and local and regional weather conditions. All these factors depend on the timing of voyages, sea levels and some luck. With these factors in mind, there is a need either to invoke sophistication in watercraft design to optimise a safe voyage or sufficient quantities of people making the journey to make up for regular losses. Both of these might be considered more likely as sea levels rose. If these conditions do not necessarily apply, then it is logical to suggest that a very small trickle of people, over a very long time, using comparatively unsophisticated craft and travelling the safest routes could have contributed most to Sahul's founding gene pool.

It is hard to believe that population pressure provided the impetus for initial migration before about 70 ky or that the first landing was an isolated event. If world population size during the Upper Pleistocene was extremely small, we can only assume that populations on the fringe of humanity, like those in south east Sunda, would have been even smaller and sparsely distributed. Or perhaps this long-held idea is not correct and populations were large enough to stay in touch with one another as well as fuel migratory bands. The regional population growth rate proposed in Chapter 2 is vastly different from those accepted as the norm but would be the size required for regular and increasing gene flow to Sahul during the last half of the Upper Pleistocene. They would also partly explain the drive behind people to move. Whatever initiated migration to Sahul, the most plausible model suggests that a slow, almost constant, trickle of people was needed to begin Australia's human story. Besides the possibility of pre-100 ky arrivals, the main migratory event saw people arrive from much farther afield, descendants of those living in the main population generators of southeast Asia.

4 *Upper Pleistocene migration patterns on Sahul*

> The gales brought us here. They picked us up from the reef out at sea, they tossed us and rolled us and pushed us up high on the sandy head and now we lie here in the sun.
>
> (Spirit Quandemooka – Oogeroo Nunuccal, Aboriginal elder and poet, Stradbroke Island, Queensland)

Introduction

Aboriginal people have differing accounts of how the first people came to Australia, where they came from and how they travelled across the continent. These events involve animal and human-like spirit beings and heroes who undertook epic journeys across an empty and often dark landscape upon which they created natural and physical features and landscapes as well as animals and the people themselves. These 'Dreaming' stories tell of how all these things were infused with life and lore and language were given to the people. They describe the long journeys, adventures and battles of the ancestral beings and how the first people encountered strange creatures. The word 'Dreamtime' is not liked by some Aboriginal people, it implies a particular time in the past. The 'Dreaming' is better because it represents no particular time: it could be millions of years ago, 60 ky ago, yesterday, today and tomorrow. The 'Dreaming' continuum in Aboriginal society is strong because it helps explain the world, where people, animals, the landscape and everything came from. Perhaps even more important it represents the special attachment Aboriginal people have to the land.

Aboriginal creation stories usually contrast strongly with general anthropological as well as scientific theories and propositions about the evolution of the continent and how the first people arrived in Australia. Apart from arguments concerning the age of the social collective memory, some of the stories concerning the first people and how they arrived here are remarkably similar to ideas developed by scientists using very different information. For example, many contemporary communities, particularly those living in northern Australia, have stories that relate how people (usually Ancestral Beings) arrived here. They often mention obscure places of origin, far across the sea and describe the craft used as well as landing places. I relate several of these accounts here:

The truth is, of course, that my own people, the Riratjingu, are descended from the great Djankawu who came from the island Baralku far across the sea. Our spirits return to Baralku when we die. Djankawu came in his canoe with his two sisters, following the morning star which guided them to the shores of Yelangbara [Port Bradshaw] on the eastern coast of Arnhem Land. They walked far across the country following the rain clouds. When they wanted water they plunged their digging stick into the ground and fresh water flowed. From them we learnt the names of all the creatures on the land and they taught us all our law.
(These are the words of the Son of Malawan from Yirrkylla, Northern Territory, in Isaacs (ed.), 1980: 5, taken from Jacob, 1991: 330).

Another story relates:

Our people are the Thurrawah family. They came from a long way off, from another land before there were people in [Australia]. Many of our people were the animals but disguised in human forms. These ancestors of mine came from a land beyond the sea, and they wanted to live elsewhere.
They decided to travel across the sea in search of new and better hunting grounds, and to make the voyage, they required the largest canoe.
(Reprinted from Jacob, 1991: 331)

Some can be quite specific about details like this one from northern Australia:

Namarrgon, the lightning man, like so many of the 'First People' entered the land on the northern coast. He was accompanied by his wife, Barrginj, and their children. They came with the rising sea levels, increasing rainfall and tropical storm activity. The very first place Namarrgon left some of his destructive essence was at Argalargal (Black Rock) on the Cobourg Peninsula. From there the family members made their way down the peninsula and then moved inland, looking for a good place to make their home.
(After Chaloupka, 1993: 56)

Scientific theories constructed by non-Aboriginal people are not always acceptable to many Aboriginal people. They feel that their close association with the land makes them special Australians; they are those who do not measure their tenure here in six or seven generations but in thousands. Aboriginal people often insist that they originated in Australia, not outside as anthropologists would have it. Some have suggested to me that, if anything, the reverse is true: people left here to populate other parts of the world. Nevertheless, there are many Dreaming stories, like those above, that describe landings from islands or lands across the sea, sometimes far away. There is also an unmistakable pride woven into these stories. It highlights the heroism and boldness of the characters, as well as the strength shown by the creation of certain features. Following the creation acts, most ancestral beings then went underground; it is believed that they still reside there and they provide meaning and a spiritual link between those who hold the stories for them and the land. For most Aboriginal Australians the land is their mother and the stories and places about the land and its creation have always constituted an integral and very important part in indigenous life by providing spiritual and social cohesion.

Problems with dating the first arrivals

Aboriginal people see the past as one continuum from the past through the present and into the future. They see nature as a totally integrated essence of environment, geography, animals, fire, water, daily living and spirituality. Non-Aboriginal people, on the other hand, are 'compartmentalised' thinkers. We need boxes to place everything in, thus breaking the links between all naturally interrelated things. Similarly, when we talk of the past we need dates because it is only with dates that we can hang the events on, which, to us, then makes logical sense.

Australian archaeology has pushed back the date for the earliest occupation of Australia a very long way. Forty years ago it was thought to be around six thousand years, by 1970 it was 28 ky and a decade later it stood at 40 ky. Until recently, the oldest occupation date hovered around this figure, mainly due to the limitations of radiocarbon dating that had been used widely until other methods became available. In the past ^{14}C dating frustrated those trying to push Australian archaeology back beyond 40 ky. Consequently, the oldest time depth for human occupation remained at between 35 ky and 40 ky for over two decades, with some naturally assuming that this date marked a hiatus for human entry rather than accepting the limitation of the dating method itself. There is no doubt that people arrived in Australia before 40 ky and this has been shown using techniques such as thermoluminescence (T/L), optical stimulated luminescence (OSL), uranium series, amino-acid racemisation, obsidian hydration and beryllium dating, as well as refined methods now applied to radiocarbon dating itself. Nevertheless, the long-held faith placed in radiocarbon dating by some has also made it the favourite dating method and in so doing it has given rise to a certain amount of scepticism regarding newer techniques. Moreover, the complicated nature of these new techniques has drawn critics, particularly from the ranks of ^{14}C enthusiasts. Many feel they are being distanced from their data by complex technology and the specialist knowledge required to fully understand, operate and manipulate new methods, interpret findings and, more importantly, identify mistakes or incorrect results. Rigorous testing and monitoring of new methods is standard. Some, however, still cling to a staunch belief in ^{14}C as the *only* reliable dating method to be trusted and that the use of any other method is tantamount to alchemy. Consequently, the situation has resulted in two camps of thought regarding the dating of the first human entry into Australia.

It has always surprised me that those involved in such a constantly changing discipline as archaeology still hammer their colours to a particular mast so firmly. Archaeological finds are more often than not serendipitous, take for example the recent discoveries of a new species of Homo in Flores. While a popular view of archaeology conjures up a rather stuffy, dusty image, the discipline is not for the inflexible and dogmatic. Similarly, the introduction of new dating techniques often draws sceptical comment from those following traditional and long-tried methods. Thermoluminescence dating, for example, has for many years been used and widely accepted by geomorphology as a reliable method for sediment dating. Uranium Series is another and without this and T/L the age of most Middle and Upper Pleistocene hominid remains from all over the world would remain unknown to us. But new dating methods are like new medicines: while they are desperately needed and always welcome, they require rigorous testing, they have drawbacks, they need to be widely accepted, they have unwanted side effects and many people are highly suspicious of them when they first come out. As with a new medicine, the main aim of a particular dating technique is usually to provide accurate and efficient results while using minimum amounts of sample. Legitimate concerns often focus on their accuracy (or inaccuracy), which suffers considerably if samples are not of a certain quality, are damaged, are sampled or handled improperly during collection, or if they become contaminated. Some techniques do not make it, such as 'cation dating' (Nobbs and Dorn, 1988). The great temptation to be able to date the obviously ancient petroglyphic rock art found in many parts of Australia, and often rejected by Aboriginal people as not theirs and predating themselves, drove development of the cation method. It was designed to provide a minimal age for the silica skins or 'desert varnish' covering most examples of this type of art (Ibid, 1988). The enigmatic 'smiling faces' of the Cleland Hills, central Australia, and the 'Panaramitee' styles in South Australia, are principal examples of this art. They often depict a naturalistic style of circles, various animal footprints as well as anthropomorphic, zoomorphic and abstract forms. It has been suggested that other motifs include the footprints of megafaunal species *Diprotodon* and *Genyornis*, extinct for at least 40 millennia, and a crocodile head (Basedow, 1914; Mountford, 1979). The latter is found thousands of kilometres from the contemporary distribution of these animals, although they did live 300 km to the northwest 65 ky ago (see Chapter 5). Cation dating, however, was found to be unsatisfactory. There is no disgrace in that, archaeology needs imagination to push the boundaries of what might be, as long as mistakes are recognised. So cation dating was always worth a try. I will return to this setting later.

Today, Australia is the driest habitable continent on earth with 70 per cent of its landmass classified as arid or semi-arid. Continental drying began 15 My ago, a trend continuing throughout the Plio-Pleistocene, particularly across the inland. In central Australia annual evaporation rates hover between 2800 and 4200 mm, summertime temperatures can exceed 50°C and rainfall is a meagre 110–200 mm, even less in El Niño years (Colls and Whitaker, 1990). In contrast, tropical regions experience rainfall that has been recorded as high as 12 m annually.

Both faunal and floral populations fluctuated in composition and quantity as the continent dried. Species gradually disappeared or adapted as the glacial/interglacial cycles superimposed on the drying acted as selective forces on them. The isolation of Australia's continental plate saved marsupials for the world to see and enjoy, as well as to show their great variety and adaptive qualities. So, when the first people arrived they must have had a shock when they discovered that Sahul's fauna was very different from that of Sunda's. The newcomers' faces must have dropped as they contemplated hopping beasts, some standing at least twice their height, and carrying young in bags on the outside of their bodies. Other animals were larger than anything on the islands, which had imposed dwarfing on many of the largest varieties such as elephants, hippos, deer and pigs. A fearless fauna unfamiliar with humans greeted the newcomers, while reptiles such as the huge goanna *Megalania* and giant *Palimnarchos* and *Porosis* crocodiles could not believe their luck with something new on the menu. Nevertheless, some parallel evolution had conferred a similarity between the animal groups on either side of the Wallace Line and it would not have taken long to switch the human mind from hunting deer to hunting its marsupial equivalent the kangaroo.

People had also landed on a continent with a series of environments, most of which they were totally unfamiliar, if we assume that they originated in the tropical or sub-tropical lands and islands of Sunda. Apart from similar rainforest areas all other ecosystems Australia offered were going to be challenges to these island hopping people. The nearest deserts/savannas are the Great Indian Desert in the northwest, the Mongolian Gobi Desert and the plains of northwest China. The snow and ice of Australia's southeast and Tasmania were totally unfamiliar and must have been as challenging as desert sands, particularly as the Ice Age door swung shut and sea levels lowered. Australia was not another Equatorial island, but one that would have certainly taxed the cultural capabilities and adaptive biology of a people used to small, lush tropical island environments.

Quaternary studies in Australia often draw those seeking to understand environmental variation and change over the last one million years into the fray of arguments concerning the timing of the entry of the first people. This is usually done through the challenge of determining what environmental and/ or biotic changes were due to natural or human induced factors. Indeed, it has been Quaternary researchers with little professional interest in the timing of the arrival of the first people that have discovered indirect evidence for entry prior to 100 ky (Singh and Geissler, 1985; Kershaw, 1985, 1986; Moss and Kershaw, 1999; Bohte and Kershaw, 1999; Wang *et al.*, 1999; Kershaw and Whitlock, 2000; Moss and Kershaw, 2000). The evidence comes from palynological studies undertaken in north and southeastern Australia. In short, a series of signals have been found by these workers which correspond to vegetational change between 100 ky and 150 ky and around 180 ky. The signals include reduction in the number of fire prone plant species with a commensurate increase in pyrophytic species and these are accompanied by an increase in the frequency of charcoal particles. One group of data come from drill cores obtained on the continental shelf in northeastern Australia. The unusual combination of floral change with increase charcoal particles does not occur before 150 ky and, therefore, has not been attributed to glacial causes. The signals have been interpreted as indicating a pattern of regular anthropogenic burning on the mainland that reduced fire-prone species and encouraged fire tolerant ones in a way natural fire frequencies would not. Is this, then, the fiery signal of the arrival of humans in northeastern Australia or perhaps only the first to use regular burning? Either way, if it is people, they arrived earlier and are exploring. Nobody questions whether the dates or the data are right, but some do not necessarily feel fire equals humans. Consequently, these data are not universally accepted as evidence for the arrival of humans, but, while desperate attempts to reach other conclusions to explain these changes are pursued, the original interpretation seems the most parsimonious.

Migration routes on the Sahulian palaeocoast

Glacial sea levels joined Australia and Papua New Guinea for 91 per cent of the last 250 ky (Voris, 2000). It includes the periods 8–115 ky and 125–210 ky. At sea levels below 60 m a long coastline stretched 200 km north of Arnhem Land sweeping out in a northeast direction then turning west encompassing the Aru Islands (Aru Hills) then north to Irian Jaya (Map 4.1). The shape of this palaeocoastline provided a net that caught all craft heading east from

118   The First Boat People

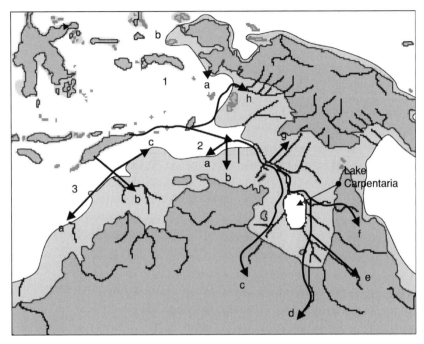

Map 4.1. Human migration into Sahul with sea levels at −75 m.

Timor or Tanimbar crossing a much reduced Arafura Sea, scooping up voyagers on a broad 1200 km front.

Even minor changes in sea level changed Sahul's coastal geography quite drastically. For example, only a five to ten metre rise drowned Lake Carpentaria. This left what could be termed the 'Bay of Carpentaria' because much of what is now the Gulf of Carpentaria was still above sea level and the Torres Strait Bridge was still in place. So, what were the migration routes open to these early seafarers? The following discussion considers some of these at a sea level of −75 m, half way between the lowest of around −145 m and modern levels.

Northern migrations (Route 1)

Landings in the far north of Sahul came from Stream A or those moving along the eastern Sunda coast. They would be drawn from the Sulawesi/Ceram corridor and, although several directions were possible, two are shown on Map 4.1, a and b. One could move north (b) via Misool Island, while the second could land directly on to the mainland (a). Their direction from here could be anywhere, but following the coast would have been the easiest way of continued movement. Crossing the Irian neck would have brought them along the northern side of the island behind the Irian Jaya/New Guinea cordillera. Both this and the previous route would take them to the eastern end of Papua New Guinea and from there down across the Torres Bridge and on to Australia's eastern seaboard. Others landing from Ceram could move northwest to the 'bird's head', but it is more likely they moved southeast because of ease of passage towards the Arafura Plain. A journey north through very rough country with high mountain ranges was not a good choice when easier southern routes were available. Along the coasts and away from impenetrable jungle would have been an easier passage, as well as exploring the rivers crossing the savannah of the northern Arafura Plain. Such entry routes do not, however, bode well for the future discovery of the campsites of these people.

Eastern migrations (Route 2)

Island hopping groups and moving through the South Moluccas from Wetar/ Timor to Tanimbar might be the most successful voyagers. This seems to have been one of the easiest crossing, which only improved during lower sea levels. After Tanimbar, they had two basic avenues open to them. The first was a short hop across to the Arafura palaeocoast that stretched west from the 'Aru Hills' towards Tanimbar. Smoke form fires on the plain may have been spotted from Tanimbar's highest point (240 m), which is conveniently situated on its east coast. Moving north along the coast after landing would have brought them to the mouth of a palaeo-river system consisting of a catchment of southern Irian rivers such as the Momats, Lorentz and Pulau (2h). These would have made a good corridor for entry, with a combined freshwater outflow attractive to these nomads.

Movement to the south would take people ahead into a huge bay described by the Arafura palaeocoast. A large drainage system emanating from Lake Carpentaria would have caused an enormous discharge into the diminished Arafura Sea, particularly during the monsoon season. The presence of this estuary would have been particularly obvious to those with highly tuned taste buds detecting the fresh water some way out to sea and drawing them into this part of the coast like a radar.

As the rafts sailed east across the diminished Arafura Sea the sailors constantly checked the direction of flight of sea and land birds. They picked up and identified floating vegetation that interpreted the nearest invisible land. Wind direction and ocean currents were regularly checked as they sniffed the air for clues to how far away land might be. Occasionally they scooped up sea water in their hands to taste it for its fresh water content. All these natural signals were fed into their cerebral computer for instant analysis that told them where they were, how far away land was and in what direction. As they approached the Arafura coastline huge thunder clouds inland and to the north boomed and they noticed the sea water become fresher. It was the monsoon season, when the fifty large rivers that filled a vast unseen lake to the east made it disgorge huge amounts of freshwater westward and far out to sea.

Entering the Carpentarian River System they would soon discover the massive Lake Carpentaria (Map 4.1). The Lake Carpentaria catchment comprised a very large river system as well as a vast network of meandering and abraded channels, extending from northern Australia (Arnhem Land to Cape York) deep into the southern half of Papua New Guinea. The catchment area was about 1 275 000 km^2 with an estimated total runoff under present conditions of around 230×10^9 m^3 annually (Torgersen et al., 1988). Lake Carpentaria would have been a target for migrating bands as a source of food and fresh water. Masses of water birds and animals are likely to have populated in and around the lake. The surrounding wide, flat black soil plain supported a typical riverine environment that must have supported abundant animal and bird life. Analyses of pollen from cores taken from the lake bed show that *Typha*, Casuarina, various species of Callitris, yams (*Dioscorea* sp.) and water lilies grew there (Torgersen et al., 1988). During those times when the Arafura Plain was exposed and Lake Carpentaria existed, the surrounding region was an open grassy plain with scattered swamps (Chivas et al., 2001).

Once at Lake Carpentaria, the world was their oyster. They could move east, up the catchment which took them into Cape York (2e and f). Alternatively, moving south, following either side of the lake, placed them on tributary systems that could take them into the continent proper for the first time, and they would be 1200 km from the Arafura beaches where they first landed. The southern boundary of the Lake Carpentaria catchment met up with the northern catchment of the Lake Eyre basin that at times of higher sea level took water 1000 km south and into the megalake system of Lake Eyre (see Chapter 5). But it is unlikely that they continued south because at times of low sea level the Lake Eyre catchment presented a lot of desert and little water.

Whether it would have taken one band or several generations to move as far as the headwaters of the Lake Carpentaria catchment is anybody's guess. Certainly, moving along a coast was always far easier for these island hoppers than through inland rainforests or over rugged mountain ranges. Papua New Guinea may have deterred exploration by early groups for these very reasons. It may have been considered unwise to follow rivers too far inland and perhaps a policy of not moving too far from the coast prevailed for some considerable time before further internal exploration of the Lake Carpentaria system was attempted.

Another alternative was that those landing at the Carpentarian system estuary could turn north, again following tributaries, and large rivers and streams flowing south across the Arafura Plain from Papua New Guinea (2g). Following these, they could cross the plain discovering two large lakes or swamp areas, Lake Ara, probably around 5500 km^2, and Lake Fura (my names) about half that size. These were filled by systems running from southern Papua New Guinea and out across the Arafura Plain. At that time the Fly/Strickland river system did not flow east as it does now but flowed directly south at a point close to the present Irian Jaya/Papua New Guinea border. From there it too flowed out across the Arafura Plain and into the northern end of Lake Carpentaria.

Keeping to the coast and moving south, explorers could either move inland (2b) towards the Arnhem Land escarpment or turn west along the coast (2a). Migrations through Timor could also end up on this part of the coast. It is obvious, however, that the permutations and possibilities for landings and subsequent movement on the Arafura Plain are almost endless.

Southern migrations (Route 3)

Use of the 'shortest' route to Sahul is a contentious issue. As I point out elsewhere, this route cannot be known instinctively. Nevertheless, some clues such as distant smoke or somehow seeing the objective from the point of departure would be useful for prompting a direct passage. But what if you cannot see your objective? This was the main drawback in making the voyage to Sahul. Those on Timor had already completed a journey of at least 16–20 km to reach there and a larger gap may not have been daunting as long as you knew land was ahead of you somewhere. The crossing from Timor would not require a more complex water craft, just the ability to carry enough water and to steer.

The continental shelf off the Kimberley is one of the largest in the world. At low sea level it cut the present sea journey from Timor to Australia by almost two thirds. So, while the crossing was always somewhat more hazardous than island hopping, it did not last long. Smoke from natural bush fires, land birds and the smell of the vast continent lying over the horizon probably drew boats on. Several possibilities arise from these voyages, but they depend on sea level. At lowest levels people would have landed on the Sahul Shelf. Movement along the coast could take them east (3c) or west (3a). To the east was the Arafura palaeocoast, in the opposite direction was the remote coast of

Western Australia. The choice of direction was important for the future of the group. Moving east would increase the chances of meeting others because of the funnelling effect of the Arafura coast on other migrations moving east. Landings on the remote northwestern coastline were either rare or unsuccessful because the voyage was longer and the dangers greater. So going west meant meeting only those who after landing had also travelled west. As the coast curved southwest past the Kimberley then south to where the Great Sandy Desert meets the sea, the more remote this coast became. In other words, two consecutive landings moving in opposite directions would, more than likely, never meet up again, those going west were doomed to isolation and, without further backup migrations, were at greater risk of extinction.

With sea levels between -100 m and -150 m, a prominent feature along this part of the Sahul palaeocoast was the Malita Valley channel which fed the Bonaparte Depression or, in those times, Bay (Yokoyama *et al.*, 2001). Those moving east along the northern edge of the Sahul Shelf would encounter the Malita Valley entrance and following it would find themselves in an enormous, calm bay. If tides along this coast were as high and as ferocious as they are today, however, the timing of entry into the bay using slender craft would have been very important and scary. Bonaparte Bay measured around 50 000 km^2 at sea levels of -150 m, but shrank to around 20 000 km^2 at -80 m. While it may have been saline at its northern end, heavy monsoonal discharge from the Kimberley plateau probably kept surface waters fresh at the southern end during summer months. The bay was calm because it was almost completely cut off from the ocean by the Sahul Shelf on its northern side. Entering the bay would draw people towards and eventually into the Kimberley region itself. A dozen major rivers running from the western side of the Northern Territory and the eastern side of the Kimberley drained a 350 000 km^2 catchment. Like Lake Carpentaria, Bonaparte Bay and its surrounding environments would have had an abundance of birds and animals attracted by the funnelled run-off from the Kimberley escarpment.

Lake Londonderry, to the southwest of Bonaparte Bay was probably fresh water fed by rivers like the King Edward, Drysdale and King George that exit on the present day northern tip of the Kimberley either side of Cape Londonderry. Measuring around 4000 km^2, Lake Londonderry would also have been a rich resource for wandering bands of hunter-gatherers. The brackish condition of some lakes may not have deterred animals, however, because the rivers feeding them always had fresh water. It is worth remembering that these large lakes and their surrounding wetlands and swamps would have provided a variety of good nesting and breeding grounds for a wide variety of birds and other animals, particularly before human intrusion. The use of a shallow draft bark skiff that easily and quietly traversed lily-covered lagoons in these areas would in quick time provide enough food for any number of bands living in the region. Therefore, they would have been a natural focus for hunter-gatherers and places where the natural carrying capacity would have been high. It is also possible that because of the rich biota these areas were natural 'stop-offs' within which bands could dwell for some time before moving on or perhaps return to after finding less abundant areas further inland. Movement across the wide exposed continental shelves with their rich resources may have been particularly slow. Inland people would have discovered the Kimberley landscapes of hot stony plateaus, something they had not encountered before. The rugged ranges and ravines, rock escarpments that ended in never-ending plains and savannas to the south would have been totally unfamiliar country to these coastal sea farers. The rugged landscapes would have slowed the rate of movement and the difficult country would soak up small bands so that movement far inland probably took many years. A quicker way to move inland was to follow the rivers around the edge of the bay, using them as corridors to the interior. The Ord River would take people directly south, while the Berkeley, King George and Durack Rivers would serve as avenues into the southwest Kimberley. Outflow of western Kimberley rivers would have also provided good access inland. From north to south the Mitchell, Roe, Prince Regent, Charnley, Isdell, May and Fitzroy Rivers all flow into the Indian Ocean with the giant Fitzroy being an excellent corridor to take people far into the interior around the southern end of the Kimberleys, and satellite imagery has shown how the old Fitzroy palaeochannel cuts across the continental shelf. Other channels flowed southeast. By following the Daly River, for example, bands could eventually end up in western Arnhem Land through the Katherine River system. Head waters of other rivers such as the Victoria lay much further south and, following this channel, people would find themselves far inland on the plains just north of the Tanami Desert, which was a harsh place during times of low sea level. Between the Victoria and Daly other rivers, such as the Fitzmaurice and Moyle, would also take people inland.

When seas rose to around -60 m, $100\,000$–$150\,000$ km^2 of continental shelf disappeared. The Malita Valley channel opened into a broad strait, while the country of the Joseph Bonaparte Plain sank beneath the ocean. The length of the Timor/Australia crossing did not change very much, however, because submergence of the Sahul Shelf left a chain of many small islands along which voyagers could eventually make their way to the mainland. Again, choice of direction at that point was important. Going east brought people to the Arafura palaeocoast, while following the island chain southwest would move people away from mainstream migration corridors.

Environments and migration on the Arafura Plain

The three main lakes on the Arafura Plain (Ara, Fura and Carpentaria) must have been prominent landmarks for explorers but none more so than Lake Carpentaria. Bands wandering the plain would constantly come across river systems, some filling the great lake, others draining it. At least 14 large rivers flowed west from Cape York and another dozen flowed north from the Barkly/Selwyn Tablelands. Groups moving across the north end of the lake would have met the Fly/Strickland river system flowing south. Skirting the edge of the great lake, exploiting wetland environments brought them into contact with many tributaries flowing to the lake and the opportunity to move up them. Water craft would have been the best way to do this, large crocs willing. The vast amount of water birds attracted to the lake would be an easy and plentiful resource, together with their eggs. Moving east would bring the groups to the eastern Cape York coast with its plentiful stocks of seafood and dugong along the coast and terrestrial game on the exposed continental shelf.

The tributaries wove across the Arafura Plain to the southern end of Lake Carpentaria offering ideal 'freshwater highways' into Australia's northern interior. These well-stocked corridors would naturally draw people along them. After reaching the headwaters of these channels, bands could fan out east, south and west. As seas rose above −50 m the northwestern corner of Lake Carpentaria was broached and it became a large bay. Northwest landings now had two choices: either cross the mouth of the harbour or move eastwards and down or through to Cape York and Australia's east coast. Crossing the bay's entrance would be easy for these experienced rafters. In this area, watercraft were the best means of transport. They provided fast and easy access along the Arafura coast, the coast of Carpentaria Bay and also for exploring inland waterways; and, no doubt, the first arrivals did not abandon their craft easily. Landings on the southern shore may have resulted in people taking an easterly route that would end in roughly the same area as others starting from the north side of the bay. The abundance of local riverine systems probably acted to spread bands within those environments. The Arafura Plain largely disappeared when seas rose above −40 m and the palaeocoast was much closer to Australia's present shape. Although narrower, the Torres Strait Bridge was a permanent feature during the whole of the last interglacial, allowing dry-foot crossings from Papua New Guinea to Australia.

At times of modern sea level, direct entry into Sahul by landing on the southern shore of the Gulf of Carpentaria was the longest way in. The journey from the Aru Islands or Tanimbar was 1400 km, or just over 1000 km to the Torres Bridge. The shortest routes to Sahul between 60 ky and 140 ky were to the coasts of southern Irian Jaya, the continental shelf at the Aru Hills and the Kimberley palaeocoast, particularly when sea levels reached between about −80 m and −150 m. The first and third are favourites as seas rose.

Internal migration strategies

Whatever the sea level above −55 m, there are a number of possibilities that may have brought people to the southern end of Lake Carpentaria. When it disappeared, the rivers feeding into it did not. As a result there was every possibility that during exploration of the coast people would discover many river entrances that led into the continent. Riverine environments provided the same menu as the coast, namely fish, shellfish, crustaceans, animals, birds, plant foods and, most importantly, fresh water running in abundance under your boat. The headwaters of many rivers flowing into the Gulf lie on the Barkly/Selwyn tableland. Water landing on the southern side of this tableland flows south into the upper catchment of the Lake Eyre basin, so, by following these, bands could easily enter this system and move safely through the heart of eastern Central Australia. A similar method of entry was suggested almost 70 years ago:

> Their routes were probably around the north and down the east and north-west coasts; from the Gulf of Carpentaria or the Sahul land up the Queensland rivers and on to the Diamantina and Cooper and so to the Great Lake (Eyre) of early times and eastern South Australia, possibly at a period when the north-east of that state was not yet arid and when the lush country around the lake was occupied by large marsupials . . .
> *(Elkin, 1938: 317)*

John Mulvaney (1961: 62) has also suggested that:

> Any migration southwards from the Gulf could easily move across to the inland drainage basin, which in a pluvial period would provide a passage down the Diamantina River or Cooper's Creek systems towards Central Australia, or down the Darling River and tributaries to the south and south-east of the continent.

Elkin's and Mulvaney's ideas are prophetic in terms of what we now know about the interglacial palaeoenvironments of central Australia, but at the height of glacial events Central Australia was not a nice place to be (see Chapter 5).

It is not having an each-way bet to suggest that considering the likelihood of a generalist economy employed by *Homo soloensis*, they would have been quite capable of moving inland along river systems as well as along coasts. We know people entered Australia prior to 60 ky, when travel in to the interior was possible, but we can only assume that coastal routes would have been favoured. If these first people were *Homo soloensis*, however, their route was probably dictated by safety concerns, rather than any economic pre-adaptation to coasts. The generalist hunter-gatherer would have taken advantage of all environments and food resources wherever they were available. This would also allow movement inland as far as safety would allow. While no archaeological site dated to before 60 ky has been found in Australia, neither has any site been found associated with *Homo soloensis* in Java, where we know they lived for a considerable time. They obviously left a very light 'footprint' even where they were numerous. It is more than likely also that many of the prints they did leave are now under the ocean, but I will take this up again in later chapters.

Movement on the continental shelf

The direction and speed of the first colonisers was controlled by general factors that were the same for any Upper Pleistocene migration, except that there was no likelihood of encountering other people in Australia, a similar situation to the first migrations into the Americas. Survival was key and this depended on the availability of food and water, the number of people involved, and getting to know the environment and its pitfalls, as well as how cunning you were. We know food and water were abundant and that people made it this far, so they must have been very resourceful and tough, but how many of them were there?

> The expectation here is that something can be learned about population densities merely by counting up all of the plants and animals in the environment. Assuming that one can delineate such plant and animal resources, the underlying idea would seem to be that carrying capacity represents the maximun [sic] population level or load that can be maintained through some idealized and usually unspecified scheme of efficient exploitation
> *(Ammerman, 1975: 223)*

The interglacial or glacial carrying capacity of northern Australia and the exposed continental shelf is beyond our capabilities to know so it is impossible to project human population numbers from using this method. The earlier people arrived, the fewer are likely to have been involved, and it is anybody's guess what the appropriate number of arrivals at any one time was. A unit of 15 people, however, would seem a reasonable maximum estimate, based on three or four rafts landing together each carrying between three and five people. The unit may have consisted of a couple of old people, two or three men (perhaps one on each raft), four women of various ages and seven sub-adults. Given arguments discussed earlier, it seems extremely unlikely that the Australian population grew from a single family. Initially, a series of landings probably brought people to various parts of the coastline and founder populations grew from these. Continued migration, however, raises a problem. People could only enter Australia along the coast where previous groups landed and where settlement may have taken place. The choices for these arrivals were either to join the people already there or to move past or replace them, but it is logical to suggest hunter-gatherers, used to moving with an empty continent before them, would not crowd together or stay in one place.

Movement could have been along the coast, but largely confined to one general area. The extent of exploration must have also depended on how secure the group felt in striking out. Several groups banding together would be safer, but that depended on numbers of people landing in one area, the antagonisms between them and predator frequencies. Groups may have spread out as a function of natural behaviour, but this would depend on how much experience of their new surroundings they had and how quickly they learned what was available to eat. If the concept of having a base was important, then exploration of surrounding areas would have been slow and tentative. In this case, movement into the interior may have been hesitant, but I emphasise that movement into arid areas would not have been a priority among people totally unused to such places. In contrast, others may have moved inland almost immediately and not returned. Speculation regarding these factors can go on almost indefinitely with no firm conclusion. Many combinations are possible depending on the cultural complexity, economic and technological capability, population size and distribution and attitudes of the initial migratory groups. It seems reasonable to suggest, however, that, if people were capable of sailing to Australia successfully, exploration of solid ground should not present too many difficulties. In order to define some possible characteristics of how exploration might have taken place, I want to refer to two basic models of migratory movement within Australia: fission or budding-off and continuous migration.

Fission or budding

The fission or budding-off process requires band growth big enough to produce a second band that then moves off without depleting the founder band. Continued forward movement using this process depends on an initial population big enough to form buds. Population growth must continue in the progenitor band sufficient to make further buds. A version of this mechanism was described almost 50 years ago. It explained how the first colonisation of Australia could have taken place and at what speed it could occur (Birdsell, 1957). In short, the model suggested:

> The most likely estimate, based upon the assumption that a horde of 25 persons represented the normal colonising unit and that the budding-off occurred when a population reached 60 per cent of its carrying capacity, gives the surprisingly short period of 2204 years of total elapsed time [to colonise Australia].
>
> *(Birdsell, 1957: 65).*

This approach lacks the element of humanity, which I have tried to ascribe to our migrants, together with what we now know about glacial/interglacial sea level change, continental shelf exposure and internal palaeoenvironmental change, all producing a very different picture from that understood in the 1950s. Birdsell's figure was derived from known band size numbers among contemporary Aboriginal groups, but these may have little to do with organisational strategies and initial band sizes 60 ky or so ago. Perhaps these marine hunter-gatherers operated 'raft-bands' so that between 25 and 50 people moved in a flotilla of several craft. Whatever it was, the fission model relies on new arrivals, therefore we have to invoke regular landings in substantial numbers to fuel the process. This seems unlikely in the Upper Pleistocene. Rather than continuous arrivals, however, all that is needed to continue the fission process is a threshold, where, after a certain number of people have arrived, the process fuels itself without requiring further migration. The rate of world population growth suggested in earlier chapters, however, would certainly be able to supply suitable numbers to serve such a model. At some point the population will be large enough to begin intrinsic growth (within Australia) without the requirement of continued migration.

If the fission process relied on intrinsic growth, mating rules and mortality rates are major determinants of how long population doubling took. Normally, however, intrinsic population growth comes only from births to existing females and continues only when young females become fertile. Using the generation gap of 16 years described in earlier chapters and an initial age of eight years for the oldest female child, the founder unit would have to wait a further eight years for its first additional person. During that time other children may have been born, although some band members may have been lost. Both old people may have died together with one or more females and some children. Unfamiliarity with animal species may have made the first bands particularly vulnerable to attack. With the loss of the two old people, two women and two children, for example, and the addition of three births, one each to the two remaining females and the teenage mother, the founder unit, after eight years, would have a net loss of three people. Out of seven original offspring, only three to five may have survived to reproductive age. Any cultural bias towards males would only serve to decrease the chances of females reaching that age. Assuming that incest is not an option, these females can only reproduce with males from another founder band, if there is one, and there is nothing to suggest that other bands did not experience a similar history. The actual process was probably infinitely more complex, however, resulting in many reproductive combinations that produced different demographic situations. All of these would have had variable consequences for the size and capabilities of future population expansion and growth. Viability of these early groups was paramount for continued migration from intrinsic sources.

There must also have been people land on this continent, live in an isolated unit, then perish for one reason or another. Each band could have had a different history, some perhaps wandering off or landing too far from others and disappearing while others prospered. It is hard to envisage the circumstances where units would be doubling in size each generation during these early stages. This is particularly so at a time when a firm foothold on the continent is being established and people are only just beginning to come to terms with the new environment. The budding-off process requires also that the bud and the original band both double in size before people move out. There are, however, various problems with this bacteria-like method of increase. If the original group continues to grow, and we must assume that it does, where do the extra people move to? They must move away, but that would mean entering territory already occupied by previous buds. Also, how long does it take for a band of 15 people to double in size? I would suggest that many generations are needed to accomplish this. I refer again to the growth rate of 0.00002 suggested for Asia at this time (57–92 ky), giving a population doubling time of 34 655 years (see Chapter 2, Table 2.4). Peripheral populations living at the edge of the Asian influence, such as those entering Australia, must have had even smaller growth rates.

The budding-off or fission method of population spread, therefore, cannot have been the mainspring for increasing the original population for three reasons. Firstly, all buds continue to reproduce, which over time causes an exponential growth rate and heavy clustering of people in initial centres of occupation that increasingly become over populated. Secondly, and in complete contrast, the growth rate of populations at the time of the first arrivals was so low that it would have taken each band at least 34 ky to produce one bud. Thirdly, it depends also on the saturation of one area before forward movement takes place: a mechanistic bow-wave model that does not fit human behaviour. There is no reason why an area should be filled before the population moves on. Inquisitive human nature also defies the possibility that a group of hunters would not explore the next valley because that must be left to a future 'bud of 25'.

With low rates of Pleistocene population growth, the fission method of continental filling would not have populated Australia to its every corner for tens of thousands of years. Such a situation flies in the face of the archaeological evidence that increasingly shows that just about all corners of the continent had been visited by 35 ky. So the first entry must have been much earlier than our archaeological evidence would suggest. Perhaps the first 40 ky or 50 ky of occupation of this continent will not be revealed till the sea lowers because settlement during that time was confined to the broad landscapes of the continental shelf. Moreover, the budding process, extremely slow during its initial stages, might have been the way Australia was at least partially populated during the first 60–80 ky. This would take a very long time to put people across large areas of the continent and/or bring them off the continental shelf. The archaeological visibility of such a process would be very difficult to detect during that long initial stage till more people arrived, probably in the late Upper Pleistocene. If the budding-off method is invoked, however, it means that Australia was fully explored by 35 ky, because no further land was available and no further extrinsic or intrinsic population growth could take place. The fission method would also produce a genetically homogeneous population if it relied solely on intrinsic growth processes originating from one or two progenitor bands. The only way the fission method could work in the time we are dealing with was by an almost constant stream of migrations that supplied the numbers needed for continental expansion.

This brings us to the next stage of inquiry. In any discussion of the populating of Australia, we have to make a choice: did it take place quickly or slowly? As Jo Birdsell pointed out, crossing Australia could have been quick but whether a populated continent was left behind, albeit sparse, is an entirely different matter. There is no reason to suspect that everybody had an immediate neighbour. Large areas may have remained unsettled until quite late, particularly marginal regions. If moving groups were followed closely by others to keep in regular contact with one another, then extrinsic population growth (additional migrations to Australia) must have been enormous because intrinsic growth could not account for it. Large extrinsic growth conjures up flotillas of rafts, but this was probably not the case either. Obviously there is a major question to consider here. If a continental population existed by 35 ky, it stands to reason that from that time on there was only a very limited growth in population or else by 1788 Australia's Aboriginal population would have been standing shoulder to shoulder at the shoreline. If it was the case that the Australian population did not increase over thousands of years, Australia would be the only place on earth where human population growth over the last 35 ky has been so static. The most likely answer to this contradiction is that as the driest inhabited continent on Earth and with 70 per cent of the land mass classified as arid or semi-arid, growth was confined to certain coastal and large riverine systems where sedentism eventually emerged perhaps 1 ky ago (Webb, 1987).

Population growth was not static, however. Initial movement must have been a slow flow of people from multiple points of entry, scattered across the old northern palaeocoast. Direction of movement from there and into the interior was subject to a complex set of environmental conditions and changes to those conditions, human choice and the ability of the explorers to adapt to inland conditions and cope with setbacks. Intrinsic population growth among these initial groups must have amounted almost to zero. Therefore, a more realistic model for the early stages of growth among these early populations depends not on the reproductive strength of the first arrivals but on a continued, limited and variable net input of people derived from migration to these shores. The implication, of course, is that extrinsic population growth in Sunda and adjacent regions is large enough to supply such migrations.

Continuous exploration

The second and most likely method of continental exploration was continuous exploration. The arrival of intermittent or regular pulses of migrants allowed them to move to other areas as a cohesive unit. Rather than wait for enough people to occupy a given region, this method calls for continuous exploration at their own pace, while keeping close to others. Isolation must have always been a hazard, however. Even with the best intentions of keeping in touch with others, neighbouring groups may have succumbed to various natural as well as predatory accidents.

In northern Australia, several large species of reptile would certainly have caused problems for small, vulnerable populations. Some species of crocodile, such as *Crocodylus porosus* and *Palimnarchos pollens* inhabited coastal, estuarine and riverine environments from the Kimberleys to northeastern Queensland as well as along the river systems that ran into Central Australia prior to 60 ky (see Chapter 5). These animals grew to a very large size and regularly attained their maximum growth potential of 8–9 m for both species and no doubt they had little fear of humans. As well as having the largest range of highly venomous snakes in the world, Australia had the enormous python *Wonambia naracoortensis* as well as the dubious privilege of having the largest land-based reptile *Megalania prisca*. This massive goanna-like monitor lizard, the largest of which probably attained the same length as the crocodiles, must have eaten its fair share of people of all ages until they themselves became extinct across the continent sometime after 50 ky ago. There were plenty of hazards for these early explorers until they became familiar with their surroundings, but the loss of individuals from the small groups was disastrous.

Occasionally, natural growth could produce enough people to form new migratory buds that then went exploring, settling in 'empty' country or mixing with other pre-existing bands. A continuous movement of people away from landing places either into the interior or along coastlines would make room for new arrivals. In this way continental exploration could occur quite rapidly. Thus, many areas could be reached comparatively quickly but without invoking the need for a large population, or one that has to saturate every square kilometre before moving on, or has the need for people to be continuously in touch with each other.

The pattern of movement that emerges from this type of 'exploration' takes advantage of those places with plentiful resources. One strategy is following riverine environments, while opportunistically venturing into more remote parts of the hinterland away from rivers when the need arose. The river then becomes a place of fallback when times are hard, disaster occurs or when the limit of foraging potential is reached. Meanwhile the main body of the colonising group continues to follow the central resource focus of the river system. Low birth rates can be accommodated in this model, whereas in the 'fission' model they need to be fairly high together with a low infant death rate. A number of physiological and reproductive factors suspected as playing a part in human hunter-gatherer lifestyles, such as infrequent menstrual cycles, more readily fit the continuous exploration rather than the fission model.

The large river systems of the north facilitated inland movement as people explored headwaters of creeks and rivers flowing into the interior. Using these freshwater highways opened up many migratory avenues via a myriad of variable sized river and creek channels. Easterly movement took people to Cape York and the east coast. Moving west would take people to Arnhem Land, and a south and southwest movement to the comparatively flat savannas of the Barkly Tableland and Selwyn Range. To reach the Barkly

Table 1 and meant an imperceptible ascent of 600 m that then gently sloped south into the Lake Eyre basin and eventually through the Simpson Savanna and down into central and southern areas of the continent. But what was the Central Australian environment like then and how did it change during glacial cycles? That is where our journey takes us next.

5 *Palaeoenvironments, megafauna and the Upper Pleistocene settlement of Central Australia*

> According to the traditions of some Australian aborigines [sic], the deserts of Central Australia were once fertile, well-watered plains. . . where to-day the only vegetation is a thin scrub, there were once giant gum-trees. . . the air, now laden with blinding, salt coated dust, was washed by soft, cooling rains, and the present deserts around Lake Eyre were one continuous garden. The rich soil of the country, watered by abundant rain, supported a luxuriant vegetation, which spread from the lake-shores and the river-banks far out across the plains. The trunks of lofty gum-trees rose through the dense undergrowth, and upheld a canopy of vegetation, that protected the country beneath from the direct rays of the sun. In this roof dwelt the strange monsters known as the 'Kadimakara'. . .
>
> Now and again the scent of the succulent herbage rose to the roof-land, and tempted its inhabitants to climb down. . . Once, while many Kadimakara were revelling in the rich foods of the lower world, their retreat was cut off by the destruction of the three gum-trees, which were the pillars of the sky. They were thus obliged to roam on earth, and wallow in the marshes of Lake Eyre, till they died, and to this day their bones lie where they fell.
>
> 'Kadimakara' – an Aboriginal Dreamtime story of the Dieri Tribe (Cooper Creek) South Australia, after Gregory, 1906: 3–4.

The previous chapter discussed some of the many variables associated with initial colonisation of the Australian mainland and set the scene for the possibilities for human migration into Central Australia. It particularly introduced the idea that people could have passed along a corridor of rivers that took them into inland Australia. The corridor was made up of a vast network of river systems that began at the abutment of the Lake Carpentaria/Lake Eyre basins on the Barkly Tableland which then flowed south 1000 km through the Lake Eyre basin and eventually poured out into Lake Eyre itself. By clinging to the river, no lateral spread took place, but rapid forward movement would occur. So, the movement of people through this system would eventually put them into vast areas of Central and Southern Australia comparatively quickly. But was the environment able to support them and what were the dangers they confronted? To answer these questions, we need to know what conditions were like in the region during the last interglacial. So I now turn to research I and others have been carrying out in the region for over a decade. In so doing, I outline a palaeoenvironmental reconstruction by looking at the general environmental conditions that prevailed between 65 ky and 150 ky ago. In addition, I include a description of the megafaunal animals that lived there in order to use them to provide a picture of the suitability of the environment for human habitation. I do this by looking at their diet, water availability in the region and patterns of reproduction and behaviour.

Background

For over 100 years northeastern South Australia has been regarded as one of the richest fossil animal deposits anywhere in Australia (Stirling, 1896, 1900, 1913; Stirling and Zeitz, 1896, 1899, 1900; Gregory, 1906; Stirton, 1967; Stirton *et al.*, 1961, 1967; Rich *et al.*, 1985; Wells and Callan, 1986; Tedford and Wells, 1990; Tedford *et al.*, 1992). The fossilised bones of Australia's megafauna, or the 'kadimakara' of Aboriginal legend, emerge from sedimentary deposits outcropping or exposed by down-cutting along modern river channels that cross the Simpson Desert and Lake Eyre region. Many of these assemblages fall within the time that humans first arrived in Australia.

During the last Ice Age, Australia lost 85 per cent of all terrestrial animal species weighing over 44 kg (Flannery, 1990a; Roberts *et al.*, 1998; Miller *et al.*, 1999). Why so many species disappeared at this particular time, after surviving previous glaciations, is a source of continuing debate. Reasons put forward include environmental change, hunting by humans, vegetation change as a result of anthropogenic burning, and a combination of all these. While some or all of these could have been responsible, the extinction pattern and its timing across the continent remains unclear. It goes without saying, however, that, if humans were involved in any way with these extinctions, their timing becomes a signpost of a human presence. Evidence from my research in the Lake Eyre basin of Central Australia may throw some light on these events and I want to outline that evidence in this chapter.

Geographical and geomorphological setting

The research area extends over 250 000 km^2 within the Simpson and Tirari Deserts north and east of Lake Eyre (Map 5.1). The most productive and well-studied fossil deposits are situated along the last 250–300 km of the Diamantina-Warburton River and Cooper Creek which empty into Lake Eyre

136 The First Boat People

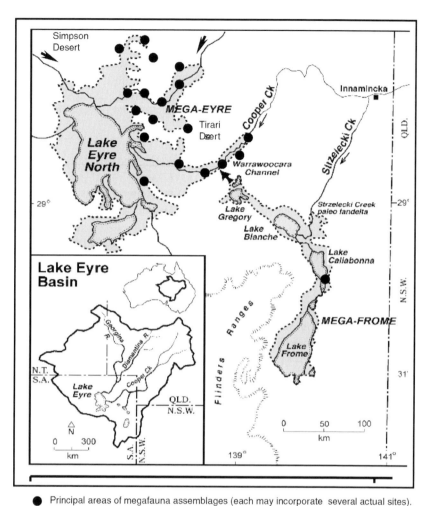

● Principal areas of megafauna assemblages (each may incorporate several actual sites).

Map 5.1. Megalake phase of the Lake Eyre/Lake Frome system at 85 ky. (After Callen et al., 1998)

and whose catchments lie in central and northern Queensland. J.-W. Gregory and his party travelled these channels in the summer of 1901–1902 and they were the first to look for animal fossils in the region. The expedition gathered many boxes of fossils that were then carried out by camel and eventually sent to Glasgow University where they were destroyed during an air raid in the Second World War.

Megafaunal remains have been found on 49 sites scattered throughout the region. Thirty of these were found by myself and colleagues during the last

14 years. Fossil megafauna is also found on Kallakoopah Creek, a 350 km anabranch of the Warburton River, that dips up into the Simpson Desert exposing Late Quaternary alluvial deposits on either bank. Other sites are found within similar deposits outcropping along numerous smaller creeks and around playa lake margins in the dune field system. The most recent, abundant and best preserved fossils are found in upper sections of the Katipiri Formation that overlie the increasingly older Kutjitara, Tirari and Etadunna Formations, the latter having a basal date extending well back into the Miocene (Magee, 1997). The Katipiri consists of differentially deposited, well-bedded, lacustrine, riverine and aeolian sediments, consisting of sands, silts and clays that represent changing environmental and weather conditions of Central Australia during the last two glacial episodes. Transported by the vast river systems that ran through Central Australia, extensive thermoluminescence (T/L) and 18 U/Th dating of these sediments over a large area have firmly established it as late Quaternary, spanning 60–260 ky (Callan and Nanson, 1992; Nanson et al., 1992; Magee, 1997, 1998). The Katipiri Formation has been divided into two phases separated at around 130 ky (Magee, 1997). The upper Formation 1 represents a series of environmental changes that span from the middle of ^{18}O Stage 3 to Stage 5 and includes a series of extensive wet periods, high fluvial activity and lake full events (Figure 5.1). These were punctuated by dry intervals but lacked episodes of severe deflation. The wettest time is between 108 ky and 130 ky, but aridity and deflation slowly increase with episodic wet periods culminating in a final brief lake-full event at around 45 ky. Naturally, channel flow decreased from 70 ky onwards with increasing aridity, finally terminating palaeochannel activity around 55–60 ky as the last glacial maximum approached. This coincides with results obtained from examination of carbon isotopes in the egg shell from the giant flightless rattite *Genyornis newtoni* that suggest the summer monsoon switched off around 60–65 ky (Johnson et al., 1999).

Glaciation initiated dune building, which eventually spread from Central Australia into the south and east of the continent (Callan and Nanson, 1992; Nanson et al., 1992). Dune deposits in the Simpson Desert/Lake Eyre region contrast strongly with underlying Katipiri sediments, with an obvious discontinuity between the two emphasising marked environmental change sometime around 60–65 ky followed by severe aridity and aeolian activity. These events coincide with the disappearance or local extinction of many megafaunal species. *In situ* fossil bone or stone has not been found in dune sediments overlying the Katipiri, so the top of the Katipiri Formation represents an environmental and faunal hiatus, marking the final disappearance of the region's megafauna. Amino acid racemisation dates obtained from *Genyornis* eggshell indicate that this element of the megafauna was an exception to the

Oxygen isotope stage	Age and error	Stratigraphic formation	Palaeo-environment
1	Present	Dune formation	Similar to present. Slightly wetter than present continuing expansion of Simpson Desert dune field.
2	12 050 ± 3140	Dune formation	Aridity and deflation. Expansion of Simpson desert.
3	24 110 ± 4930	Palaeodune formation	Poorly known, but generally drying through out – possibly slightly wetter between 30 ky and 35 ky
4	58 960 ± 5560	Palaeodune formation	Last lake-full event (-3m bsl), but not as full as earlier lakes; fluvial systems active but less sothan in stage 5.
5a	58 960 ± 5560 79 250 ± 3580	Katipiri 1(upper)	Aridity and deflation.
5b	90 950 ± 6830	Katipiri 1 (upper)	Wettest interval on record. (85–130ky). Punctuated bydry periods but without serious deflationary events. Conditions wettest in early interval (extensive, permanent, deep-water lake). Megalake phase.
5c	103 290 ± 3410	Katipiri 1 (upper)	Ditto
5d 1	110 790 ± 6280	Katipiri 1 (upper)	Ditto
5e 1	122 560 ± 2410 123 820 ± 2650	Katipiri 1 (upper)	Ditto
6	129 840 ± 3050	Katipiri 1 (upper)	Aridity and deflation
7	189 610 ± 2310	Katipiri 1 (lower)	?
8	244 180 ± 7110	Katipiri 1 (lower)	?

Source: Magee (1997).

Figure 5.1. Upper Pleistocene oxygen isotope stages and equivalent palaeoenvironments in the Lake Eyre basin.

general faunal extinction event of 65 ky. Fragmentary shell and almost complete eggs from these giant mihirungs have been discovered in palaeodune cores that mark early stages of dune building and indicating that this bird continued to inhabit the region until around 50 ky (Miller et al., 1999).

The palaeobiological environment of the southern Simpson Desert

Flora and water

The shape and distribution of Katipiri sediments indicate an extensive drainage system existed throughout the region between 75 ky and 125 ky. It formed a far more complex system, containing both main channels and abraded meander channels, than the modern drainage system, indicating a relatively wet interglacial period in this part of Central Australia (Nanson *et al*., 1988, 1998; Magee, 1997; Bowler *et al*., 1998; Magee and Miller, 1998; English *et al*., 2001). In many areas lacustrine deposits consisting of thick, interbedded layers of fine clays and silts indicate deep water stages in the large lake of the region, some of which such as Lake Eyre extended into and flooded valleys to the north and east of the modern lake boundaries (Map 5.1). Indeed, at certain times Lake Eyre was a very large water body with an estimated capacity of around 490 km^3, an equivalent of 1000 Sydney Harbours, and far greater than the 30 km^3 of the deep historical filling (Magee, 1997; DeVogel *et al*., 2004).

Generally speaking, there was almost constant water flow across the basin between 75 ky and 130 ky, fed by run-off from Lake Eyre's 1.27 Mk^2 catchment system in northern Australia. The amount of water required to fill these channels and lakes at this time seems to have been the result of a more southerly but variable monsoon incursion into northern Australia. At times the monsoon penetrated deep into the catchment, bringing high local rainfall, activating smaller rivers and creeks; topping-up standing reserves in small lakes and depressions and supplementing larger channel flows. At other times, it moved north, reducing headwater precipitation and reducing riverine flows south into the Lake Eyre region. Localised rainfall helped maintain also a broad growth of vegetation and great diversity of plant species away from rivers and lakes. Decreasing average annual temperatures maintained standing water for longer periods through lowered evapo-transpiration. All these factors contributed to the creation of a myriad of microenvironments, attractive to a variety of animals and birds. Without local rain inter-riverine and inter-lake areas might have proven less favourable to some animal species unless supplemented by soaks and springs, while main channels provided corridors of abundance.

Preliminary pollen studies of lacustrine sediments indicate many plant species found in the region today after good rainfall grew there during the last interglacial and were present when the megafauna became extinct. Poaceae, Asteraceae and Chenopodiaceae dominate fossil pollen assemblages, although few species growing outside the region today have been detected. Tree pollen is sparse but the presence of partially fossilised wood (*Ficus* sp.), rhizomorphs in Katipiri sediments and fossil tree pollen (*Callitris* and *Casuarinacaea*) suggest the presence of a variety of tree species growing across the plain (Lampert *et al*., 1989). Moroever, the fossil remains of possums (*Trichosurus vulpecula*) and koalas (*Phascolarctos* sp.) in Katipiri assemblages also provide evidence for some tree growth in the region

(Tedford and Wells, 1990). Dietary preferences of the brush tail possum and koala for eucalypt leaves and fruit and blossoms, indicate that suitable tree species were plentiful, although they were probably concentrated along rivers rather than out on the plains. While the number of regional interglacial plant species may have been similar to today, vegetation is likely to have been more abundant given higher moisture levels and lower evaporation rates. At these times the Simpson Desert may have looked more like a savannah than a desert.

Fauna

Megafaunal remains include an extensive suite of large and medium-sized browsing and grazing herbivores, large mammal and reptilian carnivores, as well as *Genyornis*. The biggest marsupial, *Diprotodon optatum*, is represented together with a broad range of other Diprotodontids. These include: *D. minor, Euryzygoma dunense, Zygomaturus trilobus, Nototherium* sp. and *Palorchestes* sp, representing a size range from a small tapir to a rhinoceros. Their activity and distribution patterns are little understood, but caloric requirements to drive large, muscular bodies must have been substantial. The locomotive capabilities of *Diprotodon* may have made it a far nimbler and speedier animal than is suggested by its appearance, something shown to be the case for certain large dinosaurs (Bakker, 1988). It follows that the more active the animal, the greater its caloric needs. Fodder consumption among large modern placental mammal species, such as hippos, rhinos and tapirs, varies from 40 to 75 kg per day. Given their body size and consequent energy requirements the largest marsupial herbivores probably required a similar intake.

A number of very large macropods such as *Procoptodon* spp., *Sthenurus* spp., *Simosthenurus* spp, *Macropus* spp. and *Protemnodon* spp. would also have had good appetites. Perhaps not all these species lived in the region at the same time, but they were certainly present at one time or another during Stage 5, because they have been found in sediments of that age. The remains of smaller animals such as possums, koalas, bandicoots and wombats, including the giant varieties *Phascolonus gigas* and *medius*, point to a wide species spectrum; moreover the combination of small with larger species represents a rich herbivorous fauna, certainly far richer than exists today.

The presence of large fish vertebra in fossil assemblages also confirms the presence of deep water bodies. These consisted of lakes and large waterholes along the bigger channel systems that produced fish reaching weights of up to 80 kg. Shellfish, freshwater crustaceans (yabbies) and snails are all present in regional assemblages. Almost all fossil sites contain significant assemblages of mammal, reptile, fish, eggshell, crustaceans and shellfish. In reality, almost the whole region is a Quaternary fossil site, representing one time frame or another. The distribution of assemblage locations indicates a widespread occurrence of species through time. Fossil sites occur at the base of or on benches marking eroded cliff lines along modern channels, as well as point bars and sand bars on the beds of modern channels. Some scatters of bone can extend a couple of kilometres. Point bars bear particularly rich bone assemblages, but much of it has travelled some way from source and almost all sites lack exact temporal control. *In situ* bone can, however, be found in gullied or cliff sections as well as on the edge of playa lakes.

Assembling population data from fossil bone is always difficult, but the lack of temporal control and a variety of taphonomic factors associated with Katipiri deposits make it impossible to estimate actual numbers of animals living in the region at any one time. Bone beds can often represent reworked deposits and contain elements from large slices of time 'pancaked' into one undifferentiated lag deposit. Fluvial activity concentrates and reworks assemblages, while later deflation and erosion may deposit bone from different time sequences or origins on to one surface that is then reworked further. Interpretation of mixed deposits thus becomes impossible with regards to particular temporal relationships and event sequencing. Instead, these assemblages represent blocks of time, but this is not invaluable in the story I have tried to unfurl. The relative abundance of bone in the Katipiri Formation, however, is hard to explain in terms of reworking alone. For example, *in situ* articulated elements, the preservation of animal droppings in the form of coproliths as well as the presence of small jaws, with a well preserved, non-eroded and complete dentition, suggests little if any reworking and a certain stability in and localisation of many deposits. All these factors were taken into account during site interpretation and the conclusions reached therefrom.

Interglacial faunal ecology and distribution

During dry periods, large quadrupeds would have congregated along large water-courses or at least those still holding water. In wet, abundant times, they probably moved out on to lush grassy inter-riverine plains. Movement across the landscape from riverine to the open plains depended on the availability of freshwater in soaks, swamps, water holes, small lakes and creeks, inter-lake channels and ephemeral streams. The relative abundance and distribution of vegetation was also a factor in their distribution. Large channels were more common in the region between 75 ky and 125 ky than they are today and were probably situated closer together, making inter-riverine corridors narrower. Moreover, with ample fodder along meandering and abraded channels, one corridor might meet up with the next, making for easy movement throughout the region. These places, together with associated reed beds and back swamp areas, would have made ideal foci for animals such as Diprotodontids. My discovery of the forelimbs of one of these creatures embedded in fluvial sediments and clays associated with larger palaeo-channels are testament to the existence of such areas and as places of entrapment for the unwary.

In contrast, large and medium-sized macropod grazers would have been more at home in open, grassy areas, visiting permanent water sources each evening and when other sources became scarce. While quadrupeds and macropods may have favoured riverine and open-plain environments, respectively, such niches were probably not mutually exclusive, with range overlap occurring as water and food resources fluctuated. Range sharing may also have occurred through differential targeting of food resources, with Diprotodontids and the largest varieties of macropod browsing on shrubs and fruit-bearing thickets.

The predators

Australia has always had few carnivores. The wide variety of herbivores represented in the Simpson fossil assemblages suggests that their numbers were large enough, however, to more than support the several species of large predator that did live there. These included *Megalania prisca*, of which there may have been several species of different sizes. All adult animals were larger than the Komodo Dragon (*Varanus komodoensis*), with the biggest probably reaching seven metres or more in length. Claims for the greatest size this animal attained vary, however.

Crush pits on *Diprotodon* long bones, 10 mm in diameter, indicate the great forces behind the enormous crocodilian jaws of either *Crocodilus porosus* or *Palimnarchos pollens* that penetrated through cortical tissues and deep into trabecular bone. Tapir to rhino-sized Diprotodontids were vulnerable on steep, muddy river banks and a 5–8 m crocodile clamping its jaws over the snout of one of these animals and then rolling would cause massive injury and instant death if the neck were broken. I would suggest, however, only the largest of these creatures would be able to kill a bull *Diprotodon* in this way. These two large species of crocodile lived right across the Lake Eyre catchment system during lake-full times. It is likely that with the exception of each other, the total lack of predators of these animals allowed them to achieve lengths in excess of nine metres (Archer pers. comm., 1998). They patrolled the entire catchment system, moving along rivers that took them from the continent's northern coasts to Central Australia, which is fine testament to the amount of water flowing throughout the system. During wet times it was possible for these creatures to move from southern parts of Central Australia to Lake Carpentaria and out to the Timor Sea. On the other hand, the Lake Eyre megalake may have provided a focus that retained its own permanent or semi-permanent populations. To stand in the arid desert environment of the southern Simpson today defies belief that these animals lived there in shady, wetland backwaters not so long ago. But the presence of very large fish, a wide variety of turtles, some as big as a coffee table, and the crocodiles testifies to the abundance of water both in the palaeochannel and lake system during Stage 5. Such animals would obviously have been a menace to humans. Flimsy craft, used to traverse these waterways, would have been particularly vulnerable, with the huge jaws able to bite them in half.

Diprotodontids, particularly smaller species, must also have been regular and easy targets for *Megalania*. Beside bulk and large lower incisors, they had little with which to defend themselves and their bulk suggests that a quick retreat was probably unlikely. The Komodo dragon (*V. komodoensis*) hunts in packs and hamstrings its prey by biting the back legs, while others attack the belly to eviscerate it (Auffenberg, 1972; Quammen, 1996). The dying creature is then trailed until exhausted. Several animals would have made a formidable phalanx and one that may not have thought twice about tackling a three ton male *Diprotodon* or a group of humans. If *Megalania* behaved in a similar fashion, it probably hunted in small groups, ambushing prey near river banks and water holes. An extended chase for *Megalania* was probably not possible, but a quiet wait in thick undergrowth or along animal trails, then a quick ambush (like the Komodo) may have been a favoured tactic. Komodos are very quick over short distances and a similar strategy of surprise and a short 50 m dash may have been enough to achieve hunting success for *Megalania*. Even single animals might have been successful, its large size negating the need for pack attacks. The open savannah would have provided prey and grass high enough to cover an ambush. If *Megalania* also possessed bacteria laden saliva like the Komodo Dragon, this would have enhanced its killing potential if it needed any. Its demographics are hard to imagine, because fossil remains are not common and a complete specimen is yet to be found (Roberts *et al.*, 2001). The teeth of these animals have been retrieved by the author from a number of places around Lake Eyre and in the Simpson Desert. If they were distributed anything like that of modern

Komodo dragons, it is possible that up to ten lizards per square kilometre lived in optimum environments (Auffenberg, 1972). It is more likely, however, that the larger *Megalania* would have needed more area, perhaps one or two square kilometres per animal or even more. It has also been suggested that Megalania was semi-aquatic and if this was the case there is little doubt there must have been a number of them scattered throughout the Lake Eyre riverine system (Molnar, 1990). Again, it could have maintained a territory keeping other lizards at bay, but the thought of two of these monsters standing on their back legs fighting, as other varanids do, is mind boggling.

Discovery of vertebrae from the large python *Wonambi naracoortensis* confirms its presence in the Lake Eyre basin. The size of just about all megafauna species is disputed by one author or another, no less so for this snake. While it has been described as having a head the size of a shovel, this has been disputed (Flannery, 1994; Wroe, 2002). Nevertheless, it was a bulky snake, although not exceeding 5m in length, and its girth was probably twice that of a modern python of equal length. Recent reassessment of its abilities conclude that it probably had weaker jaws than our modern large pythons, it lacked the ability to disarticulate its jaws and had limited capacity for lateral flexion, preventing it from constricting its prey (Barrie, 1990). It may, therefore, have largely favoured an aquatic habitat with fish making up the bulk of its diet. Again, the Lake Eyre region during periods of high lake levels and stream flow was, therefore, an ideal environment in which to find this snake. It would not have been adverse, however, to terrestrial hunting of small animals and perhaps the infants of some megafauna species, including human infants and toddlers.

The lion-sized marsupial *Thylacoleo carnifex* was the largest of Australia's mammalian carnivores and there may have been several other smaller species. It was strong, although not swift, but it may not have needed to be. Its intermembral indices suggest that it was particularly adapted to climbing and may have attacked from above as well as taking its prey to an arboreal lair similar to the leopard. Besides its strength, *Thylacoleo's* armoury included a specially adapted carnassial dentition, four convergent incisors for piercing or stabbing its prey and a pseudo-opposable thumb with a large claw for ripping or pulling food towards its mouth. The carnassial complex was ideal for slicing meat, but whether it was also used for crushing bones is disputed. Certainly the deep cut marks on some fossil bone indicate chomping by this type of dentition, but it also points to an inability of this dental configuration to slice bone in half. *Thylacoleo* was, nevertheless, a formidable creature and an additional rather gruesome ability of this animal is the suggestion that it might have strangled its prey after it dropped on it from low branches (Finch, 1982; Wells, 1985; Archer, 1984: 684). If prey included the *Diprotodon*, this hapless creature was open to attack from the water, the land and the air as it wandered the riverbanks. If trees were sparse on the plains, *Thylacoleo* may have kept to gallery forest, leaving open areas to roaming *Megalania*.

Other smaller land-based carnivores included the marsupial wolf (*Thylacinus cynocephalus*) and a large version of the Tasmanian devil (*Sarcophilus laniarius*). It is likely that devils only scavenged and, while Thylacines joined in, they only hunted smaller creatures. Why the dingo seemingly replaced the Thylacine remains unclear, but speed may have been a factor. Therefore, only slower or infant animals may have been hunted by these dog-like marsupials. All carnivores are likely to have favoured riverine corridors, but the Thylacine may have been most at home on the plains. That would have conveniently separated them from *Thylacoleo* territory, giving the two largest marsupial carnivores room to operate.

The often fragmentary and friable condition of fossil bone in the Lake Eyre region usually prevents cause of death from being determined. It has, however, been identified in a limited amount of fossils from various sites around Australia. Smaller animals display scars of slicing and cutting by carnassial dentitions, but young animals would have been particularly important in the diets of *Thylacoleo*, the Thylacines and possibly the larger Tasmanian devils. Herbivore long bones lacking epiphysial union are common in regional assemblages, suggesting moderate to high death rates among sub-adults of both medium and large species, most probably from predation. Lake Eyre's fossil assemblages show that a variety of species lived in the region between 75 ky and 125 ky and its trophic 'shape' points to a well stocked fauna throughout the system. Although a very basic estimate of prey populations eludes us using the fossil remains alone, a theoretical assessment of the size of fossil faunal populations is possible using likely reproductive biology and level of survivorship of certain animals derived from the meagre data available.

Predator/prey relationships

The feeding habits of extinct animals are always difficult to determine. So too are the hunting strategies and prey consumption patterns of Australia's extinct carnivores. What we suspect about them is based largely on comparative studies and educated guesses. For example, does the modern pack hunting strategy of Komodo dragons apply to *Megalania*? Did *Thylacoleo* act like a lion and so on? The following analysis has been undertaken to model possible predator/prey relationships as well as to reconstruct megafaunal population sizes in the Lake Eyre region during the last interglacial. From this it is possible to gain perhaps a firmer idea of the relationship between carnivorous and herbivorous species, shed light on their demographics and form a better understanding of palaeoecological conditions faced by humans moving into the region.

The analysis uses a method based on that developed by Anderson *et al.* (1985) who used almost 300 modern animals, representing 33 species of carnivore and herbivore. It uses a combined average minimum long-bone circumference, which is then transformed into logarithms and used as a basis for least-squares regressions. Measurements were taken using average minimum antero-posterior and medio-lateral dimensions of the femur and humerus. Minimum, maximum and average circumferences of both bones then allowed determination of body size for Diprotodontids and macropods. Using this method the minimum, maximum and average body weights for the two basic groups of animals could be calculated.

Thylacoleo carnifex

While being the top mammal predator, it has been suggested also that *Thylacoleo* was probably the most specialised mammalian predator ever. It would certainly not have been nice to meet in the dark or any other time. Recent suggestions propose that it was the size of a medium-sized lion, whereas previous estimations put it closer to a panther (Wroe, 1999). It had robust limbs tipped by large ripping claws and a carnassial dentition specially adapted for crunching large bones, all operated by a powerful body. This made for a ferocious animal but how many roamed the Simpson Savannah?

If *Thylacoleo* attained the size of a lion, its basic demographic profile might have been similar to that of a lion. A lion's body weight ranges from 110 kg for small females to around 185 kg for large males (Schaller, 1974). If an average combined male/female weight of 145 kg is used for *Thylacoleo* and a similar annual 50:1 prey/weight ratio, then an average annual prey weight could have been around 7250 kg. For the calculations below, however, I use the conservative range of 110–145 kg. If *Thylacoleo* was less active because of possible arboreal habits, then the lower figure might be more appropriate to estimate prey requirements. Schaller (1974: 168) estimated lion densities for the 11 000 km^2 Serengeti National Park, Kenya, at one animal every 22–26.5 km^2 or 500–600 animals, while, at the same time, pointing out the difficulties in achieving an accurate estimate. Pienaar (1969) uses the slightly higher 29 km^2 per animal or 650 animals for Kruger Park. Population densities for *Thylacoleo* would probably depend on similar environmental factors to those for lions. These include the setting (woodland, open plain, etc.), available biomass, seasonal weather patterns, breeding behaviour, hunting strategies and so on. However, it is not likely they were 'pride' animals, nor did they give birth to multiple young. The paucity of their remains also suggests that their populations were indeed sparse. Bearing this in mind and erring on the conservative side once again, if the range for *Thylacoleo* is taken as 100 km^2 per animal, to allow for a drier savanna than that of Africa, then at least 2500 animals could have lived at any one time in the 250 000 km^2 Lake Eyre region. This figure still depends on food requirements, reproductive parameters and minimum numbers required for a viable population. Therefore, an estimate of total annual prey weight for the proposed population would be in the order of 13 750–18 125 tonnes or the equivalent of 6875–9062 medium-size adult *Diprotodon*s (see below). Obviously, they hunted a wide range of macropods and other smaller marsupials and not just *todons*.

Crocodiles (*Crocodilus porosus* and *Pallimnarchus pollens*)

While the average length of 4–6 m is normal for a fully grown salt water or estuarine crocodile (*C. porosus*), an upper limit of 9 m is included in Table 5.1. The reason for this is that specimens of this size have been reported from the Bay of Bengal and Mackay in Queensland at the turn of the century. Although not confirmed, these are not too far from confirmed kills of 6.0–8.5 m animals in Northern Australia (Webb and Manolis, 1989). Moreover, the hunting of crocodiles over the last 150 years has prevented contemporary animals surviving to reach extreme proportions. Animals of 8–9 m would not have been unknown in rivers and lakes of Central Australia during the Upper Pleistocene, when their only enemies were other large crocodiles. It is also possible that they were larger in these regions because of a tendency among crocodiles to be much heavier in freshwater environments than in tidal rivers (Webb and Manolis, 1989). Crocodiles usually have only 50 big feeds per year, hence the reduced prey weight ratio compared with mammals.

Megalania prisca

Estimates of body size for *M. prisca* are derived from skeletal reconstruction, extrapolation from Komodo dragons, and combined long-bone circumference measurements on fossils and casts. It is assumed that their body weight increased exponentially in relation to linear length, similar to crocodiles. In this case it is possible that the weight of a seven metre lizard may have been

Table 5.1. *Prey weight ratios (5:1) for large crocodiles (after Webb and Manolis, 1989)*

Length (m)	4.0	4.5	5.0	5.5	6.0*	6.0*	7.0	8.0	9.0$^\#$
Weight (kg)	240	350	500	680	900	1100	1500	2300	3400
Prey ratio (kg)	1200	1750	2500	3400	4500	5500	7500	11500	17000

Note:
$^\#$ An estimated upper length for *Pallimnarchus pollens* from fossil material.
Source: * Two figures given by Webb and Manolis, 1989 in their Tables 4.2 and 4.3.

Table 5.2. *Prey weight ratios (5:1) for large lizards*

Length (m)	3	4	5	6	7
Weight (kg)	85	150	270	*430*	620*
Prey ratio (kg)	425	750	1350	*2150*	3100$^\#$

Note:
Italics – Estimated body weight of *M. prisca* and prey requirements.

twice that for a five metre specimen. Table 5.2 presents a figure of 620 kg for a seven metre *Megalania*, an estimate formulated by Hecht (1975). However, using a femoral circumference (Cf) of 275 mm and a humeral circumference (Ch) of 262 mm from a full skeletal cast of a seven metre *Megalania* in the Queensland Museum, this animal would have weighed in the region of 2213 kg. This is three and a half times the weight estimated by Hecht for the same length and similar to the weight of an eight metre crocodile. While this weight may be far too high, it is believed that this calculation is closer to *Megalania*'s actual weight than the 650 kg suggested above.

Using the five to one prey/predator ratio, a 2000 kg animal would require an annual prey weight of 10 000 kg to sustain it, or five adult Diprotodons. A three metre Komodo dragon weighs around 85 kg, therefore prey weight per square kilometre ranges from 4250 to 31 000 kg (Table 5.2). Komodo dragons weighing around 85 kg regularly ambush 250 kg feral horses and attack fully grown 550 kg water buffaloes, but they hunt in packs and share the prey. As suggested above for *Thylacoloe*, *Megalania* may not have hunted in the same way as Komodo Dragons, which also feed only once per week or fortnight. Therefore a single *Megalania* occupying a larger range would require much less game.

Komodo dragon populations average around ten animals per square kilometre in optimal savanna environments (Auffenberg, 1972). It could be argued that the dietary requirements of ten massive lizards such as *Megalania* could

Table 5.3. *Prey weight required by three top Upper Pleistocene scavenger/ predators of the Lake Eyre basin*

Species	BS (Ave.)	APWR(kg)	PS	MBR (Tonnes)
T. carnifex	110–145 kg	5500–7250	2500	13750–18125
Crocodiles*	5–6 m	2500–4500	2000	5000–9000
M. prisca	5–6 m	1350–2150	1250	1688–2688
Total		9350–13900	5750	20438–29813

Note:
BS – Average body size.
APWR – Annual prey weight range required.
PS – Population size in the Lake Eyre region (250 000 km^2).
MBR – Minimum biomass required.
* *P. pollens* and *C. porosis* combined.

not be sustained in one square kilometre of savanna, even in optimal times. Moreover, like *Thylacoleo*, the paucity of remains of this lizard in the fossil record, usually consisting of single teeth, suggests a very much smaller population per square kilometre than is found among Komodos. Perhaps a more realistic demographic figure might be one animal in 200 km^2. That would add up to a minimum population size of around 1250 lizards in the 250 000 km^2 study area. Using this density, however, requires 6250 tonnes of prey weight to sustain the regional population annually or about 3000 adult *Diprotodons*.

Both reptile and mammal predators probably hunted a whole range of species. Smaller prey, the young of larger animals and the sick and dying would all have been particularly vulnerable. Scavenging was a better alternative than tackling large healthy animals and those that could speedily retreat or deliver a powerful kick or bite. However, a large tonnage of prey was required to support the carnivore population living in the region (Table 5.3).

Table 5.4 suggests that thousands of tonnes of biomass were needed to support the carnivores. So, how much prey existed in the region? Tables 5.4 to 5.7 present data that provide some idea of how heavy some Diprotodontids and larger macropod species might have been.

The smallest and largest *Diprotodon* is from the Darling Downs, Queensland. The biggest animal from Lake Eyre was close to 3 tonnes, while a larger specimen (almost 3.5 tonnes) originates from the Darling Downs; both were probably large bulls and animals to stay well clear of. The Queensland average is around 2.7 tonnes, while the Lake Eyre average is close to 2.3 tonnes, both in excess of the weight used to determine numbers of animals in Table 5.6. It is interesting to speculate whether this average weight differential reflects

150   The First Boat People

Table 5.4. Diprotodon *weight calculated using fossil long bones from the Lake Eyre basin*

Sample range	Cf	Ch	Ct	W(kg)
Minimum	212.1	238.8	450.9	1373.2
Maximum	292.2	303.2	595.4	2933.1
Average	268.0	275.2	543.2	2283.2

Note:
Cf – Femoral circumference.
Ch – Humeral circumference.
Ct – Total circumference of femora and humerus.
W(kg) – Total weight in kilograms.
Quadruped formulae after Anderson *et al.*, 1985.

Table 5.5. Diprotodon *weight calculated using fossil long bones from the Darling Downs, Queensland*

Size	Cf	Ch	W(kg)
Smallest	216	201	1109.3
Largest	335	298	3466.8
Average	304	276	2730.6

environmental/nutritional factors between the two regions during the interglacial, with smaller animals in Central Australia and larger ones in the east. Table 5.6 shows size ranges for Diprotodontids and macropods. Averages are then used to calculate how many animals constitute the prey tonnage necessary to maintain the carnivore population (Table 5.7).

Table 5.7 shows the numbers of kills required by carnivores from each of Diprotodontids, Macropus and two short-faced kangaroo groups. In reality dietary variety would be much more balanced. In reality there would be a mixture of the above animals together with a wide range of smaller species, emus, *Genyornis* as well as birds, eggs, reptiles and possibly the occasional human. Broad predation would offset the proposed number of larger animals required perhaps by as much as 30–40 per cent. In a more realistic distribution of prey, therefore, the maximum biomass total of 20 438 tonnes could be divided between for example 8175 or 40 per cent other species, 5000 large short-faced kangaroos, 5920 *Macropus titan's* and 519 *Diprotodons*. These totals look much more believable. The relative abundance of herbivores depended on their distribution patterns. Habitat preferences and micro-environmental differences would cause

Table 5.6. *Weight estimates for two major groups of herbivores*

Diprotodontids	Weight (kg)*		
	Maximum	Minimum	Average
Diprotodon sp. (LE)	1373	2933	2283
Diprotodon sp. (DD)	1109	3466	2730
Diprotodon (av. weight)	1241	3200	2608
MACROPODS			
(*M. giganteus titan*)			
Femoral circ.	56	102	79
CLB circ.	43	85	68
Av. weight	50	93	74

Note:
* All figures rounded to nearest kilogram.

Table 5.7. *Annual prey weight required from each of four herbivore groups required to sustain the estimated Lake Eyre basin carnivore population*

Species	AW(kg)	MinB (20438t) $n =$ Animals	MaxB (29813t) $n =$ Animals
Diprotodon sp.	2608	9883	11431
Macropus titan	74	276189	402878
Procoptodons/Sthenurids*	100	204380	298130

Note:
AW – Average weight for each animal.
MinB – Minimum biomass required by carnivore population.
MaxB – Maximum biomass required by carnivore population.
* – Estimate based on average 100 kg animals from both genera.

variations in density commensurate with resource availability. Concentration of animals in well-stocked regions would have transpired and carnivores may have favoured these.

These reconstructions always raise more questions than they answer. For example, what were the limits of distribution of each family of animals, what was the carrying capacity of the region, if we could measure it, and how did it vary during glacial cycles? Were there physiological limits to animal distribution, in other words were there places they could not live?

Megafaunal breeding patterns

Mammalian megafauna probably had an adaptive reproductive biology similar to their smaller extant cousins. We know little about their physiology and reproduction patterns, although we can extrapolate to a certain extent from the biggest living marsupial and an inhabitant of the Lake Eyre basin today, the red kangaroo (*Macropus rufus*). This animal has perfected survival in arid and semi-arid conditions by adapting a breeding cycle taking full advantage of sudden resource changes typical in such environments. A female can have a 'Joey' at heel, another suckling in the pouch and a dormant embryo (embryonic diapause) in one of her two uteri. When the suckling infant leaves the pouch the embryo begins to develop, later transfers to the pouch and the female becomes sexually receptive again, a veritable production line. Such adaptive behaviour can produce one fully independent youngster every 240 days, although embryo to full independence spans 600 days. Mating continues in times of drought; young are produced but generally die in the unfavourable conditions. In prolonged drought, lasting several years or more, breeding will cease altogether, but with breaking rains a female can have pouch young again within 60 days.

The breeding cycle of the red kangaroo may be applicable to *Procoptodon*, but its much larger body mass may have required a longer reproductive cycle, thus increasing its vulnerability by decreasing replacement potential in times of stress. It may not be unreasonable to extrapolate a similar breeding strategy to other large extinct macropods, such as *Sthenurus*, *Protemnodon* and *Simosthenurus*, each having a shorter cycle than *Procoptodon*, depending upon the mass of the animal, but longer than the red kangaroo, because they were all larger. But can the red kangaroo cycle be used for Diprotodontids?

Normally, reproductive cycles of large placental mammals are longer than for those of smaller species, with single young the norm. A similar pattern is true of marsupials, so if extinct marsupial reproduction cycles were the same, then *Diprotodon* must have had the longest cycle of any marsupial. Assuming they produced young in a similar manner to other giant macropods, longer gestation, pouch and cycle times could be expected, and single young would need to achieve a comparatively larger size before achieving independence. Moreover, carrying a large, heavy young would make an adult animal vulnerable to predators, while leaving the pouch too early would place the youngster in danger. Alternatively, pouch young growth rates may have been comparatively rapid in order to free the parent of the additional weight of a growing youngster, whilst providing the young with a size suitable for some degree of protection. It seems reasonable to suggest that sufficient pouch time would be needed before a young *Diprotodon* or *Procoptodon* could become independent, probably more than the 240 days of the red kangaroo, for example. Even after leaving the pouch, they would use it as a refuge for some time. The vulnerability of young Diprotodons at this stage must have been far greater than young macropods, because of their probable lack of speed. Therefore, the length of the full reproductive or replacement cycle for these huge creatures may have been as much as 12–18 months.

The Lake Eyre basin alternated between two environmental settings during the interglacial. One was quite lush, stable, savanna-like conditions with a constant food source; the other, less so, similar to present conditions and subject to a seasonal variation in food supplies and periods of extended drought. The former would favour animals breeding seasonally and living within the carrying capacity of their surroundings, but the latter favours those capable of rapid, opportunistic breeding. A more stable regime offering regular supplies of food and water, particularly along riverine corridors, would also favour seasonally breeding animals. Diprotodontids probably favoured the relative stability of these corridors, gathering in small herds or groups for protection, and occasionally moving out across open country. If they were largely browsers, they would have avoided plains and inter-dune corridors and if they preferred generalist browsing/grazing, which optimised food availability in a variable environment, they would have been at home in both open areas, feeding on grasses, daisies and small shrubs, as well as along riverine corridors. Seasonal shortages favoured both large and small macropods with an ability to rapidly migrate in search of food and remove themselves from the area.

The long history of the *Diprotodon* in Central Australia suggests these animals had plenty of time to adapt both their behaviour and biology to at least semi-arid conditions. Whatever their behavioural patterns were, they had ensured their survival through previous glacial episodes. Migration as a survival strategy has already been mentioned, another strategy may have been a physiological adaptation like urine concentration to conserve body fluids. Another would be an opportunistic breeding pattern similar to that of the red kangaroo. Animals able to regulate their populations are much less likely to suffer population collapse during seasonal shortage and long-term drought. It is a better breeding strategy for the Simpson Savannah environment prevailing during the last interglacial, which included several episodes of extreme drought. It is also the best mechanism to help ensure survival during previous glacial events and is particularly advantageous to opportunistic, migratory species. *Diprotodons* may have adopted different breeding strategies in different parts of the continent, but the obvious advantage of an opportunistic pattern among Central Australian animals is that it does not exclude them from any Australian environment. This reproductive strategy would have been the best for megafauna living in the Lake Eyre basin during the last interglacial.

High rates of sub-adult predation, single offspring and widely spaced births would place seasonally breeding animals at a distinct disadvantage in the Simpson Savannah. Large numbers of breeding females would be needed to maintain a viable population. Moreover, the standing crop would have to be plentiful and reliable, although this is more likely to have supported a wide variety of very large as well as smaller seasonally breeding, browsing/grazing animals. If this were the case, any reduction in crop size and distribution would have brought about population collapse that would be difficult to revive in these animals. Alternatively, if Diprotodontids were opportunistic breeders, they could have withstood predation and population reductions linked to fluctuations in food resources, thus maintaining viable and/or recoverable populations even in the face of long-term environmental stress, such as that encountered during peak glacial times. During glacial periods, many areas of the continent would not support these animals, particularly Central Australia, where extreme aridity and lack of vegetation and free standing water would have made it very difficult for them. Even when populations were reduced at these times, recovery, albeit slow, was possible and indeed that is probably what happened during the nine or more previous Ice Ages. The only thing wrong with this strategy is that it made these populations vulnerable to any other pressures put on them. So, why did these natural survivors become extinct in Central Australia by 65 ky?

Interglacial/glacial megafauna collapse

The biogeography of Australia's megafauna is not well understood, but it is likely that understanding it holds the key to knowing better how and why this amazing suite of animals disappeared when they did. The long history of the megafauna in Australia shows that they had plenty of time to adapt to its climatic vagaries, perhaps using migration to better pastures during extended drought or glacial episodes. Migration and biological adaptation probably maximised survival, particularly for those living in the interior. But populations may not have been dense and there may have been areas on the continent where the larger herbivores were rare or absent altogether and, therefore, these were also places where the largest carnivores were rare or missing. Other areas could have held minimal viable populations, in other words limited numbers of animals that were never very far from collapse as a population as long as everything remained equal. Larger groups employed survival strategies during glaciations only to regroup again in interglacials. Other areas as resident groups that were always attractive and changed little even during glacial events could have held large numbers of animals. Nevertheless, all megafaunal populations must have fluctuated through time and across geographic settings. So let us take a look at their distribution across the continent and during ice ages.

It is likely that not all species of megafauna lived everywhere on the continent. Some species may not have lived in the west or in particular environments, such as rainforest or desert. Other species may have been restricted to population patches restricted to certain areas and their numbers may not have been all that big. It is worth remembering that we have similar examples today of species separation between the western and eastern portions of the continent, for example numbats in the west and koalas in the east. Some areas such as Central Australia may have been devoid of megafauna species for large spans of time, particularly during glacial events. One reason for suggesting this is that if migration was a survival strategy used between and during glaciations, then population densities in any one area would have fluctuated. Exposure of the continental shelf during glacial times coincided with very arid conditions in Central Australia, the latter triggering an expansion of animals out of the centre to peripheral parts of the continent. With Central Australia being more habitable during interglacials, animals would move back, with a commensurate increase in numbers. Therefore the continental distribution of these animals, particularly very big species, was patchy at any one time and must have suffered an overall downturn during an Ice Age.

Diprotodon remains have been found just about everywhere in eastern Australia and Tasmania but they become much rarer in the west. Apart from examples in southwestern Australia, the Nullarbor, one each from the Oakover and Fortescue Rivers and Cape Range in the Pilbara and another from Windjana Gorge in the Kimberley, they are nonexistent west of the Simpson Desert (Maps 5.2 and 5.3). A brief survey of Quaternary deposits on 12 lakes in the Great Victoria Desert I carried out with others in 2002 failed to find any trace of megafauna or any other remains including *Genyornis* egg shell for that matter. Taphonomy, however, is a renowned trickster and more survey work is required before firm conclusions can be reached. Yet our present understanding of *Diprotodon* dispersal suggests that there were indeed some areas where it might not have lived or only in extremely small numbers. On the other hand, its dispersal in eastern Australia shows that it possessed a broad adaptation to continental conditions there although it is possible that some regional environmental differences may have produced several sub-species. From the above weight estimations it seems likely that

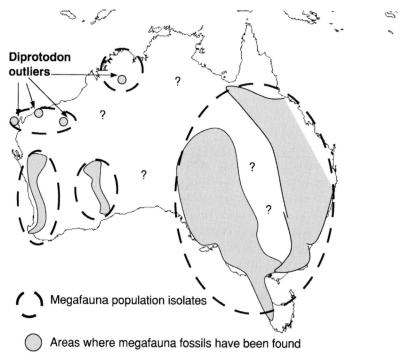

Map 5.2. Distribution of late Quaternary megafauna quadrupeds (*Diprotodon, Zygomaturus and Palorchestes*).

the eastern Australian, Darling Downs animals grew larger than those of Central Australia around Lake Eyre, which, if taken on face value, does look like a positive adaptive strategy to that environment.

If the above distributions are anywhere near correct, they are not hard to explain. Megafauna kept to areas where the most freshwater was to be found. Certainly eastern Australia bears the greatest numbers of streams and rivers, particularly in the southeast. The strength of the Central Australian presence revolved around the huge palaeochannel systems of the Lake Eyre basin. In Western Australia, however, the southwest is the only prominent population, with other remains found near more minor water sources in the northwest. Obviously the Nullarbor outlier is more difficult to explain, although likely to have focussed on a palaeochannel system that flowed prior to the last Ice Age. The north of Australia is a mystery, however. This is where these creatures should have been in their element, with large river systems flowing north into the Gulf of Carpentaria, the Arafura Sea and out from the Kimberley. These are areas where megafaunal remains should be found, although the Arnhem

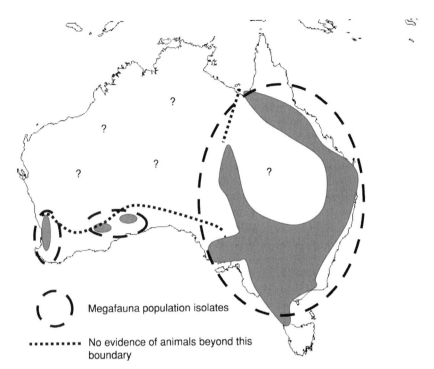

Map 5.3. Distribution of Late Quaternary megafauna macropods (*Procoptodon, Macropus, Sthenurus, Protemnodon* and *Propleopus*).

Land escarpment and the Kimberley could have posed problems for the really big species from the point of view of their rugged terrains. The lack of fossils tends to suggest that at least the larger species were not adapted to tropical conditions. If that is the case, then they would not have migrated in that direction when the centre became so harsh during glacial events.

With the distributions above in mind, it is now worthwhile introducing four important categories of uncertainty that apply to all animals and they seem to be useful in any discussion of megafaunal extinction. These include *demographic stochasticity, environmental stochasticity, natural catastrophe* and *genetic stochasticity* (Shaffer, 1981, 1987, 1990). Demographic stochasticity is the effects on any animal population from variations in birth rates, death rates and ratio of sexes. Environmental stochasticity is the fluctuation of things such as weather and food, changes in predator and competitor populations and increases in the frequency of parasites, infectious disease and other pathologies that affect the megafauna. Obviously natural catastrophes like cyclones, floods, fire, drought, as well as sudden and violent events, such as earthquakes and volcanic eruptions. On the macro scale an Ice Age could be included in this category. Genetic stochasticity refers to the balance between positive and recessive alleles in a given gene pool, or those that are beneficial and those that express in a harmful way. The former can become rare through genetic drift, particularly in founder populations containing small gene pools. Genetic drift occurs in any group that splits off from its mother group. In small isolated groups, genetic drift can eliminate positive alleles because it is operating on a rather small founder population. This can result in the expression of alleles that are harmful in some way, particularly through inbreeding which is always a possibility in small groups of animals and those belonging to dwindling populations. The genetic load is then increased or is in danger of increasing among small populations, whereas larger ones can carry an equal sized load or an even bigger one without deleterious results. If we apply these four uncertainties to certain megafaunal populations, it is possible to see what these creatures may have been up against when humans arrived, particularly those residing in small populations.

If the patterns of megafaunal demography apparent in the maps above are close to real, then Australia was divided east–west with by far the largest populations in the east and much smaller scattered populations in the west. Seemingly, they become even smaller in the north. Moreover, the whole of the Western Desert region may have been devoid of animals. So, there is megafauna from east to Central Australia, possibly with some very sparse populations in western Queensland that were extensions of the much larger populations inhabiting the Darling Downs. The western half of Australia is almost a mirror image, with a much more limited distribution of animals living only in patches in the southwest, in several places across the southern Nullarbor and isolates in the northwest. The overall population was much smaller than in the east and it was these populations that would have been susceptible to the vagaries of the four uncertainties outlined above.

These western populations were never more than minimally viable, that means they were statistically likely to continue if all things remained equal and no untoward stresses were imposed, particularly from the four uncertainties outlined above. They may have had enough flexibility to survive an Ice Age, perhaps through slow migration to better areas. Glaciations are slow processes and they may have had time to adjust, albeit that the event probably depressed their populations to their very limits. Larger eastern populations survived these events as a whole. Through their sheer numbers they were more able to survive even when some losses were incurred. The robusticity of their bigger gene pools allowed them to bounce back, no doubt slowly following the end of a glacial event. But they were not unaffected during these times and the four uncertainties must have contributed to this. Loss of numbers, fewer babies being born, drought, sudden lowering of ambient temperatures, starvation and, in smaller groups, the failure to breed and replace losses brought a cascade of events that led to a downward spiral towards extinction. One way the larger populations could have been stressed was through a process of fragmentation that particularly affected peripheral populations. Subdivision would produce small isolated groups already living under marginal environmental conditions. Disaster for these groups could follow from sudden but not necessarily catastrophic environmental changes such as a drop in ambient temperatures and/or rainfall, resulting in drying and desiccation of areas with fodder collapse. Isolation followed, as their usual avenues of contact with and escape back to the main herds, that had always been tenuous even at the best of times, suddenly closed. The small group(s) were then at the mercy of the influences of founder effect and genetic drift, both of which operate faster in animals with shorter generation turnover than humans. For pre-existing small populations, such as those in western parts of Australia, these processes were even more likely to occur because of the lack of larger populations to shelter with. This made them even more readily vulnerable to the effect of the four uncertainties and, therefore, more likely to be vulnerable during glacial events. It was at these times that the Australian megafauna population as a whole trended downwards. The effect of an Ice Age always depressed numbers through a reduction in breeding females, lower survival rates, reduced numbers of young; even one or two fewer than normal through severe extended drought would over time depress the population. Predator populations would not have been reduced till their prey animals became rare; therefore higher rates of predation may have taken place just at the wrong time, adding another stress on herbivore populations. On the positive side of the balance sheet, the inevitable collapse of predator populations then allowed for later recovery of prey species. What we have then is the large populations being reduced around their edges with some losses internally but hanging on, while smaller ones may have become locally extinct or coalesced with others when possible to maintain a minimum viable population. Now we can add the arrival of humans to this picture.

The present evidence from the Lake Eyre basin strongly indicates that megafaunal extinctions were complete by 65 ky. With the onset of very arid conditions at that time, there was less free standing water with contraction of climax vegetation and a concomitant reduction in shrubs and grasses. These changes were not sudden, but had been working up to the final drying for around 10 ky. The largest animals would have been the first affected, although it is difficult to imagine that they suddenly dropped in their tracks from starvation and dehydration. These changes were slow, but even if they were extremely rapid, occurring over centuries or even decades, the animals had plenty of time to slowly migrate to more abundant habitats as they had done during previous Ice Ages. Of course, macropods could migrate almost instantaneously and much farther than Diprotodontids within a given time period.

Aside from glacial periods, river channels still experienced some flow, providing a degree of shelter as well as possible avenues of escape and re-entry, as animals followed fluctuating water levels throughout the Lake Eyre basin riverine network. Populations of plain-dwelling macropods would have dwindled first, many escaping to better areas in scattered populations. Plains had a much lower drought threshold, with standing water drying out long before the main channels were affected. Rain in the catchment would keep main channels flowing and plain-dwelling animals could have concentrated along these. Increased competition for dwindling resources would follow, with predators experiencing a short-lived super-abundance of prey followed by serial population collapse of all species as channels then began to increasingly dry up with the gradual onset of the last glaciation.

With dropping river levels, many smaller streams dried completely, it became colder and salinity and aeolian activity increased, prompting source bordering dune building as well as dune formation out on the plains. Increasing aridity brought food shortages and further concentration of the animals that remained in the few remaining river refuges. Eventual herbivore collapse was accompanied by the loss of carnivores, with the biggest species the first to succumb. Slow migration of surviving animals would have been in full swing. Crocodiles also experienced dwindling food and water, with some retreating to upper catchment areas. Local extinctions occurred when animals became trapped in drying pools and water holes. Food sources became scarce as tortoises (*Emydura* sp. and *Chelodina* sp.), fish and water bird populations dwindled and fewer marsupials visited the river banks. Areas of local extinction widened and merged as regional drying and fodder reduction continued until whole groups of animals disappeared from the Lake Eyre drainage system.

Complete megafaunal collapse in Central Australia was almost certainly not sudden. Around 80 ky the Lake Eyre region experienced fluctuating aridity with decreased lake levels and, from time to time, widespread aridity (Magee, 1997). It is likely that animal populations never again reached the proportions that they were at that time. After 80 ky, a slow reduction in animal numbers and species variety began that gained momentum after 70 ky. A TL date of 64.9 + 5 ky (W969) from just below an *in situ* proximal section of *D. optatum* humerus from the Warburton River is the most recent evidence for this animal in the region, suggesting they may have withstood the semi-arid conditions that existed before this time. Amino acid racemisation dates from *in situ Genyornis* egg shell, collected from terminal units of the upper Katipiri Formation as well as from overlying dune sands, show that this giant flightless bird was living in the area till around 50 ky (Miller *et al.*, 1999). This is a bold exception to the general trend of regional extinctions that occurred between 60 ky and 65 ky. Perhaps *Genyornis* was omnivorous like the emu, surviving on a varied menu including lizards; but how it continued to survive in the face of extremely harsh conditions is not clear.

Evidence for humans in the Lake Eyre basin

The general consensus is that for whatever reason humans had to be a factor implicated in the megafaunal extinctions. If this is the case, then they had to be in the Lake Eyre region before 65 ky. Although dates from northern Australia point to people arriving before 60 ky, this is only a minimum entry time and a far older date for the first arrivals is assured from other evidence, such as regular burning. The earlier humans arrived, the longer they overlapped with megafauna and the less likely it is that they brought a 'blitzkrieg' down upon the hapless beasts. Megafaunal remains from Lake Mungo dated to around 40 ky show that there is at least a 20 ky overlap between human entry and the megafaunal collapse. Nevertheless, this does not eliminate the possibility that moderate hunting, firing of the land with accompanying vegetation change, together with the gradual onset of the last Ice Age all fell heavily on a delicately poised animal population. Another, more subtle, factor may have been the gradual increase in the human population over 20 ky or more that increasingly stressed animal populations so that by the time the glacial period began in earnest they were stressed to a point of no return. Such stress would have been felt differentially, however, perhaps greater in some areas than in others, depending on the direction and pattern of human migration.

The percentage of faunal extinctions in the Lake Eyre basin was regionally higher than the 30 per cent suggested for other parts of Australia, which includes modern species that left, never to return. In this case it may have been one area where megafauna migrated out, as they had always done during glacial periods, but this time the refuges they moved to were already occupied by humans. If, however, humans contributed directly to the Lake Eyre extinctions, then the timing of this event is important because it signals a minimum human occupation there before 65 ky, although their presence did not affect *Genyornis*.

162   The First Boat People

Plate 5.1. Human cranial vault from Lake Eyre basin.

Interglacial evidence for humans in the Lake Eyre region comes from two pieces of fossilised bone. The first is a small section of cranial vault found in 1988 as a surface find on a reworked lag deposit sitting on a sediments of the Lower Pleistocene Tirari Formation among megafauna fragments (Plates 5.1 and 5.2). It measures 38.4 mm by 24.0 mm, with an average thickness of 7.8 mm. It displays a shallow curve in two directions and features an inner (1.0 mm) and outer (1.8 mm) cranial table of compact bone, enclosing a layer of trabecular bone (5.4 mm). Its thickness is reduced (6.5 mm) at one end compared with the other (8.9 mm). Vault construction and the bi-directional thinning are consistent with an inferior section of the right or left temporal bone. Unequivocal identification of such a small fragment is difficult because, apart from the structures described above, it lacks other firm diagnostic features, but blood residue analysis has confirmed its human origin (T. Loy, pers. comm.). It has been examined by various experts in the field of Australian palaeontology, and the piece was firmly eliminated from any known or extinct marsupial (M. Archer, R. Tedford, R. Wells, R. Molnar, pers. comm.). The fragment has been dated by uranium series at Lucas Heights in Sydney to 132 ky±11.5 ky (LH 657) (S. Short, pers. comm.). The date corresponds to lower sections of the Katipiri Formation 1 that lie 200 m from where the bone was found.

Palaeoenvironments, megafauna and the Upper Pleistocene 163

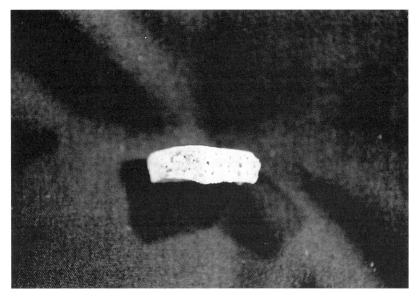

Plate 5.2. Human cranial vault from Lake Eyre basin.

With the benefit of over a decade of work in the region, direct dated *in situ* bone and T/L results from associated sediments have shown that uranium series dates are always far too high, although no consistent error of margin between the two has been found. It is for this reason that the above date is believed to be too old because of the uncertain behaviour of uranium in a bone's closed system, and therefore the date was not previously published. It was explained earlier that surface bone scatters are the norm in the region with *in situ* bone rare, so the continued difficulty of assessing the age of fossil material remains a frustrating barrier. However, rather than wait in hope that a totally reliable direct dating technique will be developed in the near future, because that is what is required, the comparative method of fluorine analysis has been used to provide at least an approximate, relative date for this bone, as well as for other fossil bone collected from the region. A comparative survey of fluorine content in megafauna bone from the Simpson Desert and Lake Eyre region was undertaken and these results, together with a set of comparative fluorine results from around the world, are presented in Table 5.8.

The Australian results in Table 5.8 are presented in Figure 5.2 and they show a clear correlation between four categories of fossil component (A, B, C, D) and the comparative age of bone and its fluorine content. Modern bone as well as fossils from around Australia are compared and they indicate a

Table 5.8. *Fluorine analysis of fossil bone*

Graph code Sample	%F	%N	%C	%H	Age
D Punkrakadarinna, *Macropus titan*	1.48	<0.10	3.89	0.20	141–183 ky (165 ky)@
D Punkrakadarinna, *Diprotodon* (*in situ*)	1.19	<0.01	1.49	0.52	64 ky
D Toolapinna West, *Thylacinus* sp.	1.56	–	–	–	>65 ky
D Toolapinna West, *Sthenurus* sp.	1.58	–	–	–	>65 ky
D OKA – 2, Sample, *Sthenurus* sp.	2.07	<0.10	1.64	0.35	>65 ky
D OKA – 2, Sample 2, *Macropus* sp.	2.90	<0.10	1.53	0.12	215–375 ky (259 ky)@
D OKA – 2, Sample 3, *Chelonid* sp.	2.65	<0.10	1.62	0.24	>65 ky
D OKA – 2, Sample 4, Crocodile	1.68	<0.10	1.65	0.36	>65 ky
D OKA – 2, Sample 5, ?Human	1.48	<0.10	1.59	0.29	>65 ky
H *OKA – 2, Human (cranial vault)*	1.37	<0.10	–	–	121–162 ky (132 ky)@
D OKA – 2 south (kangaroo mandible)	1.26	<0.01	1.53	0.38	
D OKA – 19 Bird	0.17	<0.10	–	–	Modern
D Williams Point, Wombat (*in situ*)	0.48	<0.10	1.75	0.71	40–50 ky
D New Kalamurinna, *Diprotodon optatum*	1.27	<0.10	1.91	0.35	177–283 ky (222 ky)@
D New Kalamurinna, *Diprotodon* vert.	1.38	<0.01	1.59	0.34	>65 ky
D New Kalamurinna, *Diprotodon* epiphysis	1.37	<0.01	0.75	0.72	>65 ky
D Lookout Site, *Sthenurus* sp.	1.46	<0.10	1.56	0.20	>65 ky
D Lookout Loc., *Genyornis newtonii*	1.31	<0.01	1.45	0.39	>50 ky
D Lookout Loc., *Diprotodon* rib	1.35	<0.01	1.08	0.42	>65 ky
D Lookout Loc., *Diprotodon* scapula	1.51	<0.01	0.90	0.63	>65 ky
D Lookout Loc., *Diprotodon* scapula	1.35	<0.01	0.83	0.93	>65 ky
D Lookout North, *Diprotodon* femur	1.68	<0.01	2.00	0.40	137 ky (T/L), 260 ky@
D Lake Hydra *Procoptodon*	0.55	<0.10	1.82	0.34	226 ky@
D Lake Hydra *Procoptodon*	0.85	<0.10	1.63	0.29	>65 ky
D Lake Hydra *Procoptodon*	0.62	<0.10	1.94	0.35	>65 ky
D Lake Hydra No ID (large)	0.53	<0.10	1.82	0.36	>65 ky
D Lake Hydra Fish	0.79	<0.10	1.33	0.24	>65 ky
D Sleeping Dog Cliff, *Diprotodon* (*in situ*)	0.61	<0.01	–	–	>65 ky
B Canny Dune, Bettong	0.17	<0.01	–	–	Holocene
B Canny Dune, Human	0.19	<0.01	–	–	Holocene
B Canny Dune, Human	0.17	0.11	–	–	Holocene
B Lower Cooper, Human	0.19	0.21	–	–	Holocene
MODERN SAMPLES					
A Williams Point, Rabbit	0.024	4.01	15.31	2.83	Modern
A Dune Field, Kangaroo	0.080			–	Modern
A Dune Field, Rabbit	<0.005	2.00	–	–	Modern
A Lake Hydra Northwest, Rabbit	0.059	3.66	12.48	2.14	Modern
A Lake Hydra Northwest, Bird	0.050	2.25	8.13	1.61	Modern
A Toolapinna north, cow	0.063	4.21	13.24	2.05	Modern
A Warburton River, cow mandible	0.040	3.76	11.72	1.93	Modern
A Warburton River, cow mandible	0.053	3.26	10.42	1.75	Modern

Table 5.8. (*cont.*)

Graph code Sample	%F	%N	%C	%H	Age
A Maroon Dog Bend, Dingo	0.069	2.39	–	–	Modern
AUSTRALIAN FOSSIL CRANIA					
C WLH 3	0.77	–	–	–	40 ky
C Kow Swamp 1	0.24	–	–	–	10 ky
C Kow Swamp 5	0.55	–	–	–	13 ky
C Kow Swamp 13	0.23	–	–	–	6 ky
C Kow Swamp 16	0.15	–	–	–	7 ky
C Kow Swamp 17	0.33	–	–	–	6.5 ky
C Cohuna (Macintosh)	0.36	0.22	–	–	LUP
C Keilor	0.48	<0.01	–	–	LUP
C Mossgiel	0.74	2.38	–	–	H
C Talgai	0.60	0.30	–	–	UP
PAPUA NEW GUINEA					
Aitape	0.84	<0.01	–	–	H
INDONESIA					
Modjokerto child	2.35	–	–	–	1.9 My?
Trinil 2 (calotte)	1.14	0.10	–	–	750 ky
Trinil 2 (calotte)	1.21	–	–	–	–
Trinil 3 (femur 1)	1.09*	–	–	–	–
Trinil 6 (femur II)	1.01*	–	–	–	–
Trinil 7 (femur III)	1.39*	–	–	–	–
Trinil 8 (femur IV)	1.40*	–	–	–	–
Trinil 9 (femur V)	1.06*	–	–	–	–
Sangiran 1, right maxilla	0.61	nil	–	–	–
Sangiran 2 (skull II)	1.15	0.03	–	–	800 ky[#]
Sangiran 3 (skull III)	2.93	nil	–	–	–
Sangiran 4 (skull IV)	1.33	nil	–	–	1 My
Sangiran 5, right mandible	0.56	0.01	–	–	LP
Sangiran 6, right mandible	2.32	nil	–	–	900 ky
Sangiran 8 (*Meganthropus*)	2.32	nil	–	–	LP
Ngandong 5	1.60	nil	–	–	UP
Kedungbrubus, cranium	1.76	nil	–	–	500 ky
Wadjak 1, cranium	0.58*	0.38	–	–	LUP
Wadjak 2, cranium	0.27*	–	–	–	LUP
SOUTHEAST ASIA					
Niah (Sarawak)	0.06	0.06	–	–	40 ky[#]
Tabon (Philippines)	0.06	0.07*	–	–	23 ky
Sai Yok (Thailand)	0.10	nil	–	–	4 ky
CHINA					
Salawusu, femur	0.38	–	–	–	UP
Ziyang, skull	0.79	–	–	–	UP (?37 ky)
AFRICA					
OH34 femur (Lower Bed III)	1.83	–	–	–	1 My
MIDDLE EAST AND RUSSIA					
Kiik–Koba (Crimea)	0.36	–	–	–	UP
Bisitun (radius)	0.11	0.17	–	–	UP

Table 5.8. (cont.)

Graph code Sample	%F	%N	%C	%H	Age
Shanidar 1	0.08	0.04	–	–	46.9 ky
Amud 1, rib	0.13	<0.01	–	–	25 ky[#]
Qafzeh 4, infant	0.01	nil	–	–	LUP
Skhul 5, cranium	0.05	0.05	–	–	100 ky
Tabun C1, cranium	0.19	nil	–	–	41 ky
EUROPE					
Mauer mandible	1.13	0.08	–	–	400 ky
Steinheim, cranium	1.20	0.37	–	–	225 ky
Swanscombe, occipital	1.70	0.18	–	–	225 ky
Swanscombe, left parietal	1.40	–	–	–	225 ky
Swanscombe, right parietal	1.90	0.09	–	–	225 ky
Vertesszollos, occipital	1.60	nil	–	–	186 ky
Fontechevade, parietal	0.50	0.63	–	–	120 ky
Fontechevade, frontal	0.40	–	–	–	120 ky
Saccopastore 1	1.80	0.14	–	–	80 ky
Saccopastore 2	1.70	0.12	–	–	80 ky
Krapina, layer 9, femur	0.90	0.61	–	–	75 ky
Krapina "3, skull	1.00	nil	–	–	75 ky
Krapina "3, juvenile	1.00	0.24	–	–	75 ky
Krapina "4, skull	1.00	0.36	–	–	75 ky
La Chapelle aux Saint	0.40	2.12	–	–	50 ky
La Ferrassie, foot	0.10	2.74	–	–	40 ky
Cro–Magnon	0.10	0.49	–	–	25 ky
Galley Hill	0.50	1.60	–	–	3.3 ky
Piltdown jaw	<0.03	3.90	–	–	Modern
Piltdown skull	0.10	1.90	–	–	Modern
AMERICAS					
Calaveras (California)	0.22	0.63	–	–	H
Diablo Canyon	0.19	0.74	–	–	H?
Melbourne Florida, Human	0.10	–	–	–	5–7 ky
Melbourne Florida, Mammoth	0.12	–	–	–	5–7 ky
Melbourne Florida, Horse	0.12	–	–	–	5–7 ky

Note:

F – fluorine, N – nitrogen, C – carbonate, H – hydrogen.

@ – Age determined by uranium series method (S. Short, 1988–89, Australian Nuclear Science and Technology Organisation, Lucas Heights, Sydney).

* – Average percentage of fluorine.

[#] – Estimated or average age.

H – Holocene.

LUP – Late Upper Pleistocene.

UP – Upper Pleistocene.

MP – Middle Pleistocene.

LP – Lower Pleistocene.

Palaeoenvironments, megafauna and the Upper Pleistocene 167

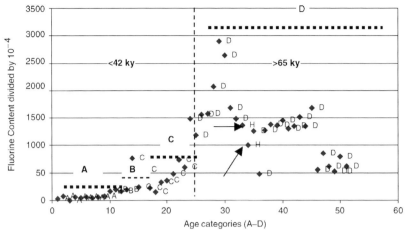

Key

A Group–Modern bone samples <200 year sold (introduced animals and recent dingo).
B Group–Holocene bone archaeological dune deposit.
C Group–Fossil human crania late Pleistocene.
D Group–Upper Pleistocene (penultimate glacial and last interglacial megafauna).
H Group–Human fossil bone (arrows) from Upper Pleistocene (Stage 5) deposits and associated with megafauna

Figure 5.2. Fluorine analysis of fossil and modern bone samples from the Lake Eyre region.

relationship between the known age of the bone (using other dating methods) and its fluorine content. Figure 5.2 is then divided between three groups that fall below 42 ky (A, B and C) and fourth (D) that is greater than 65 ky. Analysis of the cranial fragment (OKA 2) shows a 1.37 per cent fluorine content, well within the range D range (1.27–2.90 per cent) obtained from associated extinct fossil fauna found on the same site. These results far exceed those from recent specimens of kangaroo (0.08 per cent) and rabbit (0.005 per cent) used as controls. Fluorine uptake by bone has a theoretical maximum of 3.8 per cent, which suggests the OKA 2 specimen was buried in local sediments for some considerable time. The cluster of fluorine results from Katipiri aged bone show a consistent pattern of high fluorine uptake by bone across the region. Moreover the similarity between OKA 2 and other bone of Stage 5 origin strongly suggests it lies well within the last interglacial, probably towards the middle (?80–100 ky). These results are, therefore, taken to indicate that humans were certainly in this region during the last interglacial and well before the onset of the last glaciation.

The second human bone specimen consists of a completely fossilised fifth right human metacarpal (Plates 5.3 to 5.5 left specimen). It was discovered in

168 *The First Boat People*

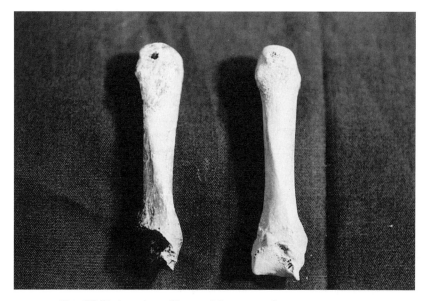

Plate 5.3. Various views of human right metacarpal.

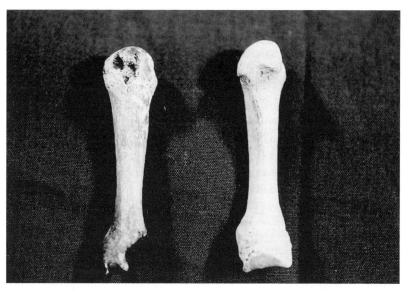

Plate 5.4. Various views of human right metacarpal.

Palaeoenvironments, megafauna and the Upper Pleistocene

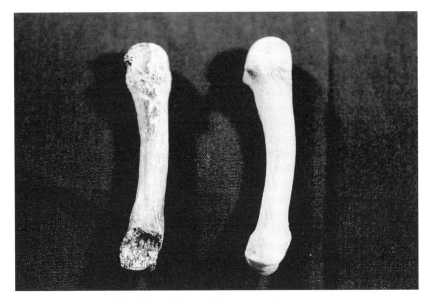

Plate 5.5. Various views of human right metacarpal.

1994 in a point bar deposit on a very remote section of the Kallakoopah Creek in the southern Simpson Desert. It was mixed with an assemblage of extinct marsupial species, including Diprotodon, Sthenurus and crocodile (*Crocodilus*?). It displays the articulation with the hamate through a quadrangular facet and another lies laterally to articulate with the fourth metacarpal. The medial side of the base has only a palmar portion of the tubercle for the insertion of the *extensor carpi ulnaris* tendon, although the tubercle for the piso-metacarpal ligament is not well developed. The dorsal surface of the shaft clearly shows the insertion line for the dorsal interosseous membrane, while the palmar surface has a line that divides this membrane from the medial insertion of the *opponens digiti minimi*. The head of the metacarpal has a medial sulcus with a small foramen. The lateral sulcus is deep. The epiphysis, although fused this might have been recent, thus indicating a young adult around 15–18 years. Overall, the bone is small and delicate with less ruggose features than usually appear on the fifth metacarpal. The sex or personal age may explain the gracility of this bone. Dimensions include: max length −51.2 mm; mid-shaft width (m−l) − 6.6 mm; head −10.8 mm (d−p) and 9.8 mm (m−l). The colour and degree of fossilisation match the extinct fauna with which it was found. This fact together with its proximity to alluvial and lacustrine Katipiri sediments strongly point to it dating to before the extinction of the megafauna 65 ky ago. Further investigation of this specimen is proceeding.

Plate 5.6. Burnt megafauna bone.

Further evidence for an interglacial human presence in the Lake Eyre region is indicated by pieces of burnt megafauna bone recovered from several Katipiri aged sites across the region (Plates 5.6 – 5.7). The bone displays significant colour change and morphological features typical of those received from high temperature burning, usually in human cremations (see reference to WLH1 in Chapter 6). These include distinctive curved cracking, white powdery surface calcination and light grey and blue/black colour gradation typical of incineration upwards of 550°C (Webb, 1990; Shipman *et al.*, 1984; Walshe, 1998). Its condition suggests it was deposited in a camp fire, eventually being covered by other coals and left for some time. The colour is also very different from the uniform black staining derived from manganese encountered on some bone fragments from sites around Lake Eyre. It has been suggested that the animal may have been caught in a bush fire. A bush fire, however, is transitory, rapid moving and normally only singes fur or, at most, chars skin, leaving deep flesh and bone untouched and the animal largely uncooked. The bone was cooked while it was 'wet' or contained within the animal, hence the cracking. The likelihood of burning right through large bone, deeply embedded in the muscle of a large animal, is even less likely in a bush fire. The origin of this charred bone is that it was the end product of somebody's meal and that must have been before 65 ky.

Palaeoenvironments, megafauna and the Upper Pleistocene

Plate 5.7. Burnt megafauna bone.

When it comes to Pleistocene bone, the discussion of burning often turns to 'accidental', whether it involves human or megafauna bone. There is a healthy scepticism among many that doubt whether *any* bone was *ever* burnt deliberately. Any scenario is proposed rather than accept that a piece of bone has been purposely cooked in somebody's fire. Suggestions range from a person or animal standing by burning trees which then fell on them, or they were standing still while a bush fire ran over them. Another is that the bone was buried and just by chance a fire was placed on top of it! Of all the places a camp fire might start, it was placed on top of that piece of bone! Amazing! Over the years I have been offered all sorts of concocted reasons why a piece of bone looks burnt rather than hear an admission that bones with the distinctive colour changes and appearance described above are the product of deliberate burning. The bones of animals and people caught in bush fires are usually not even singed let alone thoroughly cooked through to such a degree that they change colour or almost disappear in a powdery haze of calcination. In most instances bush fires hardly cook flesh and many animals caught in such fires are suffocated, not fried; often their fur is barely singed. To change the colour of a bone to blue-grey and cook it right through, it has to be placed in a maintained or long lasting fire hot enough to do the job. Moreover, calcination actually is the bone turning to ash, something not always achieved in modern crematoria.

The evidence presented above for an early human presence in the Lake Eyre basin will not be accepted by some or perhaps by many. It is tenuous, but in my view too coincidental to be dismissed altogether. I do not believe we can go on inventing excuses to dismiss all evidence however tenuous that points to people being in Australia before 65 ky ago. That practice is counterproductive and flies in the face of an archaeological history that has taught us to dismiss nothing as impossible and never put our collective heads in the sand even if what we wish not to be the case is in fact just that. I eagerly await a direct dating technique for fossil bone.

Megafauna extinctions: fire, hunting and other processes

It is widely known that Australia has its share of bush fires. At times they surround and enter large modern cities and run unchecked across thousands of square kilometres of countryside and bush. It may well have been the giant anvil-shaped smoke clouds billowing thousands of metres into the sky from enormous natural conflagrations that showed people living in Timor and neighbouring islands that this continent existed just over the horizon. The megafauna experienced and survived hundreds of these naturally occurring fires for hundreds of thousands of years before humans arrived. If fire was a direct cause of their demise, it was of a type and frequency not experienced by them before and/or it was frequent enough to change their habitat in a way that particularly affected larger species. So, was regular burning by humans one of the main factors in megafaunal extinctions?

Indirect evidence for humans in Australia long before 65 ky comes from palynological examination of core samples drilled from the continental shelf in northeastern Australia that show a dramatic increase in charcoal particles dating between 100 ky and 150 ky and which has recently been narrowed down to around 130 ky (Kershaw *et al.*, 1991, 1993; Moss and Kershaw, 1999, 2000; Kershaw and Whitlock, 2000). Accompanying changes observed in the core (known as ODP820) include a downturn in fire sensitive plant species and an increase in fire tolerant types on a scale that, it is suggested, could not be due to natural causes alone or an event that would not be recorded outside Australia. Moreover, this triple signal does not occur in the previous 2 My. The 130 ky date coincides with sea levels reaching today's levels, when people living on continental shelves would have been forced off and inland. Moving through the 2e route on Map 4.1 would have brought them to the region that was being burned according to the ODP 820 core. Similar findings were also found some years ago in the southeast of the continent, but at that time they were ignored as being far too fanciful for an early entrance of humans into Australia (Singh and Geissler, 1985). Both results have been interpreted as strongly indicative of anthropogenic burning rather than of an environmental origin. If this conclusion is correct, then Stage 5 dates for humans in central parts of the continent might be expected. Moreover, it may point to human entry into Australia at the peak of the penultimate Ice Age (150 ky), an optimum time for maximum continental shelf exposure, or perhaps even before that. Although palynological findings suggest that the 'first arrivals' fired the landscape, those people might not have been the first. Is it possible that the first people only adopted firing of the land after they arrived here? Let us step back one minute and ask: would the technique of firing the landscape have been useful on the islands of southeast Asia?

If the first Australians arrived 150 ky ago, they were not what some might call Anatomically Modern people. This begs the question, if people did arrive during the penultimate glaciation, then they must have been 'locals' from the Indonesian archipelago where humans had lived for hundreds of thousands of years. There is no evidence from Java or anywhere else in Indonesia for the use of fire by its native inhabitants (*Homo erectus* and *Homo soloensis*) at that time or earlier, and there is no evidence of humans using fire to change the landscape anywhere in the world at or before this time. It can also be comfortably argued that burning the landscape is not useful to a hunter-gatherer in a tropical region where monsoonal rainforests are largely fire sensitive and food plants could be lost together with a highly specialised rainforest fauna. Therefore, if these people arrived without the technique of burning the landscape, it must have been a strategy they developed after their arrival to suit conditions they found in other areas of Australia or to help them explore. Further, the ODP 820 core reflects burning far away from the most likely landing spots of the north and west which lay almost 2000 km away from the area of northeastern Australia from which the charcoal in the ODP 820 core originated during low sea level. Burning in this part of Australia, therefore, suggests that people were here long before, which may also indicate that they had had enough time to develop the technique between their arrival and their appearance on the northeast coast. If they brought fire with them, the first place to be burned was the continental shelf, where they remained to avoid the arid interior of the continent at that time. If people stuck to the coasts avoiding the centre, then fire may not have been needed till they began to move farther into and across the continent. On the other hand, if fire was used as a method of clearing, very small initial populations would have caused only very limited areas of burn, although the odd conflagration could have developed from a burn undertaken on a windy day in a dry and unburned landscape. These fires would be very hot, destructive and ferocious events, but nothing that had not happened previously from pre-monsoonal lightning strikes. If the first arrivals did not use fire, it can be eliminated from the list of proposed pressures humans put on megafauna. The most likely explanation would be one where humans arrived without firing practices, but over 20 or 30 ky developed the technique as seas rose and people moved inland where the vast interior of Central Australia with its savannah environments lent itself to burning on a broad scale. In this case, a fire signal lying between 130 ky on the east coast may not be the first indication of people in Australia, only that they arrived thousands of years earlier, reaching the east coast at that time.

The first people, therefore, may have had very little effect on the megafauna from the standpoint of fire use. It was not till later, when they entered central parts of the continent, became more adapted to Australia's environment and developed an understanding of how savannah environments worked that it was used on a wider scale. It was then that plant regimes began to change, altering varieties of fodder, disadvantaging some animal species, while others were favoured by the changes, particularly our contemporary species. At first, large, uncontrolled conflagrations eating up years of accumulated litter probably destroyed pockets of animals, albeit unwittingly on the part of the newcomers. This pattern would suggest that a differential distribution of these genera may have been possible, with some missing entirely from certain areas. As people moved inland, frequent burning over a long time with regular rainfall would only enhance open grassy plains and shrub lands, producing a better fodder yield that might have initially favoured some pre-existing populations of large grazing animals or even increased their frequency. Alternatively other species may have begun to be reduced by these same changes. Of course, lightning strikes had always caused bush fires all over the continent, but this as far as we can tell had never bothered the megafauna before. So, these animals may not have been affected at first by a slight increase in fire frequency when small bands of humans began infrequent burning. Moreover, it is well to remember that when Aboriginal people in some areas of the Western Desert left their lands in the 1950s and 1960s, ceasing their firing practices, animals like the Rufus Hare Wallaby (*Lagorchestes hirsutus*) were brought to the brink of extinction because they had adapted to a coexistence with regular Aboriginal mosaic burning practices (Gibson, 1986; Morton, 1990). Changes to human demography must have occurred from time to time, leaving unoccupied regions where environments reverted back to their original status, thus allowing animals a refuge for a while. Another factor to consider in these discussions is that with the distinct possibility of people moving in and out of Central Australia with changing glacial/interglacial conditions each time they entered, it would herald a new 'colonisation' of the area and any changes previously imposed on these environments would have been wiped out by long arid phases (see Chapter 8). Thus the environment would have been reset to start anew.

There are many ethnohistoric and ethnographic observations of traditional fire use by Aboriginal people (Kimber, 1983; Latz, 1995). Fire is more than the act of burning, it is a vital part of culture and a demonstration of the necessity of looking after the land. As well as 'cleaning' the country, burning initiates germination of new plants, eliminates nuisance species, such as spiky spinifex grass (*Triodia* sp.) and snakes, brings on new grass which attracts game, makes small game easier to find, aids in trapping animals and shows others and the spirit ancestors that you are looking after 'your country'. Modern fires vary in size from a few hectares to huge fire-fronts 150 km wide, affecting more than 1000 km^2 in country that has not been burned for some time. Occasionally, burning overwhelms animals, but huge losses are not normal. The ferocity and size of a fire depends on fuel build up and its combustibility, the frequency of previous fires, weather conditions, vegetational composition, wind speed and direction and so on.

Lightning and combustion from decomposition are the only natural causes of fire. In country that has not been burned for some time, fuel build up can cause very dangerous fires. This is particularly so when accompanied by high winds and in country where highly combustible plant species exist. Major conflagrations often have fire-fronts that move at speeds that would easily overtake slow-moving animals and humans (Kimber, 1983: 1). Today, the average standing crop of fuel in the southern Simpson and Tirari Deserts is too small to support regular bush fires. Even following good rains and accompanying vegetational growth, accidental burns are rare. Aboriginal people abandoned the area over 120 years ago. It is hard to believe, however, that before that time traditional burning practices included anything but limited mosaic burns, probably confined to inter-dune corridors and grassy flats. After 15 years of aerial and ground survey work covering over 250 000 km^2 as well as experiencing many electrical storms in the area, no burn scars or bush fires have been observed by me, nor, as far as I can tell, by others carrying out similar work. Local land-holders living on the fringes of the desert confirm this observation. They blame the paucity of fires on the lack of fuel build up and general sparseness of vegetation to sustain it. Therefore, if pre-glacial vegetation cover was similar to today, natural ignitions would be rare. Even when they occurred, it was unlikely that they extended very far. But during Stage 5 times, regular rainfall is likely to have produced more vegetation, including grasses and shrubs and, therefore, greater fuel build up. With more growth and fewer dune barriers, large naturally ignited fires might have occurred from time to time, but again this was something the megafauna was used to. The introduction of deliberate regular firing, however, altered fire ecology from an occasional conflagration to small mosaic fires, reducing fodder and/or replacing shrub lands with grasslands. Both browsers and grazers would have been affected by the new regime but in different ways for different animals. Browsers would experience a downturn in shrub populations as grasses took over. Fire intolerant plants would be replaced by fire tolerant types, which meant that not only were shrub lands shrinking some shrub species may have disappeared, putting animals partly or wholly dependent on them under stress. Firing of the plains not only replaced shrub and scrub with grass, it changed grassy areas into plains. Again, some grass species may not have thrived with regular burning and any animals depending on these would have missed out. The spread of grasses would make large areas vulnerable to drought conditions, where grasses die off and soils erode or blow away. In this case everyone is affected, whereas previously the small shrubs and trees would have provided some food at least for the browsers.

Overall, fire-promoted fodder growth favoured grazers, particularly those fast enough to escape fires. The large slower animals, many of which were browsers or browser/grazers, would be most at risk. During the interglacial, there were times when vegetation became increasingly scarce, as widespread aridity set in at the end of the interglacial. It seems logical that, while burning and megafauna probably coexisted fairly equitably during the initial stages of colonisation, stress slowly increased, however, during the interglacial, and this was kicked along by periodic episodes of drying in Central Australia. Larger animals would have been affected as fire regimes took hold and hunting became slightly more frequent as the human population increased. As the interglacial came to an end, fires had a greater effect on vegetation, so that animals came under severe pressure by the onset of the last glaciation.

Hunting at this time would have tipped the balance, with increasing impact on dwindling and severely stressed groups. With reduced populations, the megafauna became increasingly vulnerable to the loss of even single animals, particularly females. Their inability to replace themselves fast enough added to their predicament. For humans to achieve all this, however, there had to be enough there to play a significant predatory role. Again, in the Lake Eyre region, this had to be well before 65 ky. Persistent fodder failure would have had the same effect as the onset of aridity, with further reduction in numbers as they succumbed or moved. It is worth noting, however, that environmental change severe enough to cause widespread faunal extinction or migration may not necessarily have been detrimental to humans now totally adapted to an arid continent. It is likely that before they left the region, all water sources would have had to be dry and food stocks at rock bottom.

Palaeoenvironments, megafauna and the Upper Pleistocene

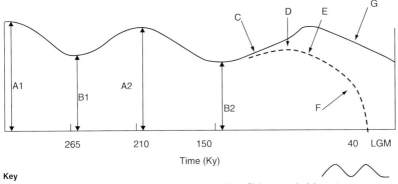

Key
Demise and recovery of megagfaunal polpulations during last three Pleistocene glacial events
Final extinction of the megafauna (ave. 40 ky) – – – – – – –
A1 & A2 – Population recovery peaks during interglacials.
B1 & B2 – Population troughs during glacial events.
C – Theoretical recovery of the population following the penultimate ice age.
C – Population recovery affected by early human activity (fire and hunting).
D – Point at which the population ceases growing and animal numbers begin to fall.
E – Effective reduction of megafaunal populations at the onset of the last ice age.
F – Final spiral to extinction as increasing fire use changes habitat and fodder regimes; hunting reduces small grops already under pressure; megafauna reproduction rates stagger under the impossible task of replacing numbers quickly enough and the last glaciation spreads aridity and diminishes rangelands.
G – Theoretical time at which the population would have decreased. Deteriorating glacial conditions reducing populations, as in the past, now increasingly act on an already stressed fauna.

Figure 5.3. Timing and processes of the Australian megafaunal extinctions.

I have constructed a model that may help document the extinction process during the last interglacial as a summary to what I have said (Figures 5.3 and 5.4). The model suggests a multiple stress view that involves firing of the landscape by humans, low level to moderate hunting, slow reproduction rates on the part of the larger species and gradual environmental degradation brought on by the last Ice Age. All these contributed in one way or another to the inevitable but comparatively slow demise of the megafauna. On top of this, I suggest that certain megafaunal species (particularly the largest types) were unevenly distributed across the continent; their populations were never large but fluctuated between glacial episodes, shrinking during glaciations and reviving during interglacials. The arrival of people in Australia around 150 ky would have had little if any effect on these animals at first because there were few people, there was very limited contact and humans did not fire the land. As seas drowned the continental shelf, so megafauna drifted back into Central Australia that was now changing from an extremely dry, arid environment during glacial times to a wetter greener one. After the sea reached modern levels, people were pushed into areas that had been far inland

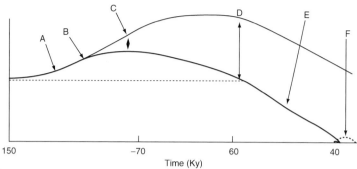

Key

A – Gradual population recovery of megafauna following penultimate Ice Age. initial human arrival has tittle effect on the population for some time.
B – Point at which human induced stresses begin to pressure normal population recovery.
C – Differential between normal post-glacial population recovery (thin line) and maximum recovery as humans spread using fire and hunting during last interglacial.
D – Megafauna population in long, slow decline suffering effects of human hunting, firing the land (for at least the last 40 ky) and the onset of the last glaciation. The megafauna become extinct in Central Australia, this time with the remaining population at a level as low as it was at the peak of the penultimate glaciation around 150 ky.
E – The megafauna population reaches a point of no return, unable to replace its numbers and suffering the increasing effects of the glacial onset, particularly aridity, and a spreading increasing human population.
F – Small pockets of animals remain for a while after the main groups become extinct. All have disappeared by 30 ky.

Figure 5.4. Final process of megafaunal extinction.

during glacial times and a certain amount of rapid adaptation to semi-arid conditions was required of them. As the environment improved during early interglacial times, there was further movement of people inland and firing of the landscape began to take place as an adaptation to inland environments, but it took some time before this practice began to have an effect on the fauna, because humans were also very thin on the ground. The effect of human movement into central parts of Australia, the introduction of limited hunting and firing held back megafauna populations from reaching the level of recovery attained in previous interglacials. Instead it reached only half the growth achieved following previous glaciations before beginning a slow downward spiral over the next 50–60 ky as humans spread, their population grew and more people arrived on the continent.

Local extinctions may have occurred many times prior to the arrival of humans, but the overall population was large enough to bounce back. Nevertheless, it is not unlikely that the fossil record does reflect the distribution of certain megafauna species across the continent and that this was indeed uneven, as well as an east–west division for some. It would put, however, the vast majority of animals in eastern and Central Australia, perhaps because of their reliance on channel systems. If this was the case, then the smaller population in the west might have succumbed rapidly when people arrived, because of the vulnerability caused by their small numbers in widely scattered herds. Bearing this in mind, it may have been possible that the large megafaunal populations, often envisaged as spread across the continent, may not have existed at least in half of it. Instead, with a number of small, patchy populations scattered across the continent and larger herds in the eastern half of the continent, it would have taken little extra pressure to bring many small local populations to the brink of extinction very easily.

Extinctions slowly gained momentum after 60–65 ky as the effects of the last glaciation increased. Movement of people together with increased use of fire and moderate hunting continued to compound the problem of the gradual dwindling animal populations but the pattern was different between regions, some being more affected than others. Remnant animal populations slowly drifted from drying central regions to peripheral areas towards the edge of the continent. That old strategy was repeated, but this time they were forced to share it with humans. Continued movement of people into Australia between 60 ky and 75 ky compounded the megafauna's problems. These people probably had better hunting strategies, more sophisticated weaponry, a more cohesive social fabric and more complex language than earlier groups. The result was the total demise of these creatures as they succumbed to the Ice Age and the widespread human occupation of Australia that was largely completed by 40 ky ago.

Is it coincidental that the megafauna disappeared in the Lake Eyre basin around 65 ky? Perhaps this event was due solely to environmental influences, or was it because of the long-term impact of humans in the region? Dates for the earliest arrival of humans in Australia hover at around 55–60 ky. Some examples include a 50–60 ky at Malakunanja II in the northern Territory, while another optical date of 53–60 ky has been found for sites in Deaf Adder Gorge 70 km to the south (Roberts *et al.*, 1994, 1998). Sustained anthropogenic burning in northeastern Queensland has been blamed for vegetational changes observed at Lyrch's Crater around 45 ky (Turney *et al.*, 2001a). In far southwestern Australia, recent redating of sediments from the Devil's Lair cave deposits put people in the region sometime between 50ky and 55 ky (Turney *et al.*, 2001b). All these reports suggest that people arrived in Australia sometime between 55 ky and 60 ky, with one suggesting that the dates obtained from Malukananja 'confirm evidence for the colonisation of Australia shortly after 60 ka and should be seen in the context of this region as having been a likely entry route for the first human movements into Sahul' (Roberts *et al.*, 1994:575). The southern end of the Lake Eyre basin lies 1700 km south of the Malakunanja II shelter, so, if anthropogenic influences were behind the Lake Eyre extinctions and if Malakunanja does mark the earliest time for people in Australia, then the speed of migration and its effects on the fauna have to be more than instantaneous, a 'blitzkrieg' event for sure and something I find extremely difficult to believe. What the above dates do show is that people had found their way right across the Australian mainland by at least 50 ky and probably as early as 55 ky. I suggest, however, that what these dates do show is that people *appear* to us at this time and I will explain what I mean by that a bit later. Nevertheless, our interpretation of time and our extremely coarse understanding of events, including environmental change and how severe it was, must continually leave us doubting our own arguments. Reconstruction of past events needs not only data but also the most parsimonious interpretation of them. Sometimes this requires a healthy dose of theory, which, however flimsy it may look, is a legitimate process and the broadsheet for laying out possibilities and collecting and re-analysing our thoughts. If people did not arrive till 55 ky, they could not have played a part in the extinction of the megafauna in the Lake Eyre region. Moreover, the ecological circumstances of the region at that time were not conducive to receiving people with little or no experience of arid environments. I might also suggest that even those with such experience would not have wished to be there at that time anyway.

It is possible that the animals may have gradually left as the environment deteriorated, only to succumb to anthropogenic pressures elsewhere. That would surely mean a 'blitzkrieg' took place elsewhere, because, not only were the migratory groups destroyed, so were the populations already living around the edge of the continent, which were probably substantial in the east. Moreover, why did these people not leave mass extinctions in their wake as they passed through southeast Asia?

Into the centre

Having proposed a process of gradual movement of people into Central Australia, I want to offer some detail of how this took place. I have already alluded to the fact that at certain times Australia's internal environmental conditions were poor and would have prevented humans from occuping the centre (see Chapter 8). Specifically, these barriers were up at 130–150 ky, 100–110 ky , 70–75 ky and 45–60 ky, when increased aridity, aeolian activity and drying water resources made living in Central Australia very difficult. So, occupation of the centre could have occurred only during 60–70 ky, 75–100 ky and 110–130 ky. If people arrived on Sahul between 60 ky and 65 ky and moved inland, they would have had to move quickly because conditions were changeable and the overall environmental trend was towards final drying, as the effects of the last Ice Age began to creep over Australia. An opportunistic ebb and flow of people may have taken place, coinciding with several short-term ameliorations embedded within the broader arid episodes. Logically, however, the present 60 ky date for the earliest people in Australia must be taken as only an indication of a much earlier entry on to the exposed Sahulian shelf: we have to allow time for these people to then move into the continent. Moreover, the 60 ky date can be superseded at any time as the archaeological record continues to produce earlier and earlier dates, a starkly obvious trend over the last 40 years. Also the 60 ky visibility we see around the continent may only be an indicator of people who leave a heavy foot on the landscape and one which has all the hallmarks archaeologists look for. So these dates may only be a signal of the arrival of those who *appear* to us in the most obvious way in the archaeological record. Those with a lighter foot mark are yet to be detected, or have they been?

A 65–70 ky frame could have seen people move into Central Australia only to leave again as the onset of the last glaciation imposed a final drying of all surface water after 55 ky. Sporadic amelioration in Central Australia's harsh conditions may have allowed people into the region for very short periods, perhaps 50 or a few hundred years at a time. Such changes may have been scattered throughout this time and would be very difficult to detect in the geological record, although we do know that they made forays to Lake Eyre around 20 000 ky at a time which the geomorphological record suggests was extremely arid.

The Simpson Desert was more like a Simpson Savanna in pre-glacial times when the Lake Eyre basin was fed by a large complex network of channels flowing from the north. The fossil record shows us that the environment supported a wide variety of animal species, from rats to rhino-sized *Diprotodons*, as well as a selection of carnivores including terrestrial and large aquatic reptile species that inhabited inland river systems. The rich trophic pyramid reflects an environment containing a wide and plentiful range of plant food sources upon which the megafauna relied and, in turn, the flora was the product of abundant local rainfall. Southerly monsoon incursions into the high catchment, coupled with regular local rainfall, fuelled a large palaeochannel system, filling enormous freshwater lakes till around 60–65 ky.

The fossil evidence suggests that there was a plentiful food supply for the megafauna until about 75 ky, then the population began to diminish, becoming extinct by 60–65 ky. If animals were largely opportunistic breeders, they were more able to withstand the gradual onset of the last Ice Age and the drier episodes during the interglacial, but importantly the Simpson Savannah also had all the ingredients to attract humans. Abundant aquatic and terrestrial food resources and ample fresh water would have attracted people into Central Australia at these times. It is suggested that a multi-factorial cause is the most likely reason for the megafaunal extinctions, at least in this part of

Australia. The process involved environmental change and human activity, in varying proportions and with local variants, but this was a process that did not begin until some time after the first humans arrived. If, as some people would argue, megafaunal extinctions were caused solely or in greater part by the presence of human activity, then people were in the Lake Eyre basin long before 60 ky.

6 Upper Pleistocene Australians: the Willandra people

> Whichever theory one prefers, it is apparent that there is a general acceptance that the early populations of South-east Asia included both a robust fossil type and a more slender (?later) form, the former being similar if not ancestral to the modern australoids.
>
> (Brothwell, 1960: 341)

Introduction

The previous five chapters of this book described processes leading to the entry of people into Australia and the impact of those people on the animals and the landscape. The discussion included growth of world population during the last 1 My, the methods and pattern of migration, demographic and genetic processes leading to the arrival of people in Australia and the environmental setting and possible consequences of their arrival. I now want to examine who these first Australians were and where they came from. Whatever the world's population was doing at that time and whoever it was that undertook these migrations across the globe, Australia was at the receiving end of the I have already hinted at the possible genetic complexities that could arise from migrations moving through South East Asia and Indonesia and how difficult these could make sorting out the 'origins' of the first Australians. This difficulty is compounded further when we consider the extremely limited fossil sample available in Australia and the fact that it may be too young to be of any real help to us.

The claim that widely differing populations arrived in Australia is not guesswork, the fossil record demonstrates this emphatically. The wide range of morphological variability among Australia's Upper Pleistocene people reflects widely differing genetic stocks among founding groups. In fact, Australia's oldest skeletal remains show a wider range of variation than modern samples. The morphological contrast between these fossils is so great that explaining it in terms of a single population with a single origin is impossible. Consequently, many researchers have proposed that the earliest peoples came from different areas within the region and this has been clearly argued in the 'Regional Continuity' model. So, in this chapter, I want to address skeletal contrasts among Australia's fossil people, as well as outline salient arguments put forward by more recent workers concerning the origin of these morphologies. To do this I use data described in an earlier but not widely known publication gathered from the Willandra Lakes collection (Webb, 1989), as well as some new findings and a rework of all the data gathered so far.

Arguments concerning Australian Pleistocene skeletal morphology

It is 30 years since a duel migration model was first proposed to explain the origin of the contrasting morphologies among fossil remains found in Australia. These morphologies were exemplified by the robustly built Kow Swamp remains from the Murray River, Victoria, and the gracile Lake Mungo fossil people from the Willandra Lakes region in western New South Wales (Thorne, 1975). The dual origin model drew also on earlier suggestions concerning the multiple origin of Australia's earliest people, but it reduced by one the 'trihybrid' or 'three migration' theory of Birdsell (1967, 1977), still popular at that time. It offered also different locations for the origin of the first Australians. Principally, these were China for the Willandra's gracile fossils and Java for robust forms, although Weidenreich (1943) had previously suggested that 'the mark of ancient Java', is seen on many ruggedly constructed modern Australian Aboriginal crania, and this suggestion was echoed by others. In 1980, an extremely robust individual (WLH50) was found in the Willandra Lakes region and it joined a range of other examples of this morphological type. It has been suggested that the contrasting gracile and robust morphologies were derived from a single population with a wide range of skeletal variability (Abbie, 1976; Macintosh and Larnach, 1976; White and O'Connell, 1982; Brown, 1987). It is no coincidence that recent exponents of this idea believe also that these people were Anatomically Modern Humans from 'Out of Africa'. There is a stark contradiction in these two ideas, however; indeed one is not at all compatible with the other, but I will return to that later.

Thorne (1977: 190) recognised that the individuals from Lake Mungo had 'no single character [lying] outside the range observed in more recent samples'. His statement was based on the characteristics of WLH1 and 3, the first, and subsequently most famous, individuals found at Lake Mungo and the formative fossils of the much larger Willandra Lakes collection assembled over following years. Right or wrong, Thorne's work brought with it a welcome freshness to the origin debate. It also built a firmer base for the arguments, because it was itself based on empirical data derived from the earliest skeletal evidence. It also challenged previous arguments about origins which were largely based on morphometric data gathered during surveys of limited numbers of living Aboriginal people removed by thousands of generations from the first migrants (Abbie, 1968, 1976; Howells, 1973). Thorne's data, which included the oldest evidence for human occupation in Australia at that time, put him in a much stronger position than his predecessors.

Kow Swamp was the first accurately dated Pleistocene human population discovered in Australia. At the time of its discovery it was believed that Australia had only been occupied for 16 ky from a date obtained at Kenniff Cave in Queensland (Mulvaney and Kamminga, 1999). The oldest date for Kow Swamp was 14 ky; thus it was close to that time, so it was natural to accept them as representing the 'First Australians'. It was also natural to feel that, because they were 'the oldest' representatives of people on the continent, their robust morphology reflected a Javan gene pool. Excavations at Kow Swamp had begun 18 months before geomorphologist Jim Bowler discovered the first of the Lake Mungo individuals in July 1968. His discovery of what later became known as WLH1 (Willandra Lakes Human 1[1]) almost doubled the age of Australia's earliest people when it was dated to between 24 500 and 25 500BP, on the basis of two radiocarbon dates on charcoal from a hearth 15 cm above the cremation (Bowler *et al.*, 1972). In 1974 the age of WLH3 found in the same area was put at between 28 ky and 32 ky by stratigraphic association (Bowler and Thorne, 1976). These dates made their effects felt elsewhere. Kow Swamp was not now the earliest fossil human evidence. The Mungo Lady discovery had shifted the paradigm immeasurably.

The discovery in 1974 of WLH3, less than half a kilometre from the WLH1 site, increased, yet again, the age for humans on this continent to around 32 ky, but these people were not robust; quite the opposite. There was and remains, however, an uncomfortable disconformity in these discoveries. Kow Swamp featured a number of robust cranio-facial and dental characteristics that were described as being reminiscent of much earlier Middle and Upper Pleistocene Javan fossils (Thorne and Wolpoff, 1981). Yet, these were present in people living 15 ky after those displaying very modern, gracile features. The gracile humans from Lake Mungo would have been better placed in a more recent temporal context. On the other hand, the robust fossils from Kow Swamp were quite rightly placed in the late Pleistocene, where their sort of morphology 'belonged'.

---

[1] A number of acronyms have been used to code fossil humans from the Willandra Lakes region. WLH is used here because it is the only suitable code that covers remains from all lakes and inter-lake areas in the system.

Besides arguing that the Kow Swamp people represented a long and semi-isolated population, with strong genetic ties to robust groups who probably entered Australia first, a convincing explanation for the persistence of their robust features was and still is difficult to construct. Since then arguments have changed as Australia's archaeological evidence has grown. Many sites from all parts of the continent have dates that exceed 35 ky (Mulvaney and Kamminga, 1999; Morwood, 2002). Moreover, recent redating of WLH1 and 3 have shown them to be 'contemporaries', both sharing a 40–41 ky date (Bowler *et al.*, 2003), although others claim a date around 60 ky (Thorne *et al.*, 1999). There are, however, good indications that people were living in the Lake Mungo region at least as far back as 55 ky, Therefore, the oldest human remains in Australia so far discovered look modern and those with the more robust morphology are a more recent arrival, if we rigidly abide by the few dated individuals we have. This makes the likelihood that the robust individuals were the 'ancient first' to arrive, with features derived from the earlier Javan *Homo soloensis* people, less likely than ever before.

The Willandra Lakes collection of fossil human remains has made a substantial contribution to our understanding of who some of the earliest Australians were. But the gracile/robust debate that has emanated from it has often ignored a whole range of individuals, concentrating instead on three well-publicised examples (WLH1, 3 and 50) to carry the arguments. I would like to now take a broader look at this collection, which has not been discussed as fully as it might within the context of the above arguments and is certainly poorly known outside as well as inside Australia.

The Willandra Hominids: a case study of an early Australian population

The Willandra Lakes fossil collection from western New South Wales is the oldest dated group of fossil human remains yet found in Australia. For over 30 years, arguments have abounded regarding the wide morphological variation seen among these fossils, what it means and if it reflects the origins of the earliest Australians. Besides Kow Swamp, most of the literature addressing early Australian morphologies unswervingly concentrates on the three individuals mentioned above. This is to the complete exclusion of at least another 130 collected specimens. Recent finds and uncollected individuals now extend the range to WLH154 and the whole collection contains both robust and gracile forms, as well as many that lie in between.

It is no accident that the three fossils above have had their fair share of publicity. WLH1 and 3 are very lightly built, in complete morphological contrast to the extremely robust WLH50. The 'wide range of variation' argument places the former two at one end of a wide morphological continuum, while WLH50 sits at the other. Apart from the unlikely implication of this argument, which suggests that we have by sheer coincidence discovered either extreme of a fossil population, with a range of variation unequalled among modern people, the likelihood of WLH1 and 50 being 'sister and brother', respectively but metaphorically, makes these arguments difficult to accept. Moreover, there has been a deliberate reluctance by commentators to discuss other remains in the collection and this has helped reinforce long-held prejudicial approaches to the origin of the first Australians. As we will see, there are many other fossils, be they all more fragmented, that display the same or similar features to the two morphological types.

The Willandra Lakes system is situated in the centre of the Murray–Darling Basin in western New South Wales (Map 6.1). It comprises a series of fossil lakes strung out along the southern half of the Willandra Creek palaeochannel, an ancestral Pleistocene distributary of the Lachlan River fed by run off from the southeastern highlands during the last glaciation. Six major and several smaller lakes make up the system that covers an area of over 1200 km^2. The freshwater lakes were filled around 50 ky and possibly experienced earlier fills. Although cycling through wet/dry phases, they remained relatively full until about 16.5 ky, when they finally began to dry up with the exception of a flush through the most northerly lake Mulurulu between 14 ky and 16 ky. The lake system provided a focus for human habitation for tens of thousands of years before this. This is exemplified by the many campsites, shell middens, hearths and human remains that litter the palaeobeach and backshore areas around the fossil lakes. The largest lake, Lake Garnpung, is about 500 km^2, which when full would have held around five million cubic metres of water.

Archaeology emerges from beneath and within late Pleistocene lunettes, long fringing sand dunes consisting of aeolian sediments and carbonate deposits formed during periods when the lakes cycled through wet/dry episodes. Dune building ceased around 17 ky ago (Bowler, 1998). Lunettes lie on the eastern and northeastern side of the lakes, formed by the prevailing wind direction in the Upper Pleistocene. They are now time capsules, because the slow inexorable building of these structures buried hundreds of camps left behind by the inhabitants who lived on the late Pleistocene beaches and sand bars around the edge of the lakes. Barring obvious intrusions into top layers, any fossil remains revealed by erosion are regarded as at least contemporary with the final dune-building phase. Continuing lunette erosion has revealed a rich pageant of human activity. The most well-known of these is the lunette on Lake Mungo's eastern shore. The archaeological data that have emerged

188 *The First Boat People*

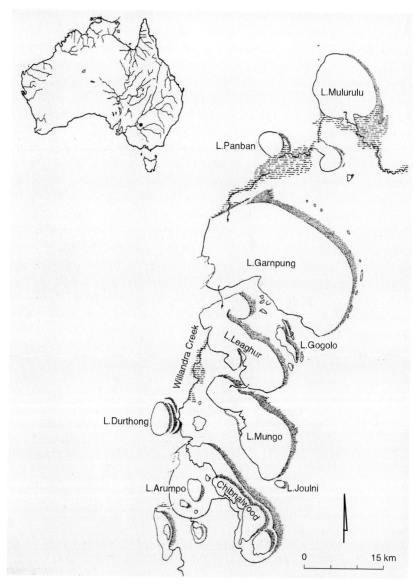

Map 6.1. The Willandra Lakes system, western New South Wales.

from it have led this lake to become a symbol of the length of time of human occupation in Australia and of course a special place not only for local Aboriginal people but for those from all over the continent. There is little doubt that the special status of this lake will always remain, even though other

Plate 6.1. The cremated cranium of WLH1.

lakes such as Arumpo and Garnpung have extensive and equally valuable archaeological deposits and are equally as prominent.

The archaeological significance of the Willandra Lakes was first made known by the discovery in 1968 of a cremation of a young female (WLH1) together with a few burnt fragments of a second individual (WLH2), on the southern end of the Lake Mungo lunette (Plate 6.1). The very modern appearance of WLH1 was subsequently described as *'ultra-feminine'* (Thorne, 1977: 190). The scientific significance of the region was underscored five years later with the discovery in 1974 of a male skeleton (WLH3) eroding from sediments further along the lunette (Plate 6.2). The cremation of WLH1 contrasted with the extended burial of WLH3, which had been covered in red ochre. The Willandra has since yielded over 200 dates gathered from exposed hearths, shell middens and the few excavations undertaken in the region. Erosion has exposed enough archaeology to show that there was a well-established, continuous late Pleistocene occupation of the area. From 1984 till recently, little archaeological research was carried out in the area at the behest of local Aboriginal communities. This contrasted with the enthusiasm of the previous ten years following the initial discoveries

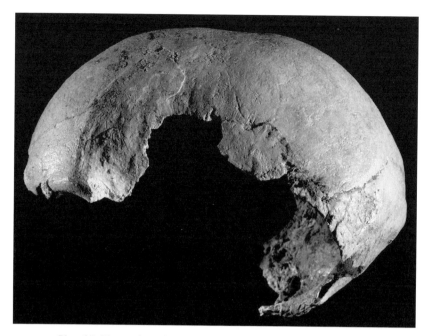

Plate 6.2. The cranium of WLH3. Note the lack of any supraorbital development.

(Bowler *et al.*, 1970, 1972; Allen, 1972, 1974; Barbetti and Allen, 1972; Jones, 1973; Mulvaney, 1973; Shawcross, 1975; Bowler and Thorne, 1976; Shawcross and Kay, 1980; Clark and Barbetti, 1982; Dowling *et al.*, 1985). Nevertheless, human skeletal remains and other archaeological materials were collected from surface discoveries during the late 1970s and early 1980s and the bulk of the collection was assembled during that time. No systematic study or description of the collection was undertaken, however, until the mid 1980s (Webb, 1989).

The age of the Willandra collection is crucial to its fullest interpretation and placement in a palaeoanthropological context and a chronology for the stratigraphic units within the Lake Mungo lunette has been constructed over the years to aid this (Bowler, 1971, 1973, 1975, 1976a, 1976b, 1980, 1983, 1986, 1998; Bowler and Thorne, 1976; Bowler and Wasson, 1984; Bowler *et al.*, 1970, 1972; cf. Barbetti and Polach, 1973). Recent reassessment of Lake Mungo's stratigraphy against a background of earlier research shows that the timing of major sedimentary sequences required updating (Bowler, 1998, Bowler *et al.*, 2003). The basic Willandra stratigraphic sequence now goes back >120 ky with the dark red lowest Gol Gol levels. Major sedimentary sequences in the lunette are: Mulurulu 13–16 ky, Zanci 18–25 ky,

Arumpo 25–32 ky, Upper Mungo 32–40 ky?, Lower Mungo 40–65 ky? and Gol Gol >120 ky. Many dates supporting the age of this stratigraphic sequence have been obtained from a variety of archaeology, including shell middens and hearths, which point to a firm occupation of the region going back almost 50 ky, although there is evidence that it is far older than that (see below). These dates not only testify to the well-established human occupation of the area, but also these people were well adapted to the semi-arid conditions. Another implication from the Willandra is that, because it is in the south of the continent, it shows that initial migrations that were not well adapted economically or otherwise to conditions in the Willandra must have entered northern Australia considerably earlier.

The vast majority of Willandra fossils were found as surface finds and cannot be dated absolutely. WLH50 is one of these (Plate 6.3A, B and C). It was retrieved as a surface find close to Lake Garnpung north of Lake Mungo but cannot be dated because it is completely fossilised. It lacks any datable organic material, although the silica skin covering its surface undoubtedly makes this fossil late Pleistocene. Attempts at dating it have produced 14 ky, obtained using a newly developed gamma spectrometer U-series method, and another date of $29\,000 \pm 5000$ BP was achieved using electron spin resonance (Caddie et al., 1987; Simpson and Grun, 1998). Some believe the latter to be a minimum date. The inability to accurately date WLH50 is frustrating, because its very robust character is extreme for a modern person, although not unique.

Attempts to date five other individuals from the series (WLH9, 23, 24, 44 and 122) have proved equivocal. Although these dates were published previously (Webb, 1989), they are no longer accepted as valid. Nevertheless, almost all individuals in the collection are heavily mineralised and were found eroding from late Pleistocene sediments predating the final lunette formation and lake drying that took place around 17 ky. This conclusion is based on considerations relating to the collection as a whole, the position of specimen recovery, bone condition, fragmentation patterns and bone surface erosion, degree of mineralisation, mineral staining and carbonate encrustation. In a recent dating reassessment of some Willandra individuals a ^{230}Th age of between 104 ky and 140 ky was obtained for WLH52 (Gillespie, 2002). The highly mineralised, red stained cranial and post-cranial fragments of this individual match the deep red sediments of the oldest sedimentary level, Gol Gol. If accurate, the Willandra collection extends across a vast time period and of course indicates that people have been in Australia far longer than we have suspected. Nevertheless, the Willandra collection is the oldest and largest representative collection of Pleistocene human remains so far discovered in Australia.

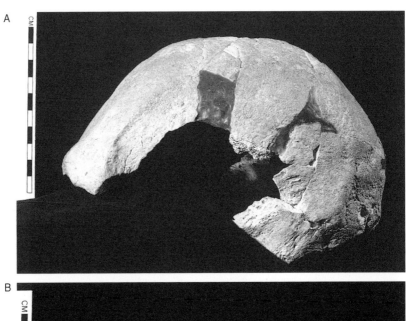

Plate 6.3. (*cont.*)

Plate 6.3. WLH50, A – Lateral view, B – Frontal view, C – Superior view.

The Willandrans: a morphological variety

The fragmentary condition of the WLH series has required the use of analyses that, while a little different from some standard approaches, could draw the most information and help define morphology. They were developed to take full advantage of the few diagnostic features available as well as to increase the number of individuals that could be examined and contrasted, many of which might otherwise be dismissed as being of little value. The data below are taken from certain common features that most commonly survive among the fossil remains from the area and can be used as morphological as well as cultural indicators. They include:

1. cranial thickening,
2. supraorbital development,
3. frontal and mastoid pneumatisation,
4. malar and glenoid fossa morphology,
5. post-cranial bone morphology, and
6. cultural inferences (cremation, tools and weapons).

These features are also important in discussions concerning regional continuity and evolutionary trends. (For a detailed anatomical description of each individual in the Willandra series see Webb, 1989.)

194 The First Boat People

Table 6.1. *Cranial vault thickness (mm) at primary anatomical points on Willandra individuals (a) and other individuals and populations (b)*

WLH No	A	B	C	D	E	F	G	H	Range[a]	x	s.d.
1	5	6	6	4	6	6	8	7	4–8	5.9	1.22
3	10	11	8	8	7	6	9	15	6–11	8.4	1.72
16	–	–	–	–	9	–	–	12	–	–	–
19	13	12	10	10	11	12	10	19	10–13	10.9	1.06
22	–	–	–	9	11	8	12	14	8–12	9.9	1.75
24	–	–	9	–	9[b]	8	10	–	–	–	–
26	–	–	–	–	10	12	–	–	–	–	–
27	–	–	–	–	14	–	–	18	–	–	–
28	–	–	–	–	12	9	–	21	–	–	–
29	–	–	–	7	7	7	7	–	–	–	–
45	–	–	9	10	13	–	10[b]	–	9–13	10.4	1.49
50	19	17	17	15	16	17	15	18	15–19	16.6	1.40
67	5	4	8[b]	9[b]	5	7	5	14	5–9	6.1	1.87
68	–	5	4	5	4	–	–	–	4–5	4.5	0.50
69	10	10	–	–	–	–	–	–	–	–	–
72	10	9	10[b]	–	–	11	–	–	9–11	10.0	0.82
73	7	–	8	8	7	8	6	–	6–8	7.3	0.88
100	13	–	–	9	11	14	10	–	9–14	11.3	1.99
101	13	–	10	–	–	–	–	–	–	–	–
124	–	8[b]	–	–	10	–	–	–	8–11	9.0	1.41
130	–	–	9	7	9	9	9	17	7–9	8.6	0.89

Notes:
A = frontal boss, B = mid-frontal (glabella/bregma), C = bregma on frontal, D = obelion (mid-sagittal or bregma/lambda), E = Parietal Eminence or Boss, F = asterion on Parietal, G = lambda on parietal or occipital, H = inion, through external occipital protuberance.
[a] Excluding inion.
[b] Estimated value.
[c] WLH50 included for comparative purposes but excluded from overall range for WLH series.

Cranial thickness

Willandran cranial thickness, using maximum thickness on any part of the superior portion of the vault, ranges from 4 to 15 mm (Table 6.1 and Appendix 2). Excluding WLH50 for reasons I explain later, 14 examples from the series have an average cranial thickness of 12.8 mm. Using primary points around the cranium, the thickest crania are WLH27 and WLH100 with 14 mm through the parietal boss and asterion, respectively. Another five individuals (WLH19, 22, 28, 45 and 101) have vaults ≥12 mm thick at one or other primary point, although another four individuals have ≥12 mm thickness on

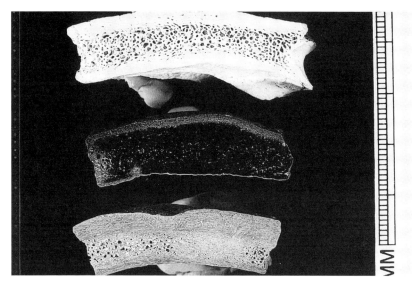

Plate 6.4. Cross-section of three cranial vaults WLH22 (top), WLH28 (middle, cremated) and WLH 63 (bottom).

at least one point on their vaults (WLH18, 56, 63 and 102) (Plates 6.4 and 6.5). At the other end of the range, there are 16 individuals with a maximum thickness of ≤7 mm or less. WLH1, 67 and 68 have uniformly thin vaults with individual ranges of 4–8, 5–9 and 4–5 mm, respectively (Plates 6.6 and 6.7). Other individuals have thin vaults also, but, generally speaking, both thick and thin walled crania have uniform vault thickness so that if the vault is thin at one primary point it is thin all over and *vice versa*.

The relative composition of Willandran vaults varies considerably (Table 6.2, Appendix 2). Seventy per cent of individuals, whose inner table can be measured, lie in the 1–2 mm range, although WLH2 has an inner table 3.5 mm thick. Normally, the outer table is thicker than the inner and ranges from 1–7.5 mm, but the greatest thickness of the two may not occur at the same place on the vault. Diploeic bone has the greatest range from 2 to 14 mm and is the major element making up cranial thickness. With the exception of two individuals, WLH2 and 68, this never falls below 40 per cent of total thickness, averages around 60 per cent and, in some cases, exceeds 80 per cent. The 25 per cent and 36 per cent of WLH2 and 68, however, is interesting in the light of their extremely gracile appearance. It might be expected that individuals with thinner cranial vaults would have a smaller percentage of diploeic bone, because the compensatory thickness of the inner and outer

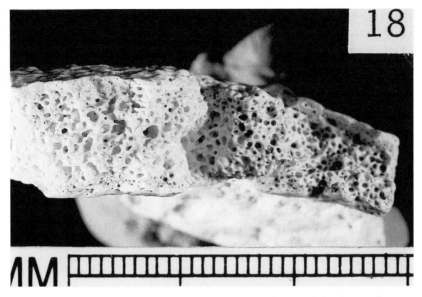

Plate 6.5. Cranial thickness of WLH18 composed almost entirely of spongy bone.

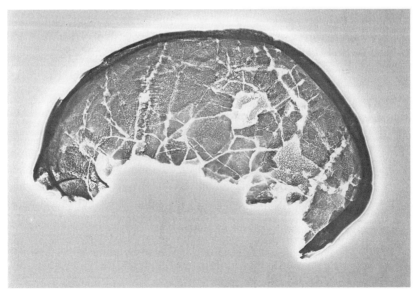

Plate 6.6. X-ray of WLH1 showing uniformly thin cranial vault structure.

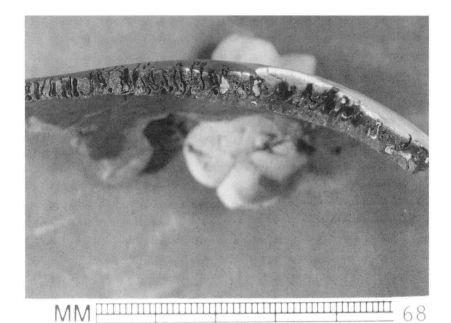

Plate 6.7. Cranial thickness along the sagittal suture of right parietal of WLH68 with calcination (white) from cremation.

Table 6.2. *Percentage of diploeic bone in vault construction*

MVT*(mm)	×	s.d.	Range
≤ 7.5	59.6	15.8	35.7–86.7
8.0–10.5	61.1	13.4	25.0–80.0
≥11.0	57.8	11.5	40.0–73.7

Note:
* MVT – maximum vault thickness.

tables will remain relatively constant, but this is not necessarily so. Table 6.2 shows that in fact the average percentage of cancellous bone as an element of vault thickness remains constant regardless of cranial thickness.

While individual vault composition is similar among people with various vault thickness, cancellous bone composition ranges from 25 to 87 per cent. At the temporal surface of the parietal and areas associated with cerebral vessels, cranial thickness is different from elsewhere on the vault and the

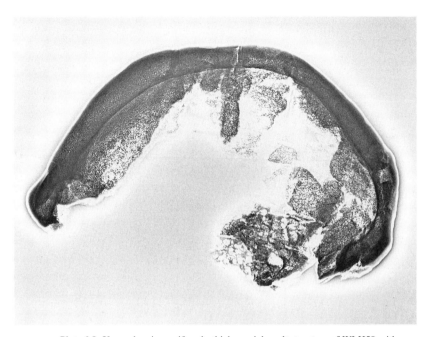

Plate 6.8. X-ray showing uniformly thick cranial vault structure of WLH50 with additional thickening at the superior occipital protuberance and at the prebregmatic region of the frontal bone where traces of vertical bone spicules ('hair on end') associated with an haemoglobinopathy.

resultant thinning occurs at the expense of the cancellous component. Normally, however, cranial composition at any given thickness remains fairly constant.

WLH50 is excluded from Table 6.1 because it is a skull of exceptional cranial vault thickness, which I have suggested elsewhere may in part be due to a pathology related to an ancestral balanced polymorphism, such as thalassaemia (Webb, 1990, 1995) (Plate 6.8). For more than a decade, however, my diagnosis has been used to bolster arguments by those who would wish the entire robustness of this specimen to be due to pathology or dismissed by others who believe the extremely thickened vault is another mark of ancient Java. It is clear that WLH50 is a very robust individual with or without pathology, because other heavily developed features are clearly not pathological. Beside any pathology that might have moderately exaggerated its vault thickness, this individual would have had a thick vault in keeping with its other robust features. Both inner and outer cranial tables are very thin compared with its overall vault thickness, so almost the entire thickness consists of cancellous bone (Plates 6.9 and 6.10). This particular feature, together

Plate 6.9. A section of the WLH50 vault (parietal) showing how it is almost entirely diploeic (spongy bone) with extraordinary thin cranial tables.

with others on the cranium, is consistent with a haemolytic condition, probably a progenitor of the genetically based polymorphisms such as HbE, HbS, Sickle Cell Anaemia or Thalassaemia, all adaptive conditions found among modern populations living in endemic malarial areas. The condition would not be unexpected if the origins of this individual or its ancestors were in the tropical or equatorial regions of Indonesia. (For a full description and associated arguments see Webb, 1995.)

The skullcaps of early fossil hominids were thicker than those of modern humans, sometimes considerably so (Table 6.3). For that reason, thick cranial walls have been identified as an archaic cranial feature. The massiveness of Zhoukoutien, Javan *Homo erectus* and *H. soloensis* crania has been discussed by Weidenreich (1943, 1951). He concluded that massive walls were associated with Middle and Upper Pleistocene hominids and that, although every so often a thick cranium is seen in modern people, calvaria became thinner as modern humans evolved. Weidenreich's measurements of *Homo erectus* populations compare with those in the WLH series, but the idea that thick cranial walls are an archaic trait has to take into account vault construction. Thick cranial vaults of archaic hominids usually comprise equal or almost equal portions of inner and outer cranial table and diploe. This is the case in both the Zhoukoutien and Javan *Homo erectus* populations, with each of the

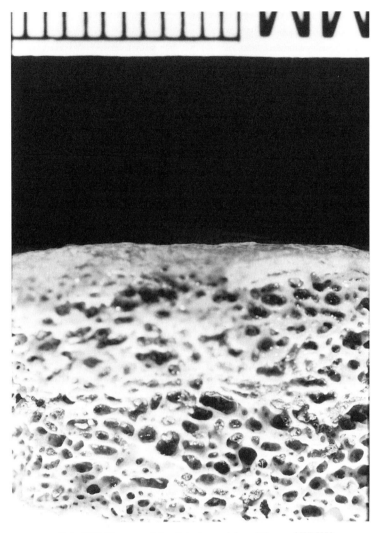

Plate 6.10. Close up of the diploeic cranial vault bone of WLH50.

tables slightly thicker than the diploe (Weidenreich, 1943: 164). While a similar pattern is seen in some individuals in the Willandra collection, the cranial thickness of most individuals is overwhelmingly composed of cancellous bone and one of these is WLH50. Robust Willandrans usually have a thick cranial vault as part of the suite of features reflecting their particular morphology. In some vault thickness matches certain individuals in the much later Kow Swamp population, but more recent Australian crania are also

Table 6.3. *Comparative cranial vault thickness ranges (mm) for various hominid populations and individuals*

Sample	A	B	C	D	E	F	G	H	Population range[a]
Australian Series									
1	5–13	6–12	4–10	4–10	4–14	6–14	5–12	7–19	4–14
2	–	5–14	7–12	6–10	8–10	9–12	7–12	–	5–14
3	5–11	6–9	6–10	7–8	4–8	9	9	–	5–11
WLH50	19	17	17	15	16	17	15	18	15–19
Homo erectus* and archaic *Homo sapiens									
4	–	3–8	4–9	4–7	4–10	5–10	4–8	11–24	3–10
5	–	–	8–9	–	7–11	9–13	8–9	14–21	7–13
6	–	–	9	–	6–14	15–17	–	21–38	6–17
7	5–16	7–11	7–10	7–11	5–16	10–18	9–15	12–20	5–18
8	–	15	16	–	–	–	–	–	15–16
9	–	7	8	–	15	18	11	19	7–19
10	–	–	–	–	6–10	10	–	–	6–10
11	–	–	7	11	11	9	11	–	7–11
12	9	13	–	–	–	–	–	–	9–13
13	9	–	–	–	11	–	–	–	9–11
14	6–11	4–8	5–9	6–11	6–11	4–9	6–10	10–24	4–11
Homo sapiens									
15	4	5	–	5	4–9	6	6	–	4–9
16	6–7	5–6	6–7	7–8	6–7	5–6	7–9	–	5–9
17	4–7	6–11	6–11	5–8	6–7	6–10	7–10	–	4–11
18	5–6	4–5	5–6	5–6	5–6	4	6–7	–	4–7
Aust. Range*	5–13	5–14	4–12	4–10	4–14	6–14	5–12	7–19	4–19
H.e./A.H.s.	5–16	3–15	4–16	4–11	4–16	5–18	4–15	10–38	3–18
H.s. Range	4–7	4–11	5–11	5–8	4–9	4–10	6–10	–	4–11

Notes:
[a] WLH50 excluded.

Samples and sources:
1 = Willandra series (WLH), 2 = Combined Kow Swamp and Coobool Crossing series (Thorne, 1975; Brown, 1982), 3 = Recent Australian Aboriginal series (Brown *et al.*, 1979), 4 = Early African *Homo erectus* (Wood, 1991), 5 = *Homo erectus* (Java) (three individuals, Pithecanthropus 2, 3 and 5, Jacob, 1966), 6 = *H. soloensis* (Ngandong or Solo), Java (Weidenreich, 1951), 7 = *Homo erectus* (Choukoutien) China (Weidenreich, 1943), 8 = *Homo erectus* (Lantian) China (Woo, Ju-Kang, 1966), 9 = *Homo erectus* (Hexian) China (Wu and Poirier, 1995), 10 = *Homo erectus* (Kabwe or Broken Hill) Zimbabwe (Weidenreich, 1943), 11 = *Homo heidelburgensis* or archaic *Homo sapiens* (Swanscombe) England (Morant, 1938), 12 = *Homo erectus* (Bodo), Ethiopia (Conroy *et al.*, 1978), 13 = Archaic *Homo sapiens* (Dali) China (Xhinzhi, Wu, 1981), 14 = *Homo sapiens neanderthalis* (Ivanhoe, 1979), 15 = Niah, Sarawak (Brothwell, 1960), 16 = British Neolithic populations (Ivanhoe, 1979), 17 = Modern Mesoamericans (ibid., 1979), 18 = Modern Europeans (ibid., 1979).

thickened somewhere on the vault, usually in the regions of lambda, asterion or inion, with a range of 13–16 mm. The general pattern of vault composition among late Pleistocene Australian crania, no matter where they are from, is quite the reverse of thick *Homo erectus* cranial vaults. Unlike erectines, Australian fossil crania have very thin inner and outer cranial tables with the bulk of the thickness consisting of cancellous bone. It may be that thick vaults are an archaic trait, but vault wall construction in many robust WLH fossils is different from earlier hominid groups, although some do retain the archaic pattern of equal thickness for the three structural features. The general drift towards gracility among modern humans has brought with it a reduction in cortical robusticity as well as a commensurate thinning of cancellous bone. In the vault, this trend seems to have taken the form of the replacement of cortical bone by cancellous tissue. Later, cancellous tissue was reduced as part of a general pattern of natural gracilisation, evolving among recent people and eventuating in a general thinning of vault walls, although this was not widespread in Australia.

Supraorbital development

Willandran supraorbital development is as varied as its cranial vault thickness. It is actually greater than indicated by metrical assessment alone and ranges from very smooth brows (WLH1, 3, 11, 51 and 68) to well-formed brow-ridges (WLH18, 19, 50, 69 and 73). Indeed, the brow of WLH68 is non-existent (Plate 6.11). Using a supraorbital module index constructed to quantify Willandran brow development has produced a range that extends from 0.3 for the gracile WLH1 to 10.8 for WLH152 (Figure 6.1 and Table 6.4). Then WLH50 stands out with a massive 19.4 module which is a larger index than the average modules for the *Homo soloensis* (Ngandong) and *Homo erectus* (Choukoutien) and quite a bit higher than the Neanderthal module (Weidenreich, 1943, 1951; Thorne and Wolpoff, 1981) (Plate 6.3). In fact, WLH50 has scores greater than individual scores for Sangiran 17 (14.8) and Ngandong XI (14.1), which has the largest brows in the Ngandong series. It is worth noting that Ngandong also has the module range for any of the compared groups. There is a large gap between the modules of WLH152 and WLH50 but an even bigger one between WLH152 and WLH1 and 3. What is interesting is the difference in the module ranges between the four fossil groups. Both erectines and the Ngandong people have high modules in all three brow elements. Neanderthals are similar, but they do have somewhat smaller modules in the medial element. In contrast the WLH series has

Upper Pleistocene Australians 203

Plate 6.11. Brow profile of WLH68 showing a bulbous forehead, a lack of any brow development, the arc of the left eye socket (bottom) and discolouration typical of cremated bone.

a very wide module range, which seems to include both archaic brows that fit within the modules of the other three samples, as well as an extremely low module score that typifies the modern brow with little or no development. These results show once again the mixture of archaic and modern features

204   *The First Boat People*

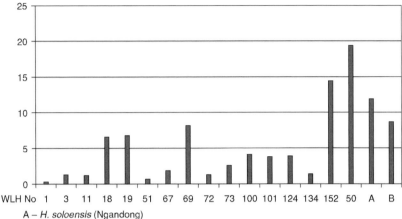

Figure 6.1. Supraorbital modules for Willandra, Ngandong and Choukoutien fossils groups.

found on various individuals in the sample, with the robust people falling within known archaic parameters.

In those Willandrans with a brow ridge, their features include a major development medially at the superciliary ridges, and the lack of lateral brow ridge extension to form a complete toral arch. When the lateral region is prominent it tends to accentuate postorbital constriction by enlarging the trigonic area (Plate 6.12). The pattern of brow development is similar to other late Pleistocene Australian populations such as Kow Swamp and Coobool Creek, but the WLH50 module is far bigger than any individual in either of these populations. Comparing the three average supraorbital measurements for Javan, Chinese and Willandran groups shows medial development in the latter with a more even toral spread in the other two together with a lateral developmental bias in Javans. It seems as though this latter feature has been lost in WLH50, although it has retained a prominent medial brow.

Those crania with well-developed brow ridges have thick cranial vaults, prominent cranial buttressing and ruggose areas of muscle attachment on their malars and the nuchal surface of the sub-occipital. In contrast, gracile individuals have little or no brow, thin, oval cranial vaults and smooth bone surfaces with little marking from muscular insertion. The lack of a brow ridge adds considerably to their gracile appearance and oval cranial shape (WLH1, 3, 11, 51 and 68). Both males and females lack brows. Conversely, while a full module cannot be properly assessed for WLH45 because of its broken brow, the two remaining measurements indicate that it cannot be high. This

Table 6.4. *Supraorbital thickness (mm) and modules for WLH Series compared with* Homo soloensis *(Ngandong) and* Homo erectus *(Zhoukoutien) and Neanderthals*

WLH No	Medial	Middle	Lateral	Supraorbital module
1	5	3	5	0.3
3	8	6	8	1.3
11	7	7	7	1.2
18	17	9	13	6.6
19	16	16	8	6.8
45	–	6	7	–
50	22	22	12	19.4
51	9	4	6	0.7
67	9	7	9	1.9
69	19	10	13	8.2
72	11	6	6	1.3
73	14	8	7	2.6
100	14	8	11	4.1
101	14	9	9	3.8
124	13	9	10	3.9
134	14	5	6	1.4
152	18	15	12	10.8
Range	5–22	3–22	5–13	0.3–19.4
x̄	13.1	8.8	8.8	–
s.d.	4.7	4.7	2.6	–

Comparative Fossil Ranges	Medial	×	Middle	×	Lateral	×	Supraorbital module range
H. erectus	12.0–19.6	14.6	11.5–17.4	14.4	10.8–14.5	12.4	4.9–16.7
H. soloensis	13.5–16.8	14.7	11.0–14.2	12.9	16.2–22.0	18.8	8.0–17.5
Neanderthals[*]	14.7–19.0	16.9	5.0–16.5	12.0	7.8–13.8	10.7	1.9–14.4
Willandra series	5.0–22.0	13.1	3.0–22.0	8.8	5.0–13.0	8.8	0.3–21.0

Note:
[*]After Wolpoff, 1999: Table 87.

is an otherwise robust individual, but probably female, a determination made by observations made on a section of its sciatic notch. Therefore, WLH3 and WLH45 clearly show that the wide range of brow ridge variation among Willandrans cannot be attributed to sexual dimorphism alone.

A univariate comparison between maximum cranial thickness and supraorbital development shows a very high correlation ($r = 0.85$) between the two (Figure 6.2). These two features cluster tightly, yet produce a scatter

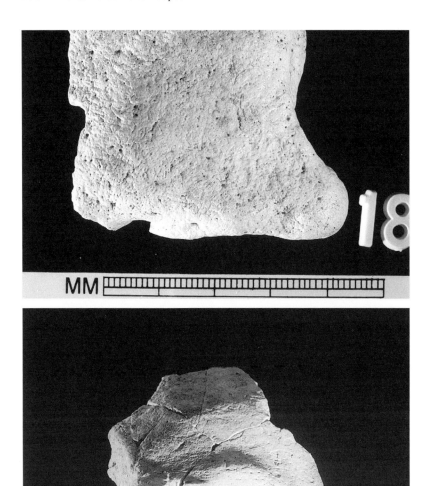

Plate 6.12. Zygomatic trigones of WLH18 (top) and WLH69 (bottom). Note also the prominent brow ridge development in the area of the superciliary ridge on WLH69.

which sharply divides gracile females (WLH1, 11, 51, 68, 134) at the extreme left from gracile males (WLH3, 67, 72, 73) in line, and robust females (WLH100, 101, and 124) from robust males (WLH18, 19, 50, 69 and 152) at the right with WLH152 set well apart from the rest and WLH50 as a very obvious outlier. To some degree the pattern reflects allometric factors, but the

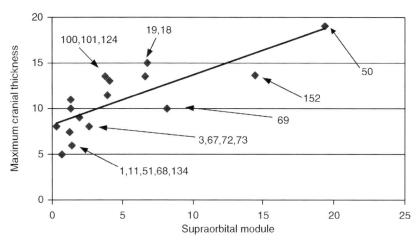

Figure 6.2. Linear regression correlations between supraorbital module and cranial vault thickness.

distribution also highlights certain shared features among the two groups, although variation between similar types is not unusual. For example, WLH69 has a very well-developed supraorbital region, although its cranial vault is not as thick as might be expected. On the other hand, WLH19 is also generally robust and it has a greater cranial thickness than WLH69. The well-developed brow of WLH152 shows that WLH50, although bigger, is not unique in this regard, and I suggest this could be a WLH50 without the vault pathology.

In summary, the supraorbital modules of three robust Willandrans (WLH50, 69 and 152) fit with the Ngandong module range, with WLH50 individually exceeding it. It also exceeds the erectine range, while two more can fit within (WLH18 and 19). Except for the most gracile Willandrans (WLH1, 3, 11 and 134), all fit into the Neanderthal ranges, while WLH50 comfortably exceeds all the fossil populations compared. It could be argued that these comparisons, drawn as they are between groups vastly separated in time and space, mean very little. They are made here, however, because it helps quantify the robust and gracile qualities of the earliest Australians that are the focus of discussions concerning regional continuity and 'Out of Africa' (Thorne, 1976; Thorne and Wolpoff, 1981; Wolpoff *et al.*, 1984; Hawks *et al.*, 2000). Another reason is that the Javan series and more indirectly the descendants of Chinese erectines have been considered ancestral to late Pleistocene Australian populations in regional continuity arguments. The pattern, however, may not be as simple as just robust and gracile groups, The middle grouping, while reflecting an overall robustness, may be an indication of the intermixing between two types which might explain why WLH50 is placed so far from the main sample (Figure 6.2). Undated samples could easily include people living later, between 15 ky and 30 ky, having a mixture of both morphologies as the two gene pools recombine among descendent generations. On the other hand, WLH50 may represent the morphology that arrived in Australia from the Indonesian region and is ancestral to later robust groups. It is worth remembering that the WLH50 calvarium was a surface find that included a section of a large right elbow, some phalanges and a few post-cranial fragments and show a similar degree of fossilisation and colour. After extensive investigation of the discovery area and repeated visitations, no further remains have been discovered. It has occurred to me more than once that WLH50 may have been originally found somewhere else and brought to the place of its discovery after being carried perhaps for years or even a few generations. It was eventually discarded, became a shallow burial naturally or might have been placed in a simple grave that eventually eroded away. Is it possible that it was fossilised when first found by Willandrans!

Malar morphology

Isolated malar bones from several individuals have been used to maximise data from the fragmentary collection. The malar has a high degree of heritability in both its shape and size, and this factor adds to its importance as a morphological determinant. Two modules for *size/length* and *robusticity* have been constructed using seven malar measurements. A linear regression comparing the two modules using a sample of eight individuals displays a cluster with three outliers (Figure 6.3). WLH2 is further separated from WLH1. While this scatter reflects the general morphological trend of the sample, separating robust from gracile individuals, the middle group lies on either side of the line, generally with graciles below and robusts above. Once again WLH50 is an outlier, strongly emphasising its robusticity compared with body size (size/length module), which is similar to others in this group. There is a statistically significant correlation ($r = 0.73$) between malar robusticity and size/length modules. Gracile outliers include WLH1 and 2, while WLH3 sits close to the central cluster between the less robust individuals and other graciles as in the previous analysis. This is due to its male status, but the size/length module reflects stature more than shape. At 173–175 cm, WLH3 is far taller than WLH1 (151 cm) (Figure 6.4A). Similarly, there is very little difference between the modules of WLH3 and 50, who have completely different skeletal morphologies; but are both male and were

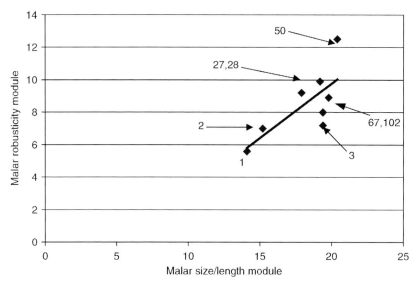

Figure 6.3. Linear regression correlations between malar size/length and robusticity modules.

possibly of a similar height, with WLH50 a little taller. Unfortunately, because of the lack of long bones from WLH50, this cannot be verified. Separation between WLH1 and 2 is probably a reflection of their female and male status, respectively, with the WLH2 malar more rugged than that of WLH1 (Plate 6.13). It is more than likely that there is a shared culture between WLH1 and 2, which is reflected in their method of ritualised cremation.

The robusticity module is a better indicator of build (Figure 6.4B). For example, while the size/length modules of WLH1 and 2 do not separate them significantly, the latter does have a more rugged construction. This includes a well-developed malar tuberosity, a thicker inferior border with well-marked, deep sulci for muscle attachment, a robust malar body, a thick marginal process and a slightly more rounded orbital margin.

Further, the marks for attachment of the *zygomaticus minor* and *major* and masseter muscles are prominent, as is the deep groove for the *levator labii superioris*. These rugged features contrast sharply with the flatter, lighter build of the WLH1 malar, strongly suggesting WLH2 is male. The robusticity module emphasises this build, placing WLH2 closer to WLH3 than WLH1, reversing the trend shown using the size/length module above. When robusticity and maximum cranial thickness modules are compared, the sample

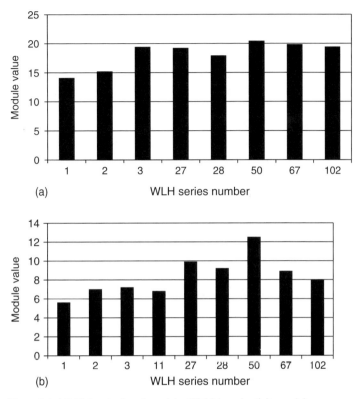

Figure 6.4. (A) Malar size/length module. (B) Malar robusticity module.

spreads out in a similar way to the previous analyses with WLH50 at one end and a cluster of graciles (WLH1, 2, 3 and 11) at the other (Figure 6.5). The inclusion of WLH11 in the gracile cluster again confirms its gracile form. The general robusticity of WLH27 and 28 is expressed but more effectively, while the outstanding position of WLH50 emphasises once again the degree of its robustness. The correlation of maximum cranial thickness with malar robusticity has a very high significance level ($r = 0.91$).

This series of analyses show a consistency of separation between robust individuals, such as WLH18, 19, 27, 28 and 50, and a gracile group represented by WLH1, 2, 3 and 11. Moreover, the fact that the male WLH3 and the probable male WLH2 belong to the gracile group shows that it is not composed solely of females. Correspondingly, there is a vast difference in male morphology, with some individuals more gracile than others.

Upper Pleistocene Australians

Plate 6.13. Comparison of the malar bones of WLH1 (left) and WLH2 (right) with its prominent malar tuberosity and overall greater rugosity than that of WLH1.

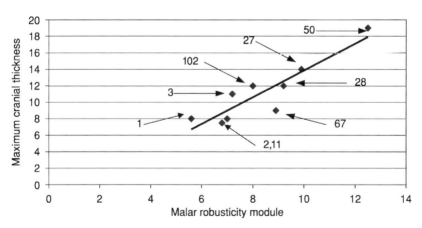

Figure 6.5. Linear regression correlation between malar robusticity and cranial vault thickness.

Pneumatisation

The fragmentary nature of the Willandra series has left few with enough bone to preserve sinuses. So, this discussion is confined to the frontal and mastoid regions only. Frontal sinuses are generally small and positioned medially, while some individuals, such as the gracile WLH68 and the very robust WLH19, have no trace of a frontal sinus. These contrast with the gracile WLH134 that has a large oval sinus with respect to its gracile brow. Therefore, sharp differences exist for this trait, even between individuals with similar morphologies. Frontal sinuses are smaller in graciles than robusts. Most of the brow in WLH3 is missing, so little can be said about the size of his. WLH1 had them but they were superior to the glabella, corresponding more to superior ethmoid than frontal sinuses *per se*. In WLH44, mastoid air cells do not extend above the root of the mastoid; they are small with the largest 3.4 mm in diameter. In another gracile (WLH47), they are large, with a 10 mm diameter. WLH24 could also be described as gracile, but pneumatisation appears to have reached the posterior root of the zygomatic arch. Superiorly, air cells stop abruptly at a point level with the supra-mastoid crest, and within the superior portion of the mastoid they measure 7 mm in diameter.

The frontal sinus of WLH50 is large by any standard and contrasts strongly with all others in the series. Its shape and size corresponds closely to that of *Homo s. soloensis* with ranges of 14–18 mm deep, 19–25 mm wide and up to 17 mm high. This contrasts with the smaller Chinese erectine sinuses at 4–15 mm deep and 8–25 mm wide that are smaller and centrally located (Weidenreich, 1943: 167; 1951: 252). But the rather thin and solid brow ridge and post-orbital sulcus common to most Chinese erectines may have restricted frontal sinus development. Weidenreich (1943: 165) confirmed this pattern '*It seems to me as if the tendency in pneumatisation were less pronounced in* Sinanthropus *than in* Pithecanthropus. . . Therefore, there are two differences in the frontal sinuses of Choukoutien and Ngandong people: the former are small and are centrally located in the interorbital region while the rest of the supraorbital region is solid' (Weidenreich, 1943: 165).

Robust Willandrans, such as WLH18, 19, 22, 69 and 101, have a different sinus construction from WLH50. The frontal sinus of WLH101, for example, is divided into oval, cell-like structures, similar to, but larger than, those normally encountered in the mastoid region. They spread posteriorly rather than laterally, reaching deep above the root of the frontal crest as they do in WLH69. The frontal sinus in WLH22 is similar, placed above the frontal crest but occupying a larger area. The largest frontal sinus is that over the

Plate 6.14. Frontal sinus of WLH50, at top of picture.

left orbit of WLH50 and estimated to be 6.8 cm^3 (Plate 6.14). It is difficult to assess how this development affected superciliary ridge development in this series. In those cases where it occurs, the sinus nearly always lies a considerable distance from the ridge and not directly within it. Even in WLH134, which has a prominent superciliary ridge, the sinus is situated infero-medially. The one exception to the general pattern is the frontal sinus of WLH23, which is centrally located and consists of a rather cavernous area almost directly below the superciliary ridges. The latter are not well developed and it seems that even when the two features are in close approximation, the ridges are not as well developed as might be expected from the size of the sinus.

Frontal sinus development in the Willandra series is so varied that superficially it looks as though there is no pattern to it and brow-ridge development is independent of sinus development and position. Surprisingly, in those with negligible brow-ridges, the frontal sinus is placed centrally directly behind, above or below the glabella, if it appears at all. The complete absence of sinuses in gracile individuals with little or no supraorbital region is not unexpected, but this explanation obviously does not apply to robust forms with well-developed brows.

Robust individuals display extensive mastoid pneumatisation. Again, WLH50 leads the way, extending posteriorly to the occipitomastoid border, superiorly to the parietal notch and anteriorly above the supra-mastoid crest. The air cells are uniformly large; the largest 14 mm in diameter. Because of missing bone the medial expansion of air cells in WLH50 cannot be assessed. The air cells of WLH27 penetrate the cranial floor beneath the right sigmoid sinus, extend to the occipitomastoid border as well as above the supra-mastoid crest, as they do in WLH50. It is often difficult to determine how far air cells penetrate superiorly within the temporal wall. Their subtle change in shape and size, which becomes smaller with increasing distance from the mastoid, enable them to blend with normal cancellous bone, making it difficult to distinguish between the two. Nevertheless, they probably reach the parietal notch in WLH27 as they do in WLH50. In WLH101, pneumatisation almost reaches the posterior root of the zygomatic arch, while superiorly it extends to the occipitomastoid border, with the largest cells measuring 7 mm in diameter. In the robust female WLH45, the pattern is similar to that of WLH101 but the cells are small, measuring only 1–2 mm in diameter.

Frontal and mastoid development in Willandrans depends to a large degree on general allometric factors. Larger, more robust individuals have greater bone mass, larger mastoid processes and well-developed, high brow-ridges, so they will have more pneumatisation than those with lighter cranial structures.

Glenoid fossa

Only five individuals in the series (WLH19, 44, 72, 100, 101 and 124) have a complete right or left glenoid fossa that can be measured. The sample is small, but several points are worth making. There is a wide size range with length varying by around 50 per cent and breadth by 30 per cent (Tables 6.5A and 6.5B). Depth has a very large range, with the depth of WLH72 over three times that of WLH44.

The Willandran length slightly exceeds that of Chinese erectines, but is markedly less than in the Ngandong group. The Willandran fossae are, however, much shallower than either Javan or Chinese fossae, although none of the robusts has them preserved for comparison. I suspect, however, that they would have been comparable with the other two groups. Willandran glenoid length slightly exceeds that of the Chinese, but is less than that in early Javan erectines and far smaller than in *Homo soloensis*. Willandran depth just overlaps the Chinese range and contrasts strongly with the deep and long Javan fossae. The Willandran length/breadth indices almost encompass both Chinese and Javan groups, while their depth/length and depth/breadth indices are far smaller.

Table 6.5A. *Willandra Series glenoid fossae, dimensions (mm)*

	Dimensions			Indices		
WLH No	Length	Breadth	Depth	L/B	D/L	D/B
19	16.4	24.2	7.0	67.8	42.7	28.9
44	23.2	19.0	4.0	122.1	17.2	21.1
72	21.3	26.3	12.5	81.0	58.7	47.5
100	24.2	27.0	10.0	89.6	41.3	37.0
101	21.0	26.5	8.0	79.3	38.1	30.2
124	15.0	22.5	6.0	66.7	40.0	26.7

Table 6.5B. *Comparative glenoid fossae ranges (mm)*

Ranges	Dimensions (mm)			Indices		
Sample	Length	Breadth	Depth	L/B	D/L	D/B
WLH series	15.0–24.2	19.0–27.0	4.0–12.5	66.7–122.1	17.2–42.7	21.1–37.0
H. s.	28.0–36.0	23.0–31.0	13.0–18.0	100.0–123.0	46.4–64.3	56.5–64.3
C H.e.	16.0–21.0	23.0–27.0	11.5–15.0	72.0–78.3	68.3–83.2	46.0–65.2

Notes:
C H. e. – Chinese *Homo erectus.*
H s. – *Homo soloensis.*

The shape and size of fossae among gracile Willandrans suggest there has been a considerable reduction in depth since the Middle and Upper Pleistocene, while length and breadth have remained similar. Depth includes examples of very small, shallow fossae (WLH44), providing a marked contrast to Middle and Upper Pleistocene Asian groups. In contrast WLH72 is three times the depth seen among gracile individuals. This once again highlights the wide morphological variation that exists in the sample and emphasises two groups with outstandingly contrasting morphologies. Even the extreme depth in the Willandra is, however, only at the shallow end of Asian erectines.

Craniometric data

Little significant metrical information can be gleaned from the Willandran series, but the frontal curvature index of WLH19 is 18.9, only slightly higher than the mean of 18.3 for the combined Coobool Creek/Kow Swamp

Table 6.6A. *Mandibular metrical data (mm)*

WLH No SH	SH	SB	CH	CB
1	23	14	–	12
3	33	15	33	15
11	28	13	22	12
20	24	14	33	15
22	33	18	36	15
27	34	16	–	–
73	30	14	27	14
102	–	–	31	16
103	–	18	–	16
121	–	–	–	17
124	–	–	26	14
152	–	–	38	18[a]
Range	23–34	13–18	22–38	12–18
X⁻	29.3	15.3	30.8	14.9
s.d.	4.5	1.9	5.4	1.9

Notes:
SH – Symphysis height, SB – Symphysis breadth,
CH – Corpus height, CB – Corpus breadth.
[a] Estimated value.

Table 6.6B. *Comparative mandibular metrical ranges (mm) among three fossil and a modern Australian population*

	WLH	Kow Swamp[a]	CC[b]	HMV[a,b]
Symphysis height	23–34	36–39	33–45	25–43
Symphysis breadth	13–18	15–19	13–20	12–20
Corpus height	22–38	28–36	29–38	22–37
Corpus breadth	12–18	13–20	13–17	11–21

Notes:
CC – Coobool Crossing, HMV – Late Holocene Murray Valley Sample.
[a] – Thorne, 1976.
[b] – Brown, 1982.

populations, inclusive of those deemed to be head bound, but WLH19 shows no sign of the frontal flattening typical of cranial deformation. Mandibular metrical data from the Willandra series are limited but symphysial height contrasts strongly, barely overlapping with a combined Kow Swamp/Coobool Creek sample (Tables 6.6A and 6.6B).

As well as being completely outside the female range of Coobool Crossing, WLH1, 11 and 20 are almost five, three and four standard deviations (s.d. 3.14 mm) smaller than the Coobool mean (37.2 mm), respectively (Brown, 1982). Moreover, WLH1, 11 and 20 are over four, two and three standard deviations (s.d. 2.48 mm), respectively, smaller than the mean for a comparative and more recent female population from the Murray Valley (32.5 mm). Corpus height of WLH11 is almost four standard deviations (s.d. 2.43 mm) below the mean (31.2 mm) for Coobool females and two standard deviations smaller than recent groups. The small size of WLH11 supports conclusions drawn from analyses of its supraorbital development, cranial thickness and malar morphology that put it into the gracile grouping. Both males in the gracile sample (WLH3 and 22) have a symphysial height two standard deviations (s.d. 2.62 mm) below the mean for Coobool males (39.1 mm). Gracile Willandrans have very small mandibles compared with other late Pleistocene populations and lie at the bottom end of the modern range (Table 6.6B).

Postcranial remains

Humeri

Only 14 individuals have humeri suitable for measuring, yet even among these the anteroposterior and mediolateral diameter ranges are quite large. Minimum circumference measurements lie at the robust end of the late Holocene range (Dongen, 1963). The 56.6 mm measurement obtained for WLH45 is interesting, because it puts this female well within the range of modern Aboriginal males and almost two standard deviations above the female mean (48.5 mm). This result is consistent with other robust features described for WLH45. The robusticity index for the Willandran humerus is high, but the sample consists of only seven bones; a very small number on which to base any firm conclusions. The range of humeral length is 305–361 mm, well within and at the upper end of the Holocene range (276–366 mm) (Dongen, 1963).

The humeral cortex of WLH110 is extraordinary thick, measuring 7–8 mm across and the diameter of the medullary cavity is 7 mm (Plate 6.15). This extraordinary feature is not attributable to any obvious pathological processes and except for the thickened wall the bone is completely normal. The few pieces of skeleton from this individual do not allow further investigation of overall robusticity. Robust individuals are more likely to have thick cortices than those with a lighter skeletal form and this has been noted in the very robust Shanidar 4 Neanderthal femur and is a common feature among this group (Trinkaus, 1984: 267). Indeed, thickened cortices accompanied by

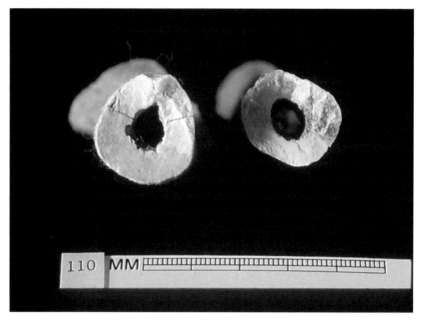

Plate 6.15. Thick humeral cortex in WLH110.

narrow medullary cavities are now widely recognised as being one of the suite of derived characteristics among erectines and Neanderthals (Antón, 2003). Humerus II from Zhoukoudian shows a similar form with the narrowest diameter occupying only 22 per cent of the entire transverse diameter of the shaft (Wu and Poirer, 1995 : 59). The authors note that thick walls and narrow medullary cavities are typical among *Homo erectus* long bones with 'almost no medullary canal at midshaft' in the erectine tibia (Wu and Poirer, 1995: 57–61). The medullary cavity in the right humerus of WLH110 occupies only 30 per cent of the total transverse thickness, something also reflected in its left arm (33 per cent). It is hard to believe that this thick cortex has nothing to do with the derived characteristics associated with phylogenetic origins and that its cranial form would not have been a robust type.

Femora

The diameter of the subtrochanteric and mid-shaft portions of Willandra femora are placed at the robust end of the late Holocene range. The similarity, however, between femoral and humeral remains in this regard may be a reflection of taphonomic processes preserving robust bone over gracile forms. The mediolateral diameter of the mid-shaft exceeds the modern range by about 4 mm, suggesting a more transversely flatter bone. The platymeric indices, however, lie well within the modern range. Again, some generalisations seem valid, even though the sample is restricted. The first is that the Willandran robusticity index is almost exactly the same as that for modern Australian femora. Secondly, the female WLH45, once again, looks robust by having a femoral index 0.8 below the male maximum and only 0.1 below the maximum for females. Thirdly, estimated femoral length suggests a wide range of stature, from 149 cm in WLH1 to 177 cm for WLH67.

An outstanding feature of some femora is very thick cortices around midshaft and lateral to the *linea aspera*, where thickness ranges from 3 mm for WLH3 and WLH121 to 12–16 mm on certain parts of the shaft of WLH107. This exceeds by far the 8mm thick cortices observed in one Neanderthal femur (Trinkaus, 1984). Cortical bone composition is solid, consists entirely of compact bone and lacks trabecular networks or pathology of any kind. There is a small, positive association between cortical bone thickness and cranial thickness; but, because of the poor condition of the remains and the invariable lack of one or other of these components in testable individuals, this cannot be taken farther than simply being a general impression.

Tibiae

WLH6 consists only of a right tibia, which is the smallest in the series, lying at the lower end of the range for all tibial measurements. Prominent lines of muscle insertion suggest a male individual with an estimated height of around 158 cm. Besides being short, this bone is narrow in both mid-shaft dimensions, particularly when compared with recent tibiae (Plate 6.16). Other Willandran tibiae are larger in both mid-shaft dimensions, although the right tibia from WLH45 is only slightly so. Hyperplatycnemia occurs only in the right tibia of WLH67. Like the humerus, the sample indicates a wide range of stature among the group.

Willandran cultural indicators

Cremation

In the initial WLH1 report the following reconstruction of disposal was offered:

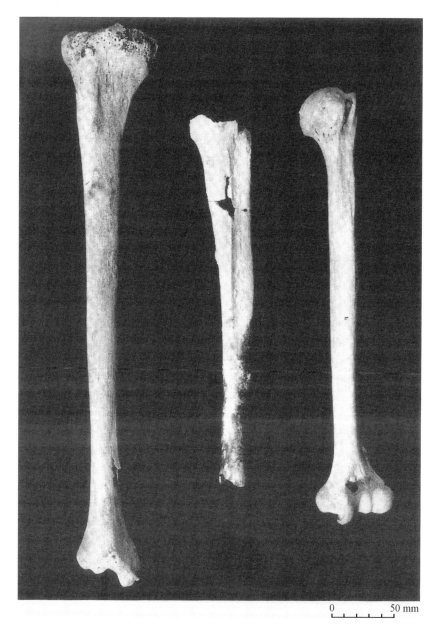

Plate 6.16. Comparison of WLH7 tibia (centre) with a modern example (left).

The individual was cremated as a complete and fully fleshed cadaver, though the pyre was insufficient to achieve full incineration. There was a total and thorough smashing of the burnt skeleton, particularly the face and cranial vault. The ash and smashed bones were gathered together and deposited in a conical hole either beneath the pyre or immediately adjacent to it. *(Bowler et al., 1970:57).*

It is significant that when the Mungo Lady (WLH1) died she was not just thrown out as dinner for the next passing *Thylacoleo*. She was given a special burial, indicating that her people and their culture demanded more than just a hole in the ground or a cursory disposal of some kind. Differential colouration of adjacent cranial bone fragments also indicated that the pieces were in different parts of the fire. In other words, she was partially cremated, her bones taken from the pyre, smashed further and the fragments placed in a purposely excavated place in the centre of a fire for renewed burning. The whole thing was then covered with sand, which later became impregnated with calcium carbonate leached from surrounding beach sands. This complex burial process implies ceremony and ritual and is the oldest in the world (Bowler *et al.*, 2003). Positive identification of deliberate cremation, however, can be problematic. To identify WLH1 as a cremation Thorne (1975: 186) suggested four categories of bone alteration:

1. slight scorching and blackening but without destruction of the surface,
2. burning that removed surface bone features with slight glazing that left small patches of dark grey mottling and small patches of whitish surface,
3. thorough burning with powdery surface and blue-grey to white colour, and
4. intense colouration with the bone surface a vivid white colour and split or crazed.

In short, cremated green fleshed or defleshed bone has a colour range from dark black/deep blue to light grey, including white powdery calcination after exposure to very high temperatures, usually greater than 800°C (Buikstra and Swegle, 1998). These colour changes are often accompanied by distinctive transverse cracking, particularly on tubular bone and thorough carbonising through the bone with charring of broken edges; these characteristics were discussed for bone from Lake Eyre in the previous chapter (Plate 6.17). The various blues and greys have been noted by others as indicative of two phases of burning at different temperatures (Shipman *et al.*, 1984). Calcination is also indicative of high temperature burning, either in a large, constantly maintained fire over a period of time or placed in the centre of a very hot fire. This would indicate deliberate rather than accidental burning.

The WLH1 burial is not unique. Other examples of burnt bone with similar colour changes, calcination and cracking occur in WLH1, 2, 9, 10, 68, 115,

Plate 6.17. Curved cracking, calcination and colour changes on tubular bones of WLH115, all typical indicators of the bone having been cremated in a high temperature fire.

121, 122, 123 and 132, suggesting cremation or deliberate burning of human bone was fairly common practice, particularly among gracile people of which all these are examples. Other burials eroding out in the region have features suggesting that they have also undergone high temperature burning and these have been categorised as cremations. One of these individuals, the female WLH68, which consists only of calvarial fragments, is extremely gracile, lacks any hint of a brow and the cranial vault is 4–5 mm thick, but partial fusion of the sagittal suture indicates an adult. Its colour ranges from blue-black to shades of grey with calcination (Plate 6.18). The thoroughly burnt edges of the fractured bone show it was broken before being cremated. This individual not only looks like WLH1 it has even more gracile features.

Seven others in the series (WLH6, 22, 24, 63, 93, 120 and 126) have burn marks on some part of their surfaces, but these are regarded as 'burnt' rather than 'cremated'. They may have been partly cremated; but, because they do not show the characteristic colour variation and damage described above, they have not been regarded as deliberate cremations. It is likely, however, that this distinctive method of disposal was not restricted to graciles. Two others (WLH28 and 63) display the characteristic features of cremation, and have

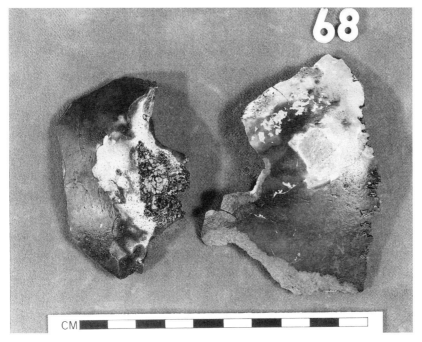

Plate 6.18. Cremated cranial sections of WLH68. At left is a frontal view of the frontal and supraorbital region. Colour changes include a range from black to dark blue, grey and areas of white calcination. The right bone is a section of the right parietal with the sagittal suture along the right margin showing partial fusion with the left parietal.

the distinctive bone colour changes of grey and blue/black seen on other cremated examples, but these are robust (Plate 6.19).

Bone smashing

Taphonomic processes fragment exposed bone but there are indications among some Willandrans that certain bones were fractured before burial. The neat pattern of fracture, producing straight, even edges indicates that some bones were not fresh when this occurred. Smashing may have taken place months or even years after death, when dry bone would shatter causing complete fracturing. Much of the fragmentation is on long bones and the crania of robust individuals, including WLH16, 18, 19, 69, 106, 107 and 110. These have thick to very thick cortices, some of which are broken into very small pieces. Normal taphonomic forces in such an environmentally stable

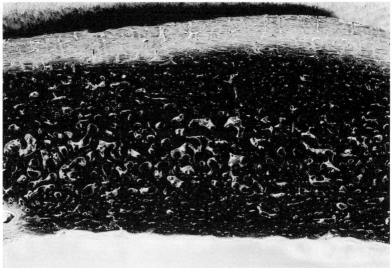

Plate 6.19. Heavily charred pieces of the cranium of WLH28 (top). The charring has penetrated right throught the thick cranial vault (bottom). Note also the substantial proportion of diploeic bone comprising 80–90% of vault thickness structure.

region do not usually cause such breakage patterns and it would require a strong, deliberate blow to fracture such thick cortical and cranial vault bone. Often, broken edges have been completely rounded by erosion and covered by a carbonate layer. Thus the bone was broken a long time before the deposition of carbonate, some of which took place well before 20 ky ago, and certainly long enough to bevel the edges of the bone before that took place. Internal stability of the lunette in which it was buried for so long is also unlikely to have caused post-depositional breakage. Support for this conclusion comes from the preservation of a number of burials that have emerged over the years, such as WLH3. It had thin cortices and after 42 ky of interment was extremely friable, causing bone fracture and loss during excavation, but it had retained perfect anatomical position until that time.

Other burial practices

Besides cremation, several other burial practices are worth noting. These include the extended burial of WLH3 that was discovered only 500 m from WLH1. These two completely differing styles of interment are, in themselves, a fascinating contrast in funerary rites at a very early time in Australia's human story. Whether males and females received different types of interment is not known, because WLH2, which is believed to be male, has been associated with the WLH1 cremation. The similar ages of WLH1, 2 and 3 suggest several methods of internment were being used by the Willandrans so long ago as flexible cultural practices. Bone smashing may have been another aspect of mortuary custom as a secondary burial practice following exposure of the body and defleshing by birds, carnivores or by time. The bones were then left to dry and later they were gathered together, smashed up, then deposited in a grave or pit of some kind.

Observation of eroding burials over 25 years or so has shown that upright burial was also practiced. The body was placed in a grave in a sitting or crouching position, so that when the grave eroded the cranium was the first part of the body exposed (WLH135 and 154). It would be nice to know when these various burial practices were introduced and by whom. All that can be said is that the earliest Willandrans used a number of complex burial procedures that included crouch, extended and cremation, with possible secondary smashing of bone on some occasions.

Ochre use and ceremony

A large amount of ochre was sprinkled over the body of the middle aged man WLH3 after he was placed partially on his side with his hands clasped in his lap and knees slightly flexed. It signals a prominent additional cultural indicator in the burial practices of this group. The nearest source of ochre is over 200 km northeast of Lake Mungo in the Barrier Ranges (Johnston and Clark, 1998). The ochre used in this burial must have been either traded from the ranges into the Willandra for this particular purpose as well as other uses, or supplies were sought by the Willandrans themselves and they made the 400 km round trip to fetch it. The existence of trading could imply cultural complexity among these as well as others inhabiting the region 40–45 ky ago. The construction of such networks and trading of goods requires trading centres with reciprocal exchange mechanisms, suggesting a certain degree of 'settled' and structured society in the region, which implies a much older time frame for the first people there.

Ochre could also have been used for body decoration as part of ceremonial activities or for unknown secular purposes. Rock art sites are unknown in the region, mainly because of the lack of suitable surfaces, such as rock walls and rock shelters. While art has not been found in association with the Willandrans, undated petroglyphs as well as painted motifs are found in the Barrier Ranges (Mootwingee), the possible origin of the ochre used for the WLH3 burial. The difficulties of dating this art prevents us from knowing whether the Willandrans or their contemporaries were instrumental in its production when collecting their ochre; after all they did go a long way for it. So they may have left inscriptions as a calling card to the spirit beings who produced the ochre and who may have played an important role in the ceremony and ritual of their burial practices. What we can say is that ceremony and ritual were definite aspects of their life and complex social behaviour existed around the Willandra Lakes a long time before the last glacial maximum.

The use of ochre is not the only indicator of ceremony and ritual. The premortem loss of both mandibular canine teeth in WLH3 has been described elsewhere (Webb 1989, 1995), but it is worth mentioning again to bring together different elements of Willandran cultural complexity. The resorption of the alveolar bone around both mandibular canine sockets has caused shrinkage of alveolar bone both in height and thickness mesio-distally. This has resulted in a ridge between the inter-proximal surfaces of the lateral incisor and first premolar bilaterally (Plate 6.20). Moreover, the incisors and premolars either side of the empty sockets now lean inwards as though trying to fill the gap left by the missing canines. These changes suggest that the canines were lost many years before the individual died, although canine teeth have the longest roots of any tooth and are not usually lost under normal circumstances. An injury or blow to the face severe enough to remove a canine would also damage or remove teeth either side, and this has not occurred. Therefore, bilateral loss of the canines is even more unlikely, particularly with an identical pattern of socket resorption. Moreover, canine loss is not normally observed among skeletal populations with the odd exception of edentulous individuals and those with extreme forms of occlusal

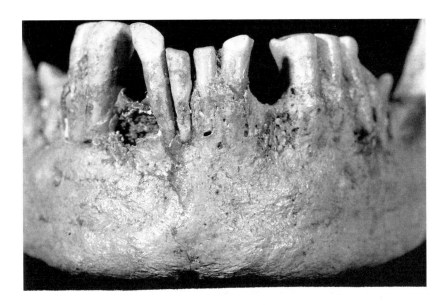

Plate 6.20. WLH3 mandible showing both canine teeth missing, resorption of the alveolar bone, and a compensatory lean of adjacent teeth towards the gap. All these characteristics strongly point to tooth avulsion some decades before the death of this man forty thousand years ago.

attrition; neither of these causes is the case here. The removal of canines is unknown in Australia or anywhere else as far as I am aware, so I am at a loss to know why these teeth are missing in WLH3, except by their deliberate avulsion perhaps during an initiation ceremony.

The lower, central incisors of WLH22 were also lost *pre mortem*. Complete resorption of both sockets has occurred, suggesting that they were lost many years before death. Nothing remains of the sockets themselves, which have been completely filled with new spongy bone. Advanced resorption points to a considerable time lapse since their loss or removal and has led to a narrowing of mandibular width. Two other individuals (WLH11 and 20) also have *pre mortem* tooth loss, but this is confined to the molars.

Traditional tooth avulsion among male and female native Australians, normally as part of initiation rites of passage, was usually confined to one or both upper incisors (Campbell, 1981). The removal of lower incisors was also practised, particularly among central Australian people, so the loss of these in WLH22 suggests that this initiation ritual goes back a long way. Ritual and ceremony are strongly implicated in the burial procedures of these

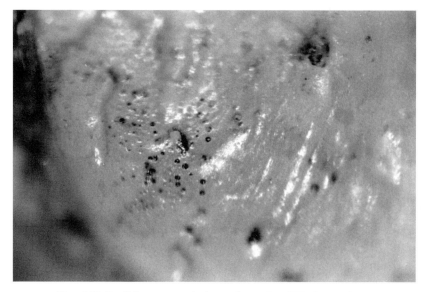

Plate 6.21. Parallel grooves on the lower, first molar of WLJ3. The lingual to buccal orientation of these scratches point to vegetable material, possibly containing silica phytoliths or small sand grains, being drawn across the surface of the tooth.

people, so it is not unlikely that other ceremonies existed in the culture, including passage through various age grades.

Material culture and dental attrition

Another feature of WLH3 is its unusual dental wear pattern, which contrasts with those of recent Native Australians with the exception of some individuals from the Murray River. The later introduction of a larger vegetable component into the diet as well as the use of grindstones added more abrasives to the food and resulted in predominantly flatter and even patterns of dental attrition among recent tribal Aboriginal groups. The pattern in WLH3 is not flat or even; it is very uneven, with peculiar attrition on the lower incisors and molars, particularly the left M_2 and M_3. The buccal side of these teeth is sloped so much that it almost reaches the level of the socket. Magnification of the polished occlusal surfaces shows striations cut bucco-lingually into the enamel and dentine (Plate 6.21). These grooves and the unusual slope of the tooth surface suggest that vegetable or plant fibre was repeatedly drawn across and down the teeth and that hard quartz grains or phytoliths in the material cut these grooves in the tooth surface.

This dental wear pattern suggests that this individual was shredding vegetable fibre for teasing out and making into nets, bags or baskets. What is also interesting is that this type of manufacture was, therefore, not necessarily left up to women. Nets are more than useful when living by a lake, not just for fishing but trapping birds and waterfowl. People may have used bags made of vegetable fibre when collecting shellfish (*Unio* sp.) that they brought back to their camps for the family meal, the remnants of which now form the myriad of middens that can be seen eroding from lunettes and other sand features around the fossil lakes in the region. Processing of plant fibre by those living close to large bodies of water would not be unusual. But to find evidence of this activity as a skeletal marker on a 50–60 year old man, buried beside an inland lake, is a fascinating insight into and a poignant reminder of his life that would have been totally lost without the skeletal evidence. Moreover, the man lived in the region 25 ky before the lake finally dried up and that is *really* an echo from the past.

Severe osteoarthritis in the right arm of WLH3 is also worth mentioning in regard to material culture. A detailed description of this pathology has been published elsewhere, but I mention it here because of its cause (Webb, 1995). The articulated ends of the bones of the right elbow have almost been destroyed and then fused (ankylosed) together. This severe form of osteoarthritis is usually caused by heavy and repetitive use of the elbow complex in a particular manner. That activity included quick extension of the elbow and rapid pronation of the lower arm. Long-term use of the arm in this way often results in almost total removal of the articular cartilage on the head of the radius, which then grinds into the joint surface of the distal humerus removing the articular cartilage on that bone. The set of traits observed on WLH3 have been termed 'atlatl elbow' from observations made in the elbows of North American native people who used the spear thrower (Angel, 1966; Ortner, 1968). Although WLH3's arthritic condition may have been exacerbated by an infection of the humerus, the main protagonist was a particular pattern of stress on the right elbow consistent with using a spear thrower. Another identical example of severe osteoarthritis with joint destruction and ankylosing of the right elbow also occurs on a second male person from the Willandra region recently examined. However, in consideration for local Aboriginal sensitivities, no further description can be made other than to identify it as WLH152 and say that it is not a gracile individual.

Our earliest evidence for the spear thrower in Australia is now dated to 19 ky, although it was probably in use in Europe (at La Gravette) 10 ky earlier (Walsh and Morwood, 1999). The Australian date for this weapon was made on a fossilised mud wasp nest overlaying a 'Gwion Gwion' (Bradshaw) figure painted in the Kimberley region of northwestern Australia that holds this weapon. This date is a minimum date, but the Willandra evidence may indicate the presence of this implement in Australia at a considerably earlier time. Gwions are also famous for holding another Australian icon, the boomerang. The oldest date for this implement, however, is 23 ky, obtained from an archaeological site in southern Poland where one made from mammoth tusk was recovered (Valde-Nowak *et al.*, 1987).

Bone points have been located at seven sites in the Willandra. Examples include those sharpened at one and both ends. It is not known whether they were used for manufacturing other weapons (i.e. pressure flaking), preparation of skins, sewing, net construction, personal decoration or even magic or sorcery purposes, but they are made from both animal and human bone. None has been dated so far.

Willandran cultural complexity

Cultural indicators among the Willandrans indicate they had a complex society with a rich ceremonial life. There are two different burial practices in use at least by 42 ky. These are accompanied by tooth avulsion, possibly resulting from some sort of rights-of-passage ceremony. Ochre was not only being used in burial ceremony, but it is likely that it was also used for body decoration. Burial practices and tooth avulsion suggest use of symbols and complex ritual activity pointing also to the use of complex language in order to transmit tradition from generation to generation, communicate ideas and explain ritual and ceremony, as well as make sure ceremonial processes were carried out properly. The dental attrition on WLH3 points to the use of the molar tooth complex in shredding plant fibre, probably to make string or cord for weaving into bags, nets and bindings. Hearth size and contents patterns also suggest a family structure and a differential hunting strategy. From this evidence there is little doubt that they possessed social structures and networks, family groups and may have been organised into bands. They may also have traded with others, an allowance made possible through an organised lifestyle and the fact that they had been occupying the area (as were others) for some considerable time before 42 ky. Finally, the pathological elbow of WLH3 points to the use of sophisticated weaponry, namely the spear thrower. Whether they also had shields, hafted axes and stone points on spears cannot be confirmed, but the undoubted skill and social complexity of these people would not have prevented this from being the case.

The Willandran material culture, ritual and ceremony and burial practices strongly point to Australia as being anything but a socio-cultural backwater in the latter stages of the Upper Pleistocene. The Willandrans had become fully adapted to their surroundings, which consisted of lakes, riverine ecosystems and semi-arid backcountry. They had already developed an adaptive social system that allowed them to live in many different environmental settings. Their adaptive lifestyle was underpinned by a very lively and spiritual existence focussed around the half dozen freshwater lakes and no doubt similar adaptations had taken place in other regions of Australia. No-body at this time had a 'coastal economy' unless they lived near the coast. Everybody else were generalist hunter-gatherers adapted to wherever they lived. The social and cultural wherewithal of these people were their main armament, if they had their culture they had their ticket to live anywhere, a trend to be found elsewhere in Australia and Tasmania.

Nanwoon, Tasmania

The last statement requires that I briefly mention another gracile individual discovered in 1987 far from the Willandra, in southwest Tasmania. It consists of a dark stained but very slender occipital bone (Plate 6.22). It is small, light and delicately built with thin cranial walls (1.0–1.5 mm) and shows little if any impression from muscular attachment in its sub-occipital region. Maximum thickness anywhere on the bone is only 6.3 mm, but the average thickness is around 4 mm. The oval appearance and delicate construction of this bone suggest that the entire cranium was very rounded, with little buttressing, and had an accompanying small face, jaws and teeth. The striking similarity between this bone and that of WLH1 suggests that it is also female and may be from the same 'gracile' gene pool. The importance of this individual is its presence in southwestern Tasmania in the late Pleistocene. The bone is >20 ky, but the lack of further work at Nanwoon at the request of Tasmania Aboriginal people has prevented a firm date being established. The Nanwoon area was abandoned by Aboriginal people around 13 ky in the face of changing environmental conditions in that part of southwestern Tasmania (Cosgrove, pers. comm.). These people were living in a very cool climate at that time. They utilised cave systems and rock shelters situated at the base of escarpments in the southwest of the island and enjoyed a far clearer view of the hills and valleys before them, as they gazed out across a savanna-like environment, than is the case now. The earliest dates for people in Tasmania put them there about 35 ky ago. This was a time when icebergs floated around the south coasts and icy winds blew off a glacier on the central west highlands. The cultural, economic and, possibly, biological adaptations required by these people to live in this place were far different from those enjoying the balmy north of the continent or even those occupying the Willandra.

By 35 ky ago the Australian people were in place. They occupied the largest island in the world to its every corner. They lived in a vast range of different environments successfully. Others might join them later, but the

Plate 6.22. Nanwoon occipital bone showing a hole of indeterminate origin on the right side.

exploration was finished. What happened next was an exploration of the mind and the development and intensification of cerebral culture as Australia headed deeper into an Ice Age.

7 *Origins: a morphological puzzle*

Arguing against the evidence: 'Mungo myths'

After assessment of Australia's oldest fossil human remains in the previous chapter, I feel a few old chestnuts should be addressed before going further. These involve the morphological appearance of individuals in the Willandra Lakes collection and the validity of the sex and age assigned to primary examples.

The fact that there are many gracile individuals in the Willandra series vastly weakens the argument that WLH1 and 3 are unique archaeological finds, occupying a position at one end of a single range of skeletal variation. By its very nature, a range of variation usually follows a bell curve. This would place only a few gracile people at one end and correspondingly few robusts at the other. The regularity with which starkly contrasting robust and gracile individuals are found in the Willandra, however, makes the one range of variation arguments untenable, because it is unlikely that so many individuals at either end of such a morphological spectrum would be discovered purely by chance, while the types in between remain comparatively few.

Working in this field, one often hears whispers on the ether concerning certain fossil remains. There is one about WLH1 which says that this individual should be called 'Mungo Girl' rather than 'Mungo Lady', her sub-adult status thus explaining her gracility. More often than not such whispers are started by those having little or no experience of examining the remains, and in certain cases have never set eyes on them. Let me reassure all those who would suggest that a sub-adult status is more appropriate for WLH1, nothing could be farther from the truth. The eruption of third molars, a partly fused suboccipital synchondrosis and cranial sutures together with full metaphyseal closure of long bones firmly places WHL1 as a young adult female person (20–25 years?), and certainly not sub-adult. Similarly with WLH3, it has been suggested that *he* is really a *she*. Sexual dimorphism does not alone account for gracility. Some individuals have been called female only because of the delicacy of their cranial structure. It is worth noting that, if only the cranium of WLH3 had been found, its construction would, without any doubt, have placed it as female. The presence of a pelvis, however, confirmed its male status. These arguments extend to suggesting that all gracile individuals are female, so what we are seeing in these two morphologies can all be explained by sexual dimorphism. I would point to many other examples of gracility in the collection and ask the sceptics to explain all these away as female, one end of a range or sub-adult. Besides its pelvis, the method of interment, its pattern of tooth avulsion and the likelihood that this individual used a spear thrower do not indicate a female status for WLH3.

The 'Gracile' people

There are eight basic skeletal features that characterise Australian gracile individuals. These are outlined below, followed by the Willandra Lakes Hominid that displays these features.

Gracile morphology is characterised by:

1. an oval and fully expanded brain case (WLH1, 3, 29, 68 and 135);
2. a very full frontal squama which lacks supraorbital development and well-developed zygomatic trigones (WLH1, 3, 11 and 68);
3. parietal bossing (WLH1);
4. thin cranial vault bones (WLH1, 9, 10, 11, 29, 68, 115, 122, 130, 135);
5. small, delicately constructed mandibles, glenoid fossae and smooth malar tuberosities that indicate a small dentition and upper jaw and a delicate facial bone structure (WLH1, 11 and 44);
6. a lack of strongly developed muscular markings, particularly on the sub-occipital region and malar (WLH1, 3, 11), and
7. small stature (WLH1 and 6).

Cultural markers include:

1. ceremony and ritual,
2. the use of ochre,
3. defined burial practices,
4. manufacture of a wide range of material culture,
5. sophisticated weaponry and tool use,
6. family and social cohesion.

If allometry is the reason for gracility among Australia's early inhabitants, it is logical to assume that the people were small. From the few post-cranial remains available, it is apparent that many graciles *were* small with height estimates for WLH1 and 6, 149 cm and 158 cm, respectively. However, stature estimate for WLH3 is between 172 and 177 cm, showing he is much taller and naturally possesses a larger but still very gracile skeleton. Besides

WLH1 and 3, many other Willandrans are extraordinarily gracile, such as WLH11 and 68 who have no equal among females from later robust populations. From these observations it seems likely that allometry is the reason for gracility: smaller people have small and delicately constructed bones and in the Willandra there are male and female examples of this.

There are two important points to make here. Firstly, those who would wield a calliper to prove a dogmatic interpretation for personal reasons, adhering to inflexible craniometric interpretations can dismiss the overall gracility of individuals like WLH1, 3, 11, 51, 68 and 134. Essentially, it is a difference of emphasis of one form of evidence over another and the use of statistically reconstructed morphology against a broader interpretation of observed phenotypic quality, subtlety of construction and shape itself. The latter includes an assemblage of features difficult to quantify, as well as the size of the post-cranial skeleton, which while poorly represented or missing in most individuals is rarely mentioned or considered as a reflection of the wider skeletal form. The overall shape of partial or fragmentary skeletons is often statistically difficult to construct in a way that makes very much biological or evolutionary sense. As a result, direct comparison of certain parts of the WLH1 and 3 skeletons on the one hand with representatives of the robust gene pool have only compounded problems of interpretation. For example, most of the upper face, maxilla and some sections of the cranial vault are variously missing from WLH1 and 3. Nevertheless, while the cranial thickness and craniometrics of WLH3 may lie within the lower end of the robust range, there is an almost complete absence of supraorbital development that has no equivalent in any male cranium in any Australian skeletal collection that I have observed. If this is not taken into consideration, overall morphological difference between the two is lost or can be ignored. It is worth remembering Birdsell's observation proffered for genetic studies, but equally applicable to morphometrics:

It is one of the scientific vogues these days to use the computer to calculate various forms of 'genetic distance' between series of populations. The methods are statistically complex and varied in form, but all fail in terms of expressing evolutionary significance. At best these statistical devices are measuring differences between populations and not biological relationships. *(Birdsell, 1979: 419)*

Secondly, there is no doubt that wide morphological variation is present in the Willandra collection, but this is not the case among contemporary individuals of the same genetic population. It is natural to expect such variation in a fossil population scattered over a minimum of 20 ky, particularly in Australia. In terms of variability, we are not talking about an endemic population that has evolved over hundreds of thousand of years and displays a narrow morphological range as a consequence, as has occurred outside Australia. This continent had no Middle Pleistocene population from which its Upper Pleistocene morphology evolved. It rather quickly built its population out of many migrations from a variety of origins outside Australia.

To establish if a wide morphological variety existed at any one time is difficult because dating of Australia's human fossils has problems, one of which is that for one reason or another radiocarbon people are sceptical about Uranium series dates. Therefore, we must not forget the enormous time depth of the collection and the temporal differences that must exist between certain individuals. This is usually ignored, however, when it is suggested that the gracility of WLH1 is unique and that it occupied a position at the extreme end of a single range of variation. To prove this there needs to be a comparison between WLH1 and its contemporaries. The only contemporaries are WLH2 and 3, so there we have two males and a female, all of whom display the same sort of morphology with some minor differences explained by sexual dimorphism.

Gracile origins

So who were these generally small and lightly built people and where did they come from? Gracile traits similar to those outlined above have been observed on the Niah cranium (?40 ky) from Sarawak (Brothwell, 1960; Bowler et al., 1972; Birdsell, 1979). They led Brothwell (1960: 341) to suggest that 'the differentiation of a more lightly built physical type [among modern humans] may have taken place at a much earlier period than has so far been anticipated' and that if this were so it would give 'weight to the opinion of Birdsell (1941) that "Negritos" were present in Upper Pleistocene times and spread through Australia from more northerly territory'.

The fragmentary remains of seven adults and one child, dated to 28 ky, have been recovered from cave deposits Batadomba Lena in the centre of Sri Lanka (Kennedy and Deraniyagala, 1989). Although incomplete and fragmented, they display gracility. One mandible has a symphysial height (25 mm) between that of WLH1 (23 mm) and WLH11 (28 mm). Another interesting feature is a thin cranial vault (parietal range 4.5–6 mm), which is similar to the occipital bone from Nanwoon in Tasmania. Interestingly, the Sri Lankan remains are also burnt and calcined, possibly from cremation, as well as being associated with ochre. But are they the Negrito people or their ancestors?

Populations of small stature generally have a gracile skeleton and an average height of around 150 cm (Flower, 1880, 1885; Smith, 1912; Slome, 1929; Schebesta, 1952; Cappieri, 1974; Omoto, 1985). *Homo floresiensis* was very small and obviously contemporary with the Willandrans, although living far away in Flores. While these people have nothing genetically to do with the Willandrans, the phenomenon of diminutive stature was operating on humans, albeit descendants of *Homo erectus*. Could it have been operating also, in a style of parallel evolution, among more modern peoples? Unfortunately, we know almost nothing of the phylogeny and geographical origins of such people, because of the general paucity of human skeletal remains from areas to the north in which they might have originated. This prevents us from knowing who else could have come to Australia, what they looked like and who might have been the progenitor migrants to these shores. One thing is for sure, and that is the island and rainforest environments that stretched from mainland southeast Asia, through island southeast Asia and into Papua New Guinea, in both glacial and interglacial times, would have been the type of conditions in which these smaller people would have thrived. The island phenomenon of stature reduction observed among many large mammals and possibly responsible for the stature of *Homo floresiensis* may also have operated on other people occupying islands in southeast Asia and beyond.

Essentially, this discussion draws us towards the inevitable question: how long ago did we develop the phenotypic variability found among modern human populations? Indeed, were small, gracile Willandrans examples of the morphological adaptations that were taking place among anatomically modern humans by at least 50–60 ky ago? Indeed, we might ask similar questions concerning other phenotypic characteristics, such as when did the epicanthic fold appear or the wide variety of human skin colouration, hair type as well as many of our other physiological and morphological specialisations and adaptive strengths? Of course, the presence of Negrito or Negrito-like people at that time presupposes that differentiation of types had taken place among modern humans, presumably through adaptation to certain environments and certainly before 60 ky ago. Such people must have occupied those environments for long enough to adapt in their unique way. For Out of Africa arguments, this had to take place between the time they left the Middle East (approx. 92 ky, Qafzeh 9) and their subsequent arrival in Australia. To invoke a Negrito population existing before 40 ky ago, one must naturally assume that most other 'racial' varieties had also emerged.

The idea that people of small stature moved through the Australian region in the late Pleistocene is not new (Bijlmer, 1939; Keesing, 1950; Birdsell, 1967; Howells, 1970, 1973). Thirty years ago Birdsell (1967: 147) was only too aware of the complete lack of . . . archaeological evidence in either Asia or Australia to document the passage of Negritos . . . Later, he went on to describe WLH1 as . . . somewhat reminiscent of the skull from Niah which certainly is to be classified as Negritoid in derivation . . . (Birdsell,

1979: 421). Besides skeletal remains, it is difficult to know what type of 'archaeological evidence' would signal Negritos, but some examples of Willandran gracility tend to indicate the presence of such people in Australia at a very early time. As I indicated above, it could be that this presence signals a move towards specialised adaptation among anatomically modern humans. But, if this is so, then we have a specialised adaptive morphology arising from a generalised African one within migrating groups and taking place in a maximum of 40 ky. In my view this is a little hard to accept.

Recent populations of short stature people are more widespread in southeast Asia than in any other part of the world and most inhabit island refuges. Three groups (Mincopies, Onge and Garawa) live on the Andaman Islands. The Aeta, Agta and Dumagat live on Luzon island in the Philippines and the Ati and Batak on Negros and Palawan Islands, respectively (Omoto, 1985). Similar groups live elsewhere in the Philippines and southeast Asia, but are not obviously related to these groups. Mainland populations include the Semang on the Malaysian Peninsula, the Senoi on Sumatra and several groups, such as the Tapiro, Yali and Goliath people who live in the remote mountainous regions of West Papua and New Guinea (Molnar, 1985). Small stature was not unknown in Australia with a population living in parts of the northeastern rainforest (Birdsell, 1967, Howells, 1970). Small stature and phenotypic uniformity, however, is not as rigidly circumscribed today as it might have been once. Numbers have been reduced and mixing with surrounding populations has tended to blur original population boundaries and reduced height differentials. Moreover, even among these populations, there is stature variation. For example the Efe people from central Africa have a mean height of 142 cm for men and 135 cm for women, whereas Negritos are a little taller with a range of 142–152 cm (Diamond, 1991).

It has been suggested that Negritos emerged during the Upper Pleistocene from divergent groups of Proto-Malays who entered and became adapted to rainforest environments (Omoto, 1985). Such body proportions are probably adaptive to rainforest conditions, but this explanation does not account for small stature among people living in arid regions such as the Kalahari (Diamond, 1991). Reduced amounts of growth hormone have been implicated in the production of small stature, but diet may have also played a part, perhaps as an adaptive mechanism to cope with limited caloric intake. For example, African pygmies recognise that the rainforest does not provide enough for their nutritional requirements. But this brings into question the whole notion of these people being adapted to such ecosystems. Also with the feast and famine regime endured by most hunter-gatherers, small stature should be a wider phenomenon than it is. Reduction in body size among large mammal species isolated on islands, such as elephants and hippos, may also occur among similarly isolated island living human groups. The reasons for small body size remain uncertain, but, whatever the selective forces were, they existed among people inhabiting southeast Asia probably a long time before 50 ky.

The migration of Negrito or Negrito-like people to Australia or the development of this body form in Australia is not significant. A warm or tropical environment is a shared factor among small people and this might have favoured the emergence of this body form several times in different groups across the region. If that were so, no direct ancestral link between such populations is required but a lot of time is for it to take place several times. Under these circumstances short stature could have arisen anywhere 'given the environmental condition and a sufficient time depth, say, 20 000 years or so' (Omoto, 1985: 130). So perhaps it occurred in Australia. The implication is, however, that we require at least 20 ky of occupation previous to the known time of occupation in the Willandra. So, a Negrito-like morphology within Australia may point to a local adaptation and the gracility of Willandrans is only a signal of the emergence of this locally derived characteristic that was also taking place elsewhere. We should not lose sight of the biological and demographic consequences of the large time frames being dealt with here. If such people developed in Australia, the process would require isolation from others over some considerable time and presumably within a rainforest environment. This seems highly unlikely. I suggest that this type of adaptation only happened once, external to Australia and was brought here. The people then found the Willandra region during their explorations. Although the Willandra Lakes have always been a very long way from the nearest rainforest, we should not forget the Kalahari where small stature is also found. Is it possible, then, that the gracile Mungo people had an epicanthic fold on their eyes?

The 'robust' people

Robust individuals include WLH18, 19, 27, 28, 45, 50, 69, 73, 100, 101 and 154. They have a number of robust features but others (WLH16, 22, 63, 107, 110) have only one or two robust features, including thick cranial vaults and cortical bone. Robust characteristics include:

1. angular cranial vaults, sometimes with a flattened frontal squamae (WLH18, 50);
2. thick cranial vault bones (WLH18, 19, 22, 28, 56, 63, 100, 101, 154);
3. prominent brow ridges (WLH18, 19, 50, 69, 154);

4. rugged muscle attachment areas on the cranium and elsewhere (WLH19, 28, 50, 154);
5. a pronounced nuchal crest (WLH19, 28, 50);
6. large jaws and teeth;
7. pronounced prognathism;
8. large, broad faces with rugged malars and prominent zygomatic trigons;
9. post-cranial robusticity including heavily built long bones often with thick cortices (WLH107 and 110).

These features have been used in regional continuity arguments to link early Australians with Javan hominids. One of these, a thick cranial vault, has a range among Upper Pleistocene Australians far wider than any other population, archaic or modern, for which I have been able to find comparable data. Also, Australian vaults are built somewhat differently from Javan *Homo erectus* or *Homo soloensis* groups, both of whom usually have thick inner and outer cranial tables making up two thirds of the thickness, with another third consisting of a moderate amount of cancellous bone between. This is typical of erectine and 'archaic' *sapien* vaults. Thick Australian vault bones display a variety of composition, but usually with the vast majority of depth consisting of cancellous bone with only thin inner and outer tables. I will return to this feature again later.

Another trait used in regional continuity arguments is the well-developed supraorbital region noted among Australian fossils. While rugged brows occur in many Willandrans, they display a medial thickening in the superciliary region, whereas *Homo soloensis* show development laterally in the zygomatic trigone region. The Australian pattern is also different from that of Middle and Upper Pleistocene Chinese hominids.

The degree of frontal pneumatisation among robust Willandrans varies considerably and seems to have little to do with supraorbital development. Some individuals have prominent expansion of frontal and mastoid air cavities, similar to early Javan fossils, while others have little or none, which follows the Chinese pattern. Weidenreich (1943: 165), however, was 'far from inclined to consider the presence or absence of frontal sinuses decisive criteria for the classification of types', which is understandable caution. The broad range of sinus morphology among the Willandra series, in relation to other anatomical observations, suggests a mix of characteristics from earlier regional hominid populations. The complexity of the processes could have brought together as well as given rise to a series of characteristics which, because of the comparatively few fossil remains available, tend to continually confuse rather than sort into obvious patterns of geographical derivation.

Robust origins

the resemblance of some specific characteristics to the morphology common in the Solo [Ngandong] sample is so marked that it is difficult to deny an evolutionary relationship in the Australasian region, a point suggested by Weidenreich several decades ago *[Wolpoff, 1980: 330].*

These words have been echoed by many anthropologists for over half a century. The statement adequately sums up regional continuity arguments relating robust Australian fossils to those of Upper Pleistocene Java. It reflects also the rugged construction of certain Australian fossils, and to avoid 're-inventing the wheel' they are not examined here. Few that have nailed their flag to regional continuity arguments have pulled it down particularly on the question of 'an unquestionable Indonesian element in Aboriginal ancestry' (Wolpoff, 1999). Briefly, however, regional continuity arguments relating to Australia point to the exceptional robusticity displayed in many otherwise modern-looking fossil crania. This robusticity is usually tied to Indonesian Pleistocene populations who migrated here, notably *Homo soloensis.* It also presupposes, although, because of the lack of any other evidence, it is not always overtly stated, that these individuals entered Australia before gracile forms. The obvious logic of this argument rests on the proximity of these populations to Australia and their long existence in the region. Whether they did migrate here has so far been difficult to prove because we lack properly dated robust fossils of the correct age. Some are sceptical that we will ever find out if this is the case, if for no other reason than most Australian fossil remains are plagued with the phenomenon that they cannot be dated for one reason or another. Nevertheless, even when the relevant archaeology is found (and I have no doubt that it will) many will, no doubt, find something wrong with the dating technique used or some other excuse to discredit it. I refer here to the ongoing debate in Australia between those who are 'pre-50 ky arrivalists' and those who are not.

The archaic appearance of many Willandran robust crania and the very thick cortices observed on some long bones makes a close relationship between these individuals and the much more gracile individuals found here very unlikely. When considering their collective robust characteristics they are difficult to explain as a set of derived characters from anywhere other than Java. Moreover, there is no evidence for a local drift by modern graciles towards an archaic robustness. Such a drift would also fly in the face of the overall trend towards gracility by modern humans during the last 30 ky. Besides those in the *Homo soloensis* people of Java, robust features are not part of regional skeletal populations during the Upper Pleistocene. Arguments based on the drift to robustness with the onset of the last glaciation also have little support when similar trends are lacking among populations living under the same conditions in other areas of the world. Taken in concert, the robust skeletal traits outlined above continue to point to an Indonesian origin, albeit most do not conform exactly to all the Javan traits.

It is proposed that the differences between the Willandran robust crania and those from Ngandong, such as the pattern of brow development and cranial thickness composition, are due to factors of isolation, genetic drift, founder effect, local adaptation and mutation that operated on the small number of people involved in the 50–100 ky journey to and colonisation of the Australian continental shelf (see next chapter). It is also likely that it reflects an equal variety of colonisation strategies and differential movement of groups into and through the continent during those 50–100 ky they lived here before the arrival of more gracile people, perhaps around 60–70 ky ago, with whom they gradually blended. There may also have been a small contribution to the robust gene pool brought in by other migrations that arrived during that long time period, bringing a more modern appearance to the group and representing a completely unrelated gene pool. Like many others, I eagerly await fresh evidence that will throw new light on the stale arguments about the origins of the robusticity found in Australian Pleistocene populations.

While origins always take centre stage, equally important is the reason for the persistence of robustness in Australia as far as the Holocene. One possible answer is that, given the genetic complexities of migration, an archaic or conservative morphology is more likely to have continued as a result of movement out of the Indonesian region without an accompanying mixing of genes from other areas. In other words, if Australia's progenitor gene pool carried with it a predominantly archaic set of features and migration from other regions was limited or had not begun, then skeletal robusticity was unlikely to change or be reduced very much. Persistence of type would take place as descendants moved into the continent and, in some cases, into virtual isolation.

A wide morphological variety

All gracile individuals belong to early, if not the earliest, humans that inhabited the Willandra. The gradual onset of the last glacial maximum forced robusts out of Central Australia, where they had been living, and some moved into the better watered Willandra. The result was a variable morphology, consisting of both extremes. Later the majority consisted of intermediates, giving rise to the wide range of variation taking place after 20–25 ky and throughout the Holocene, although a robust morphology persisted albeit somewhat reduced in its overall rugosity. These processes also produced the variation of morphology from the Javan ancestral traits, whilst preserving a general robustness. It could also be argued that the wide range of morphological variation among earlier people is a reflection of such variety among a *single* population. It is extremely difficult to believe that a single, small population could exhibit such a broad range of skeletal size and shape in this part of the world at that time, bearing in mind that no other population in the region displays this. Indeed I cannot think of another population with which such a wide variety could be compared. Argument for such a wide range of variation existing here in Australia at that time also flies in the face of well-accepted paradigms that propose population centres have the greatest morphological variety, while edges, as Australia surely was, display conservatism.

There can be little doubt that many discrete migrations may have landed on this continent during the late Pleistocene and this occurred increasingly during Holocene, with one bringing its pet dog with them. Each brought its own particular morphological contribution that lay within the wide range of morphological variation represented by earlier founder groups. New arrivals moved as discrete groups within Australia's largely empty continent and for many thousands of years extremes of type could be maintained because they lived in different areas. Each group had its own exploration history; some became more adapted to one environment over another. They later came together through a variety of convoluted factors. These included:

1. additional migrations of variable size and landing sites;
2. glacial and interglacial environmental changes, which altered various parts of Australia differentially, particularly Central Australia,
3. variable rates of exploration;
4. extended use of continental shelf by different groups of migrants before entering the continent;
5. movement into and adaptation to particular environments; and
6. marine transgression that affected some coastal regions more radically than others, as well as eliminating huge areas of land that the earliest people must have occupied.

All these processes were extremely slow, sporadic, took place at different times and had greater effects in some places than in others. Groups may have avoided each other at first or one may have eliminated another. Eventually they were drawn together by a slowly increasing population and reduction in the amount of available habitable land as the last glacial maximum (LGM) took hold of central portions of the continent. The result was that at the LGM, the population sported a wide range of morphological variation that was wide enough not to be affected by further migration from outside, because the extreme parameters encompassed all comers. So, morphological varieties arriving with new migrations were easily incorporated into the breadth of variation already in place. In fact, any addition only strengthened and added vigour to what was here.

Upper Pleistocene genetic mixing in Australia took place not in equal amounts, but in stochastic bursts and in different proportions of robustness and gracility, as well as a large helping of intermediate morphologies. At various times new arrivals brought in a variety of shape and size over the last 60–70 ky. They penetrated the continent in different ways that added further complexity to the Australian morphology. Besides these events, local adaptation was taking place, emphasising and favouring some traits over others. This set the scene for why we see a morphological variety so much greater here, in a genetically peripheral setting, than elsewhere. Variety should be expected in places where migrations originate (places of high genetic pressure) and genetic variety is at its greatest; but, because of the input from Indonesia, Australia has been an exception to this.

I have suggested in previous chapters that at this time humanity was beginning to blossom biologically as well as demographically. This included the development of great adaptive variety that slowly emerged as phenotypic variety across the world. It also saw the emergence of people with cultural and biological wherewithall, enabling them to live in a wide variety of habitats. Over thousands of years, Sahul, as an end-of-destination, received many 'population samples' in different frequencies as they emerged from southern Asia and these must have then undergone founder effect.

It does not seem unreasonable to suggest that, if two contrasting morphologies were present in Australia before 50 ky, subsequent mixing would produce descendent populations, at least 3000 generations later, that incorporated a range of variation almost spanning the original two. It could also be argued that the finite size range of the non-pathological human cranium would put everybody on the planet into a single range of cranial variation for the last 200 ky or more, but that would not make them all part of the same population or give them the same origin. Once a gracile and a robust population began to mix, individuals displaying degrees of these forms could emerge at different stages of the mixing process. These may have established new populations elsewhere which, in themselves, could have initiated morphological change that proceeded at different rates in different parts of the continent. Other groups may through the founder effect have retained a bias towards one or other original type, which through isolation or dominance forged a persistence of their respective morphology in some areas. This kind of process, with its limited amounts of material, would complicate the fossil picture even further. The whole thing is made infinitely more complex if there was a continuation of internal mixing of the differing gene pools, while further migrations introduced additional phenotypic variability to the continent.

On a subjective level, there is no problem with having two very different morphological varieties enter Australia at different times. Indeed, this might be expected if human variation was blossoming in neighbouring regions due to the mixing of migrations and population growth. The relatively scanty fossil record would make it very difficult to separate these two morphologies definitively if their respective morpho-metrical ranges overlapped somewhat in the first place. The uncertainty of sexing fossil crania alone, particularly fragmentary individuals, is well known, so identifying such overlaps becomes almost impossible. Flexibility of interpretation is extremely important here. The sex of a number of individuals in the Willandra series does not wholly account for their osteological gracility. On the other hand, femaleness has been attributed to some only because of the delicacy of their skeletal construction and proportions. Sexual dimorphism is only at its most obvious when extremes are met. Morphological contrast is difficult to observe between rugged females and gracile males that lack skeletal parts most useful for sex determination. Therefore, while accurate sexing is essential in order to clarify these aspects of the argument, the very fragmentary fossil remains of Australia's earliest people rarely allows it. Using pelvic morphology, I have shown an overlap among the Willandrans involving individuals WLH3 (male) and WLH45 (female) as well as between females (WLH1 and 45). There are major differences in male morphology also and these are exemplified by the contrast between WLH2 and 3 and others such as WLH18, 19 and 50. The contrast between certain individuals of the same and different sexes is so great that it is impossible to believe that WLH68 (very gracile female) and 19 or 50 are 'sister' and 'brothers', or that WLH3 and 19 are 'brothers', or that WLH1 and 45 are 'sisters'.

With the huge leap in our understanding of the time people have been on this continent, that has taken place during the last 45 years, there can be little doubt that archaeological evidence for the earliest occupation of Australia will be pushed back well beyond 100 ky in the next 40. With the time of first entry continually being pushed back, the usefulness of the Kow Swamp and Coobool Creek populations, as well as individuals such as Keilor (13 ky) and Cohuna (12 ky?), for determining the origins and affinities of Australia's founding is diminishing considerably. They will always retain an importance, however, for understanding subsequent trends in skeletal biology in Australia, as will the much later examples of Mossgiel (>4.6 ky), Cossack (<6.5 ky), Green Gully (6.5 ky) and Nitchie (6.8 ky) among others. At least 1750 generations separate the earliest Kow Swamp people and WLH1 and 3 and nearly 3000 separate Kow Swamp from those who lived in the Malakunanja rock shelter 45 ky before. Direct links between these people are tentative indeed. Further, we do not know whether there was a continuous ancestral link between founder groups and Willandra people.

A people-empty Greater Australia may have greeted more than one colonising group. A viable population may have taken several attempts to establish itself before humans eventually clung permanently to this continent. When it did occur, however, we might expect that over the thousands of generations that followed enormous demographic and genetic changes took place, while the continent imposed its own adaptive themes on the population. It is likely that certain skeletal features that arose among later people reflected such mechanisms. The proposal that there is a direct affiliation between the earliest and end-Pleistocene groups ignores the intervening internal biological, genetic and demographic history of the continent, as well as evolutionary changes taking place externally and which shaped the morphological diversity of continuing migrations into Australia. No argument can sustain the proposition that Australia's earliest arrivals featured a single population containing the morphological extremes displayed among our fossil hominid groups. If two groups did exist at the same time, they must have maintained a considerable degree of genetic isolation to have maintained their respective appearance. It has been proposed that a 'very small body size appears to be a successful ecological adaptation by which people can coexist for a long time with neighbours quite different from themselves' (Brues, 1977: 261). This might explain an apparent coexistence of the two morphologies among the Willandrans, but once again I emphasise the time frame involved and the possibility that they were not necessarily contemporary in the Willandra, even if they missed each other by several hundred years or even a few decades!

Considering a morphological contrast

Those interested in the most logical answer to the morphological contrasts among the earliest Australians can only draw the conclusion that the robust origins can only originate in the Indonesian archipelago. Indonesian genes arrived here bringing the dominant morphology that had thrived there for hundreds of millennia. As yet, it cannot be proven that the first arrivals were *Homo soloensis* or contemporary types, but the earlier people arrived here, the more likely it is that those populations were derived from this stock. People had been living next door for almost 2 My and were crossing water gaps that moved them in our direction almost 900 ky ago, so the likelihood that they were going to be the first to enter Australia and at an early time is the most parsimonious conclusion that can be drawn.

The problem of sorting out relationships between robust and gracile Willandrans is, however, difficult when the broader picture is considered. What happened in the Willandra is what happened in the Willandra, and it may not be applicable to other parts of the continent. There is every possibility that the Willandra has skewed the picture of the earliest people, just because of the preservation and archaeological environment found there. The archaeological visibility of the Willandra is almost unique compared with other parts of the continent, but conclusions drawn about the relationships and subsequent history of people inhabiting this very small area have, for too long, been automatically taken to apply to all of Australia. It is as though we have been subliminally encouraged to extrapolate the human evolutionary issues arising there as the way it was everywhere. Therefore, I suggest that the robust morphology was dominant in Australia till much later migrations brought in a variety of more gracile forms related to populations moving through or derived from those in southeast Asia and elsewhere. Whether these, in turn, reflect the arrival of anatomically modern people (from Africa or elsewhere) remains debatable. I suggest that the robust/gracile debate would not have arisen if a similar site to Lake Mungo had been discovered in Central Australia, because the more gracile forms did not enter that part of the continent till the late Pleistocene. By then, however, earlier discrete gene pools reflecting particular morphological types had blended into a heterogeneous form with a wide range of morphological variation reflecting adaptations to broad environmental conditions across the continent. Nevertheless, robusticity remained prominent among Central Australian as well as among some Murray River peoples, but it was not dominant everywhere.

Who were the 'gracile' people then, anatomically modern humans (Out of Africa), descendants of earlier Chinese populations or something else? There is little doubt they are anatomically modern humans. Their skeletal form and their associated cultural indicators support that very strongly. Their resemblance to late Pleistocene fossils such as Liujiang (?60 ky) and Ziyang (?37 ky), who are undoubtedly modern humans, is, however, erroneous. These late Pleistocene Chinese people could be descendants of anatomically modern humans who had arrived in China some time before. Subsequent movement of their descendants out of China and down into southeast Asia eventually led them to Australia. That would tie the two regions together skeletally but not in the same way as previously suggested with Liujiang as a direct descendant of Mapa people. In these circumstances the graciles of Australia and China reflect not a Chinese link but a broader movement of modern humans across the face of the globe. No mystery in that. But for those who strictly adhere to the Out of Africa hypothesis there can be no mixing of anatomically modern humans and earlier 'archaic' peoples from Indonesia or we would see it reflected in the mitochondrial DNA, I suppose.

Gracile morphology did not exist right across the continent either. Indeed, it is likely that it was restricted to certain areas in the south and east of the continent, and the Willandra was one of these. Elsewhere the story was different, with less gracile groups playing a similar role to the graciles of the Willandra. The extent of the gracile spread would have depended on how many of them there were and how their populations were distributed. It is likely that there were not great numbers of these people; in fact, there may have been comparatively few and their migratory pattern is likely to have been confined to areas where they were most adapted, such as coasts and rainforests. I can hear many saying already that there was not much rainforest in the Willandra. No, but coastal environments were reflected in the major river systems that these people no doubt explored and used to take themselves inland. Rainforests were the last bastion of these groups, where those of small stature lived and continue to do so today.

Interestingly, it is the comparative loss of gracility in favour of robustness that marks the late Pleistocene. After that gracilisation takes off again during the Holocene. This trend suggests an initial loss of gracile genes in almost all populations, without replacement, and consolidation of a morphology favouring robustness. Holocene populations then received genes from further migrations entering Australia carrying heterogenous morphologies that were biased towards gracility. This helped maintain a general robustness among late Holocene populations in southern Australia, while in the north gracility is predominant, probably due to the 'front-line' effect of being in the area that would receive the new migrations.

Making direct comparisons between people separated by vast time periods is impossible because of their movement both into and within Australia for at least 80 ky, the isolation that may have occurred and the subsequent genetic mixing at different frequencies in different parts of the continent. Genetic drift and founder effect could occur on such a vast continent and these swept through areas at varying frequencies and at different rates. The ebb and flow of environmental change altered patterns of habitation. Thus, the rate of genetic mixing through thinning and depopulation, particularly in marginal regions such as Central Australia, took place during glacial cycles. Amelioration of conditions allowing reoccupation of these areas could have brought previously separated people together, perhaps for a very long time. Consider, for example, two groups living on the northern and southern boundaries of the harsh desert region of Central Australia during the last glaciation.

Origins: a morphological puzzle 249

Separated by an extremely inhospitable and waterless environment, they would have been kept apart, perhaps for tens or even hundreds of generations. An occasional amelioration in glacial conditions may have allowed the occasional foray into the regions between, but generally speaking each would develop along its own genetic pathway determined by length of isolation, founder composition and local selective pressures. The result is likely to have been two groups who began to look somewhat different from each other.

As if this picture is not complex enough, these mechanisms would have been working on broader levels, over wide geographic regions, incorporating populations made up of small gene pools. Such processes would have been capable of maintaining a high enough frequency of certain extreme phenotypes, to produce, in some places, a population that largely displayed robusticity, while another was biased towards gracility, although this would depend on the original population composition. To compare and contrast individual fossil crania that may have been subjected to these very complex processes tends to confound the process of understanding the history of human variability in Australia using cranial analysis.

The question has been posed that, if two such morphologically diverse populations gave rise to modern Aboriginal people, why did the same process not take place outside Australia? Surely, gracile people moving through and encountering robust populations in Java would have triggered the same processes of mixing and outcome, producing an 'Aboriginal' morphology there. It is obvious that the processes leading to the development of the Aboriginal morphology did not happen outside Australia and are unique to this continent. Of course, it is more than likely that the migrational patterns in and around the Indonesian archipelago may not have been quite the same as those operating in Australia. Some of those who arrived here may not have travelled through or even been near Indonesia. Is it possible that, for one reason or another, a large part of the Indonesian *Homo soloensis* population had already migrated to Australia, leaving few individuals behind. Consider for a moment, that there may have been a comparatively small population of these people living along the Indonesian archipelago at the time they began to leave. From what we have found so far, these hominids quite suddenly disappear in Indonesia or at least none of their fossils has been found that appear to be mixed or date to later that 50 ky ago. Mixing of subsequent migratory genes from southeast Asia would not then have taken place in Indonesia, but here in Australia!

We also have to consider the special circumstances determined by Australia's geographical and environmental setting. These may have been the vital ingredients that produced the unique Aboriginal morphology. They include Australia's desert and temperate environments found nowhere else in the region. These invoke in very special ways the effects of adaptation, which favoured a certain body shape and size. It includes the relative isolation of the continent from its neighbours as the sea assumed modern levels after 10 ky for the first time in 120 ky. This also gave rise to the special circumstances surrounding the history of Tasmanian Aboriginal people who were cut off from the mainland for at least 14 ky (Lambeck and Chappell, 2001). Another factor was the requirement for any would-be inhabitant to cross wide oceanic gaps between the continent and elsewhere, invoking founder effect. The vastness of Australia, an area of 7.5–10 million square kilometres, depending on sea levels, is found only in Asia. Australia's space could isolate groups from one another and promote the mechanisms of drift and founder effect. All these were compounded by the environmental swing caused by the glacial–interglacial cycles and which operated over the time humans have been in Australia. All these factors altered the circumstances and the odds of forming viable groups and populations.

The heat generated by the gracile/robust debate far outweighs its importance. I have never quite understood the 'terror' felt by some at the thought of having two quite different morphologies existing on this continent at the same time. Perhaps this was prompted more by being different than by being logical. The gracile component of Australia's precursor populations was, no doubt, one of many morphological varieties that entered the continent during the formative stages of Upper Pleistocene human migration here. Logic still dictates that a robust form of human arrived here first. Its population was small and its subsequent rate of growth was very small. Ill-adapted to Central Australian conditions, they lived on the exposed continental shelf until environmental conditions changed early in the last interglacial. Slowly they moved inland towards the centre where they began to adapt over a long period to the more arid conditions they found there.

Extreme gracility can be explained by the arrival of modern people of small stature. Their appearance sharply contrasts with fossil remains of other larger and more robust people. Like many other modern human variants, such as the epicanthic fold, little is known about when or why short stature evolved among modern peoples Short stature, however, appears on this continent at a very early stage, suggesting it did not evolve here but came in from outside and is found among people living far from rainforest environments. The gracile people were, nevertheless, modern humans and probably a coastal people that used riverine corridors to explore inland. There, they met the more robust groups who began to vacate Australia's centre as it became drier with the onset of the last glaciation.

The fossil record shows the presence of a very gracile type that largely disappears. It is followed by the broad Australian population having a wide range of skeletal variation, something to be expected if gracile genes did not continue to enter Australia. Moreover, if gracile people contracted to rainforest environments, through pressure from other more populous or stronger groups, that may also appear as though this morphology 'disappeared'. This would be particularly so if the chosen place or its refuge was yet to be archaeologically investigated.

In other areas of Australia, dominant groups eventually absorbed graciles. Intrinsic population growth and the continuing arrival of a more middle of the road morphology compounded this process, something often overlooked in these discussions. Therefore, those who would argue that the early Australian population had a wide range of morphological variation are right – it did. But they have to accept that this was the product of the eventual amalgamation of several distinct morphological types, presenting its own distinctive appearance that was derived from different geographical regions external to the Australian continent. The one that would become so dominant over most of the continent was that sporting the 'average' morphology, neither very robust nor very gracile. The slow arrival of these people in the late Pleistocene and throughout the Holocene gradually contributed to a balance that saw extreme forms disappear, although the appearance of robust and gracile forms continued into the late Holocene, with robustness dominating in some areas from time to time.

Although it does not occupy the Aboriginal mind very much, the main anthropological question in Australia continues to be, where did Australian Aborigines come from? Aboriginal people have told me many times that they are not migrants and that they were created by the ancestral beings that created everything else in the Dreaming. My reading of the fossil remains and emerging evidence of the length of time that they have been here makes it abundantly clear that they are correct. The well-known morphological appearance of the Aboriginal Australian is derived from adaptation to this continent, possibly through two glacial episodes, but it also reflects the variety of founding and subsequent groups that set foot on these shores a very long time ago, perhaps as many as 150 ky ago. One thing is for sure, early Aboriginal people were definitely not freshly Out of Africa.

8 Migratory time frames and Upper Pleistocene environmental sequences in Australia

I want now to put the human migration story into a possible sequence, tying it to the sea level and environmental changes related to the penultimate glacial cycle. Humans lived everywhere in Australia by 45 ky and had straddled the continent 10 ky earlier (Turney *et al.*, 2001a, b; Pearce and Barbetti, 1981; Groube *et al.*, 1986; Bowler, 1987; Roberts *et al.*, 1990, 1994, 1998; Smith and Sharp, 1993; Flood, 1999; Grun *et al.*, 1999; O'Connor, 1995; Mulvaney and Kamminga, 1999; Turney *et al.*, 2001a, b; O'Connell and Allen, 2004). That signals the presence of a well-established, viable population, adapted to a wide variety of environments, ranging from islands off Australia's Kimberley coast to central New South Wales, from the Swan River in southwestern Australia to northern Queensland. Even Tasmania was occupied by at least 34 ky (Mulvaney and Kamminga, 1999; O'Connell and Allen, 2004). Although largely cut off from the mainland from 135 ky to 43 ky, there were brief connections of the eastern sill at 76 ky, 68–62 ky and 46 ky, but the elevation of the narrow Bassian land bridge was only around five metres, which made it susceptible to flooding and the dangers of storms common in the area (Lambeck and Chappell, 2001). So, if people arrived on the south coast after 63 ky, they had to wait till 43 ky before they could continue their explorations. They did not take long because we know they were there in the far southwest of Tasmania by 34 ky, or, in Ice Age terms, in the far southeast of Sahul. It marks the earliest evidence for habitation of Australia's cold environments south of latitude 40° and is the earliest evidence for humans living in the high latitudes of the southern hemisphere. At that time, Tasmania's southwest consisted of open savannah with patches of temperate rainforest. During the Ice Age, a glacier formed in the central highlands, icebergs flowed past the south coast, and snow and icy rain would have regularly blown across the hill slopes and valleys of this harsh, cold and very windswept part of Sahul. Tasmanian conditions were in marked contrast to those enjoyed by people living on the balmy Arafura Plain far to the north. So the first Tasmanians mark one very important step in the human occupation of Australia. They show that people were adapted, both culturally and physically, to a variety of climates across the continent, including one that can only be experienced in northern and central parts of Asia. So when using a 60 ky

Migratory time frames and Upper Pleistocene

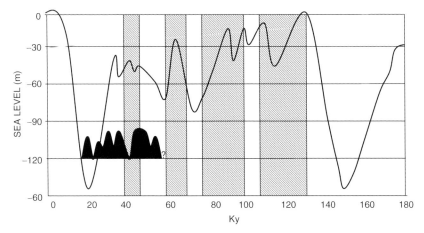

Figure 8.1. Upper Pleistocene megalake phases of Lake Eyre (hatched) compared to sea level change. Willandra lake full events of the late Pleistocene are shown against the last phase of the Lake Eyre full events.

old entry time for the time of first entry does not leave much time for such broad adaptation to take place.

Cyclic exposure and drowning of Australia's continental shelf acted on human groups within Australia in a similar way to the 'migration pump' described in earlier chapters that shifted people on and off continental shelves and moved those outside Australia. The cycle worked like this. As sea levels dropped, the continental shelf became exposed, presenting new areas for habitation. It also extended our shores towards islands from where migrants originated. With the return of higher sea levels, our shore line receded and people were simultaneously pushed off the shelf and forced inland. At the same time, exposure and drowning of continental shelves were linked to events happening far inland.

During low sea levels, there was a kind of transposition of habitable land. Generally speaking, when sea levels dropped below −70 m, Central Australia became too arid for any permanent habitation but large parts of the continental shelf were exposed. In turn, when they rose above −70 m, the loss of continental shelf was replaced by an opening up of Central Australia through increased local precipitation in the interior and increased water flow in Central Australian river systems. The mechanism controlling this was the southern monsoon incursion and the degree to which it penetrated the continent, which fluctuated over time (Figure 8.1). During low to very low sea levels, the monsoon shifted north out of the Lake Eyre catchment; in other words, it was operating above 16° south. At other times it swung lower, perhaps as far as 22° or even below, dropping massive amounts of water in the Lake Eyre catchment. At other times it moved variably between these latitudes but keeping far enough south to maintain flows in the Lake Eyre system. It switched off around 65 ky only to appear briefly at 40 ky and then finally moved north out of catchment reach once again as the Ice Age tightened its grip.

When moving south into the continent, the monsoon not only filled the inland waterways but transformed deserts into savannahs. Therefore sea levels and the depth of penetration and strength of the monsoon operated a vast environmental gate that basically swung open during high sea levels and shut during times of low sea level. It was also a habitation gate controlling human access to Australia's heart and opening up a range of habitats that produced all the ingredients any hunter-gatherer required: animals, water birds, fish, shellfish, vegetable foods and plenty of fresh water. Animals and people could inhabit vast savannah-like plains and travel the long river systems deep into the continent as if on 'freshwater highways'. With the loss of monsoon rain, aridity ensued with rivers and creeks drying and animals migrating out of Central Australia.

The search for an accurate view of the timing and depth of Upper Pleistocene sea levels has produced a number of slightly different diagrams as methods have become refined (Aharon and Chappell, 1986; Shackleton, 1987; Chivas *et al.*, 2001; Lambeck and Chappell, 2001). Most agree, however, that modern levels existed only at around 115–125 ky and again from around 10 ky. We have then a picture of normal sea levels lower, or much lower, than they are today. With the exception of the time period between 115 ky and 125 ky, the initial campsites of any arrivals before 10 ky are now submerged. The former date, therefore, marks a very small window of opportunity to find the camps of the earliest people that might have entered this continent. The only other time when the sea was at modern levels was around 250 ky, but these were not optimal times to make a crossing, the shortest of which measured several hundred kilometres. It is likely then that we will never discover when and where people landed if they arrived at times of low sea level, because all landing sites are now drowned. Moreover, with a hostile inland environment at these times, particularly for a people quite unfamiliar with desert and arid savannah conditions, a coastal existence would have been preferred to moving inland until amelioration of inland conditions and rising sea levels pushed people off the palaeocoasts. Long periods of habitation on the continental shelf would also now be invisible and a population favouring these large areas would now be undetectable. Inland campsites of those moving away from coasts as seas rose will always reflect a later arrival than was the case. Movement inland by the first arrivals is more easily understood if the landing place and direction from which they came is known and how they might have reacted when entering a people-empty continent full of unfamiliar environments and even less familiar animals.

In order to discuss movement of people within Australia, I want to outline some possible stages of the process with particular reference to environmental conditions in Central Australia during the period 60–150 ky. To do this I will draw on my research in the Lake Eyre basin, which I have detailed in Chapter 5. I use the period 60–150 ky as a parsimonious time frame, for the first arrivals and divide it into four time frames, which correspond to changes in environmental, climatic and sea level conditions and oxygen isotope phases. They include 130–150 ky (Stage 6), 106–130 ky (Stage 5d and e), 75–106 ky (Stage 5a, b and c), and 60–75 ky (Stage 4). I describe each of these stages from two perspectives: environmental conditions and human activity.

130–150 ky (Stage 6) – the environment

This time frame is, in reality, longer, spanning 130–180 ky when sea levels averaged -75 m with a maximum drop to around -145 m at 150 ky. Voyages at this time were optimised, particularly in shallow waters where islets and quays emerged offering convenient stopping off points. If migration depended on low sea levels, the best time for crossings was between 145 ky and 155 ky. Sahul was at its biggest at this time, increasing to around ten million square kilometres. Conditions in Central Australia were harsh with extreme aridity, widespread aeolian activity, dune building, desiccation, deflation and very cold winters. Any water would have evaporated without replenishment at the beginning of this period and the extensive riverine system that crossed the region was dry due to reduced local precipitation and a northward excursion of the monsoon away from its upper catchment in northern Queensland. Vegetation largely disappeared under these conditions and cold, dry, winds increased sand movement and dune formation forcing most animal species out to peripheral areas where water and fodder were more plentiful, a pattern repeated from previous glacial events.

The Simpson and Gibson Deserts resembled the Sahara more than at any other time. Environmental conditions in the Kimberleys were also inhospitable, with widespread aridity and dune formation in some places. While internal conditions were poor, Sahul's continental shelf was a much more attractive place to be. Small but permanent glaciers covered the southeastern high country and probably one third of central and southwest Tasmania. Barrier reef islands became coastal hills, wide savannah plains stretched out west of the Great Divide and extended 50 km south from the cliffs now fringing the Great Australian Bight. At this time, the massive Lake Carpentaria straddled the exposed Arafura Plain that joined Australia to Papua New Guinea. The westward extension of the Sahul Shelf beckoned, as it reached out towards the eastern end of the Indonesian archipelago, and Sahul's northwestern shelf stretched almost two thirds of the way to Timor.

130–150 ky (Stage 6) – the people

People could have entered Australia between 140 ky and 160 ky from Indonesia via Timor and/or other islands to the east with seas at their lowest −145 m. Landings could have occurred anywhere along the exposed palaeo-shoreline of the Sahul Shelf, but the only people arriving at this time would have been *Homo soloensis* as they were the only humans around at the time. There are no other possible candidates except perhaps *Homo floresiensis* and modern humans had not appeared. *Homo soloensis* had an average brain capacity of around 1200 cm^3 and a distinctive set of robust cranial features. These included broad faces, well-developed cranial buttressing, thick cranial walls, prominent brow ridges, flattened frontal bone, thick occipital torus above a prominent muscle insertion area, supra-mastoid crests, angular tori and marked temporal lines or crests inferring large jaws, big teeth and probably a generally powerful body.

The fossils of *Homo soloensis* were discovered during the 1930s in central Java on the Solo River system that traverses Java (Weidenreich, 1951; Santa Luca, 1980). It is not likely that they had a specialised 'coastal economy' or were people that depended on the sea any more than any other environment. The fact that all the fossil crania of these people have been found along the Solo river system suggests they probably had an adaptive economy and were quite at home in riverine environments. After landing they would gradually move across the Arafura Plain and/or other exposed areas of the continental shelf, following drainage systems emptying northern Australia and lakes on the Arafura Plain. One of these flowed from Lake Carpentaria, an enormous body of water fluctuating between 30 000 and 70 000 km^2 or about the size of Lake Victoria, Tanzania. It formed when sea levels were at or below −53 m and existed between 125 ky and 165 ky. The lake discharged west, possibly through a number of rivers that flowed across the Arafura Sill. The enormous outpouring of freshwater, particularly during the monsoon season, could have been detected out at sea, as the ancient sailors tasted the sea to determine where and how close land might be. Once locked into the fresh water discharge, it would have been easy and very natural for them to follow it to an outflow estuary. Following these they could then make their way to Lake Carpentaria, where the surrounding ecosystem and resource base would have supported an abundance of animal and bird life. Exploring the lake shores and surrounding region, they would have discovered the Fly/Strickland system feeding into the north end of the lake from Papua New Guinea. Another 30–40 major rivers flowed into the lake from an enormous catchment in Australia. On the eastern side, the catchment drained an area extending across two-thirds of Cape York down as far as the western side of the Atherton Tableland. The southern catchment encompassed the Barkly Tableland and in the west it drained eastern Arnhem Land.

Landings far to the west would have met large river systems draining the Kimberleys and Arnhem Land. They flowed out across the continental shelf as extensions of the present-day Pentecost, Ord, Victoria, Daly, Adelaide, Mary and South and East Alligator Rivers to name only the biggest rivers in the region. On the west coast, the massive Fitzroy River would have been a prominent land mark and source for exploration of the southern Kimberley. The forbidding Central Australian conditions would have deterred the movement of these people inland, as they were used to lush, tropical island environments, and it would have been natural for them to have concentrated on exposed continental shelves and hinterland regions behind coastlines where resources were similar to the coasts they had left behind and were readily available. Unfortunately, coastal campsites in this region as well as any that were placed as far as 200 km inland from the coast are now drowned. Short ameliorations in the generally harsh glacial conditions, perhaps lasting a couple of hundred years or so, may have prompted opportunistic forays into places that were not drowned later under rising seas. They could have taken people far inland but never very far beyond the Kimberleys themselves, because, while food and water may have become more available during these times, the overall harshness of Central Australia was always present. Finding archaeological sites associated with such sporadic events, however, is always going to be very difficult.

By 130 ky the sea had risen rapidly towards modern levels. The rise included one of around 45 m in 5 ky between 130 ky and 135 ky. Such a rapid return would have claimed the shallow continental shelf at such a rate that the loss of land would have been experienced well within one generation. The rise in sea levels affected many varieties of animal, as habitats changed or disappeared altogether. These events had happened before during previous transgressions, but this time humans were in the landscape. Initially, the effect of hunting on fauna would have been minimal and confined to very small areas where the landing had taken place and because of the small numbers of humans involved. Any firing of the land could have started very large fires in areas that had only experienced irregular natural ignitions before. I cannot envisage *Homo soloensis* using fire to clear the land; it is certainly not something that they would have found useful in the rainforests of Indonesia. I can, however, see how it might have been developed within Australia some time later as a useful tool for clearing the land and hunting. The first people, therefore, may have had very little effect on the megafauna from that standpoint. It was not till later, when they ventured farther into the continent or pushed out around its east and western edges and humans and megafauna shared savannah landscapes that the former 'discovered' the beneficial use of firing techniques in such landscapes. So, the first impact on fauna would have been almost negligible. Later, with the use of fire, it began to take effect, but very slowly. With the loss of the continental shelves and the subsequent movement of people inland, the impact moved to a new level. It was then that plant regimes began to change, resulting in changes to the fodder regime with reduction in certain types. Any large, uncontrolled conflagrations started by the new arrivals probably resulted in destroying pockets of animals, albeit unwittingly on the part of the people. When considering these events, it is well to remember an earlier suggestion that megafaunal species may have been unevenly distributed throughout the continent, with some occupying isolated pockets. Some may have been confined to certain regions, with very limited or no members inhabiting other regions. Those with minimum viable populations would have been affected much quicker by the appearance of humans than others with larger populations and more representative animal species. In this case, fire would have had a differential effect, with burning in some areas having a greater effect than in other places.

100–130 ky (Stage 5d, e and into c) – the environment

Between 125 ky and 130 ky, sea levels rose from -65 to -35 m, rapidly drowning Lake Carpentaria and creating the Gulf of Carpentaria. The rate of rise continued so that modern levels were reached around 120 ky. As a consequence, Sahul was divided into Papua New Guinea (PNG), Australia and Tasmania, continental shelves were inundated and Sahul shrank by 25 per cent. These events would have pushed people and fauna off shelves and into Australia proper in a comparatively short time. However, sea levels soon began to drop again, reaching -2 m by 110 ky. They would not reach modern levels again for 100 ky. Voyages to Australia were longer at this time and consequently more difficult, but there were more options open for exploring the continent. Sea levels began to fluctuate after this time dropping to -40 m by 105 ky only to rise again to -20 m by 100 ky.

The high sea levels of 125 ky marked a return to wetter conditions in the centre of Australia. Although there is evidence for rapidly fluctuating wet/dry episodes there, the dry periods were not severe and lacked the widespread deflationary processes that took place during the glacial maximum (Magee, 1997; DeVogel *et al.*, 2004). A wetter climate brought good times for the megafauna and it prompted another migration back into central parts of the continent after a long absence. Recolonisation of central parts of the continent by megafauna brought with it an opportunity to increase populations and occupy new habitats. Besides being free of humans, animals were attracted to the landscape because of the extensive wet period that brought to life the vast palaeochannel system crossing the inland. Local as well as catchment rains maintained flow, encouraging widespread vegetation growth. The wet period continued for the next 20 ky and Central Australian lakes filled, providing large reserves of freshwater in numerous lakes and streams. Lake Eyre grew to almost four times its present size (9500 km^2), covering 35 000 km^2 and containing 430 m^3 of water, almost 15 times that of the deepest historic fill (DeVogel *et al.*, 2004). These Upper Pleistocene levels have left tell-tale beach ridges and thick lacustrine deposits from a time when thousands of animals were being attracted into the region (DeVogel *et al.*, 2004; Bowler *et al.*, 1998; English *et al.*, 2001).

The Central Australian wet originated from a change in monsoon rainfall patterns that moved south, bringing heavy rain to the northern half of Australia and to the Lake Eyre basin catchment. They pushed deep into catchments, feeding rivers that ran south as well as bringing regular precipitation across a vast area of the inland. Palaeodrainage systems crossed Central Australia and one of the biggest formed an unbroken chain of rivers and streams stretching from the Barkly Tableland to central South Australia. Water beginning as monsoonal rain in northern Australia drained south through 2000 km of channels, eventually emptying into Lake Eyre and neighbouring lakes. Around 125 ky, Lake Eyre began one of several 'megalake' phases that formed a gigantic, permanent, deep-water body containing enough water to fill over 1000 Sydney Harbours, and it remained roughly this size for 15 ky. It exceeded all its present boundaries, flooding palaeovalleys to its north, in the Simpson Desert, and east, now largely occupied or dissected by the modern Warburton and Cooper river systems.

The riverine and lake systems maintained by regular local rainfall provided an environment more like a 'Simpson Savannah' than today's Simpson Desert (see Chapter 5). The area supported a variety of animal species, all exploited by the broadest selection of carnivores that Australia had to offer in the Upper Pleistocene. The rivers and lakes contained fish up to 100 kg, turtles of all sizes, some as big as a large coffee table, and eight metre freshwater and estuarine crocodiles. The rich trophic pyramid living in the region at that time reflects not only the wide range of animal species, present but the bountiful plant community upon which the vast majority of animals browsed or grazed. During the drop in sea levels between 105 ky and 110 ky, environmental conditions in Central Australia became much drier with limited deflationary episodes.

105–130 ky (Stage 5d, e and into c) – the people

With a return to modern sea levels, those living on the continental shelf were pushed inland. The accompanying change to wetter conditions in Central Australia made it possible for them to move farther inland instead of clinging to the coasts. Movement into the continent proper and exploration of central regions became possible for the first time. The possibility of further migrations arriving at this time may have prompted the move or it might have been purely that they were inquisitive. Alternatively, any new arrivals could have moved around those already in place and continued on.

With sea levels forcing people off the continental shelf and inland, and by moving upstream along rivers flowing north into the Lake Carpentaria basin, they would have eventually reached the upper catchment on the Barkly Tableland. Once there, they could pass into the tributaries flowing south through the upper catchment of the Lake Eyre basin. Many animals and water birds would have been attracted to the waterways. The gallery forest wound along vast stretches of channel providing refuge areas for a wide variety of species and excellent environments for wandering bands of humans to hunt as they moved down into the basin from the Arafura Plain.

Because of the small numbers of people involved in this process, intracontinental exploration would have been slow. Initially, people would have moved across the emptiness of Central Australia very gradually. It was an area that would quickly absorb their limited numbers. With little natural increase in the overall population size, groups were few, remained small and may have become scattered across the 2–2.5 million square kilometers area. Some might have reached peripheral areas of Central Australia, bordering on the Great Victoria Desert and Nullarbor, where poorer and more arid conditions existed and where there was an absence of large palaeochannels. Prevailing conditions deterred movement in those directions till a later widespread downturn in Central Australian environmental conditions forced people out of the centre and pushed them towards the southern coastal regions.

75–106 ky (Stage 5a, b & c) – the environment

As sea levels rose from -40 to -20 m between 100 ky and 105 ky, the southern monsoon incursion weakened and moved north. Less water fell in the Lake Eyre catchment, streams and lakes dried and vegetation became sparse once again. Harsh conditions prevailed throughout Central Australia and anybody living there would have moved to better pastures. Over the next 25 ky, sea levels fluctuated but the trend was downwards. Lake Carpentaria existed briefly around 87 ky, Tasmania was separated but a broad link with New Guinea was maintained. While sea levels gradually dropped, the monsoon moved south again around 100 ky and intermittently hung around till around 65 ky. The inland palaeochannel system flowed throughout this time, filling Lake Eyre around 80 ky and so began another megalake phase that with some fluctuation continued beyond the end of oxygen isotope Stage 5. Vegetation flourished and this was the longest 'good time' that the area would ever experience. The sea level and monsoon patterns over this period suggest that they worked independently and even though sea levels continued to fluctuate, the trend was downwards. The southerly monsoon incursion remained for some time and animals migrated back into Central Australia establishing their populations for the next 25 ky, but more people than before followed them in.

75–106 ky (Stage 5a, b & c) – the people

By 106 ky perhaps 40 ky had passed since the first landings, plenty of time for *Homo soloensis* to become biologically and behaviourally adapted to what was a rather benign Central Australian environment. The reader will notice I have avoided using the words 'cultural adaptation'. The reason is the obvious lack of culture associated with *H. soloensis*. Whatever their cultural world included, little of it was left behind by these people in their homeland during their tens of thousands of years, tenancy there, so I assume that its signal would be almost nonexistent here in Australia. In the face of this and the fact that they probably lived for a great deal of their time on the continental shelf will always make detecting their presence and finding their living areas exceptionally difficult.

The effects of a new environment as well as the usual results of isolation, drift, adaptation, selection, founder effect and mutation may have altered the original morphology of these people, so that, while robust traits persisted, some aspects of the cranial architecture may have changed. So too the environment may have selected for certain traits that enabled these sub-equatorial people to live in semi-arid conditions. The possibility that they drifted back and forth, in and out of Australia's heart for perhaps 40 ky probably played a part in these processes. The demographic profile and the history of the population could have also altered their appearance somewhat from the ancestral Javan form, even though a general robusticity was maintained as an advantage in changeable environmental conditions and the vagaries of, at times, an uncertain diet. Physiological adaptation may also have begun to play its part in allowing greater flexibility for living in semi-arid environments. People were no longer 'new arrivals' or 'migrants' they were 'Australians', with as many as 2500 generations behind them. Of course, this does not mean new arrivals may not have continued to land on Sahulian shores throughout this time, but who these people may have been if it were not more *Homo soloensis* is difficult to know, unless it was migrations from farther afield, perhaps southern Asia. The amount of space in northern and north-central parts of Australia provided more than enough room to absorb these early arrivals and, moreover, with the constant low sea levels, certain amounts of continental shelf had reappeared to boost accommodation prospects. Again, they would face the old problems of becoming isolated and extinct.

Over 40 ky of human habitation now began to impact on the megafauna. Initially, during the first 10–20 ky years, it was negligible but by 105 ky a number of factors began to affect them. Of course the megafauna were not used to humans, their mobility and their hunting. Eventually they also had to cope with more humans arriving and spreading out across the landscape, albeit very, very thinly. Animals and people came together by mutual attraction to food resources and watering holes. The abundance of such places in peripheral regions of the north presented more choice of where to be, so it kept animal/human contact at lower levels. In Central Australia places of mutual attraction were river systems, where permanent water could be found, rather than out on the plains. So the geographic setting limited choice, for example to obtain water, and that would increase the frequency of contact between the two and so the stress of any human predation was increased in such areas.

It was probably when humans began to explore away from the coasts that they first used fire to eliminate impenetrable undergrowth and thicket. Burning the land, however, may have started large conflagrations that began to affect larger animal species. The greatest impact would have been felt in marginal or negatively changing environments. Vegetation changed commensurate with burning, it favoured grasses and pyrophytic species, as well as producing large areas of open forest and savannah. Environmental change associated with glacial fluctuations had always been a problem for the megafauna, but it was now compounded by the presence of humans. Now, changes to fodder regimes and alteration of habitat brought about by human firing favoured some species over others, so the human impact on megafauna began. These processes have to be kept in perspective, however. People did not live everywhere, they lived in comparatively small patches in the north and, where possible, some northern central parts of the continent adjacent to palaeochannel systems and lakes. Not all of Australia's megafauna was stressed; in fact the vast majority of animals still had not had contact with the new predator.

Sahul formed once more, as a retreating sea exposed continental shelves joining Tasmania to the mainland and producing the broad Arafura Plain. Lake Carpentaria appeared again around 75 ky, as sea levels dropped below −50 m and it largely remained till around 12 ky. The scene was set once again for easy crossings from Sunda to Sahul.

60–75 ky (Stage 4) – the environment

Around 75 ky, monsoonal incursions moved north again and Central Australia dried. This marked the beginning of 4 ky of aridity and deflation. People living in Central Australia would have found it increasingly hard to remain and many may have moved out to wetter areas, leaving the place almost empty. It is more than likely that the megafauna responded in a similar way, so that again the pattern of animal and human demography was closely linked. Both slowly migrated out of drying areas, seeking water, but humans followed animals for food. The Toba volcano exploded at the beginning of this stage, which must have put some urgency into migration away from that area and set back the movement of others (see Chapter 3). The noise would have been heard in Central Australia.

After a 5 ky break, the southern monsoonal incursion switched on again (70 ky), and Central Australia becomes habitable once more, with people and animals returning. The palaeochannel system flows again but with less vigour and a more variable flow than before. From 50 ky to 70 ky, sea levels fluctuated ten metres between −70 and −80 m, the lowest for over 65 ky, and exposing large areas of continental shelf. The opportunity arose once again for Sahul bound migrations to arrive by optimising their chances of successful voyages. The coastal profile at this time is similar to that of 135 ky, although the ocean gap between Timor and Australia is somewhat wider.

Following a brief lake-full period around 65 ky, Lake Eyre contracts and is dry by 58 ky. This is preceded by the drying of smaller rivers and lakes, and main channels become a chain of water holes. Previously lush vegetation thins with reduction of those species favouring a wetter environment. By this time all megafauna had become extinct or moved away from the region with the exception of *Genyornis*. Another Ice Age was approaching and Central Australia felt it first as usual. Eventually all but the hardiest vegetation disappears as the arid phase intensifies and this time the megafauna would not return. The environment finally dried out after 58 ky and continued to be dry for almost 20 ky as the Ice Age approached. Vegetation disappears and blowing sand takes over.

50–75 ky (Stage 4) – the people

This stage is short but very important in Australia's human story. Encouraged by falling sea levels, the first groups of modern humans enter Australia between 70 ky and 75 ky. It is not likely that they arrived earlier because the earliest evidence for them is in the Middle East around 90–99 ky, so it is likely that they arrived in South East Asia some considerable time after that. I suggest 70–75 ky also, because it allows for a 20 ky movement across Asia into South East Asia and across to Sahul. Secondly, sea levels (-50 to -80 m) optimise successful crossings between 70 ky and 75 ky, the best since 85–90 ky. Even so, the capabilities required to cross from Sunda to Sahul at -80 m would probably get people across at any time. We must not underestimate the capabilities of these groups to sail across ocean gaps of 200–300 km. A bamboo raft with its built-in watertight storage facilities within the compartmentalised stem structure of this grass makes an ideal craft capable of carrying both food and water for quite long journeys. Thirdly, the eruption of the Toba volcano may have encouraged roaming bands to move on quite quickly in a southeastly direction, opposite to that of the enormous ash and gas cloud fanning out to the northeast of Sumatra.

The comparatively arid conditions operating in Central Australia between 70 ky and 75 ky are likely to have deterred the new groups from moving inland. If they did arrive at this time they may have come into contact with populations of the earlier people, as they retreated to wetter regions in northern Australia. Encounters between the two possibly resulted in one avoiding the other, with the recent arrivals moving into unoccupied areas and exploiting niches with which they were most familiar. Coastal regions may have been favoured as these had been left largely empty, particularly in the east and south, because the original groups had stayed in the north or along inland river systems. Moreover, the later groups probably had a more specialised economy, perhaps one based on the coast as suggested by Bowdler (1977). Besides using regular burning, they may also have possessed a more sophisticated technology that included nets, fishing spears, spear throwers, throwing sticks (boomerangs?), hafted weapons, shields, more sophisticated watercraft and the skills necessary for making fish traps and regularly exploiting offshore islands. The sea would have been their highway for migration throughout island southeast Asia and Indonesia with coastal movement facilitating ease of passage, after all sailing a coast is easier than tramping though a rainforest. Coastal voyaging is also faster; it moves people quite large distances relatively quickly and it is a good way to find river systems flowing from inland. Exploration of these would be a natural adjunct to their coastal movement, because riverine resources replicate those on the coasts. Eventual movement up the large river systems such as the Murray/Darling system took them into the interior of southeastern Australia (Bowdler, 1977: 223).

A drying Central Australia offered nothing for these people so used to living close to water and they were ill-adapted to semi-arid/arid conditions. So, finding lakes like the Willandra and Tandou/Menindee systems attracted settlement sometime between 55 ky and 65 ky. Moreover, the gracile morphological appearance of many individuals in the Willandra series indicates little or no genetic mixing was taking place between these people and more robust groups before that time, because they had largely missed each other by occupying different landscapes. But now these they were poised to meet up in one of many such contacts that increasingly took place across the continent.

Between 65 and 70 ky, another amelioration in the Central Australian environment allowed robust groups to return, which took them further away from the newcomers. By this time they had an advantage over their new neighbours in that they favoured the inland regions because of their adaptation to semi-arid conditions following a long history of occupation in the region and because their generalist economy was adapted to that environment. The newly arrived modern groups continued to favour coastal as well as riverine areas. Thus the separation of the two groups was maintained. Moreover, the limited size and scattered distribution of the population helped maintain this separation. With another wet phase beginning in Central Australia around 65 ky and the palaeochannel system flowing again with monsoonal run off, Lake Eyre entered a brief megalake phase, although not as extensive as before, but by 60 ky it was again dry, never to became as full again. This was also the last time that humans and megafauna shared the savannah of the Lake Eyre basin.

The different migration strategies of the modern and robust groups were determined by the culture and economy of each, together with environmental conditions and individual adaptations to different landscapes. If the first people did not land in Australia until 60 ky, transcontinental crossings using 'freshwater highways' would not have been possible and the migrants would have been totally unfamiliar with the arid conditions that prevailed there. In that case, coastal migratory routes would have been favoured, not necessarily because of their possession of a 'coastal economy', but because that was the only way open to them. The centre would have been far too unpleasant to contemplate entering it.

<60 ky – humans and the megafauna

Megafauna had previously withstood the constant environmental, climatic and sea level changes, albeit that they were not left unscathed by these events (see Chapter 5). But now there was the addition of living with humans. Initially the impact of extremely small human groups keeping largely to the exposed continental shelves and the absence of regular or extensive burning practices brought little additional stress to these animals. Later they became tied together by a common search for basic resources and shared similar environments in response to changing climatic conditions. Later, however, animals and humans were thrown together in their search for plentiful water and food supplies on the continental periphery as Central Australia dried and they both left the area (45–60 ky). By this time the long-term use of fire, even if limited to certain areas, had begun to slowly change the Australian landscape. These changes had a cumulative effect that began before the onset of the last glaciation and continued into it. Megafauna began to come under increasing pressure from the resulting changes to vegetation and fodder regimes that accompanied the burning as well as minimal amounts of hunting by an increasing human population with whom the animals were being closely associated in the dwindling resource areas.

Large animals were incapable of replacing themselves fast enough, because they were tied to single births and long gestation periods. These factors added appreciably to their plight. They were now losing individuals in greater numbers than ever before and some species began to decline to dangerously low levels. The effect of even minimal hunting and limited burning by humans on populations already under stress from the four uncertainties of demographic, environmental and genetic stochasticity and natural catastrophe (introduced in Chapter 5) compounded their problems, causing an exponential downturn in populations right across Australia. Human movement out of the centre caused an effective increase in their population size in the better environments and this impacted on any megafauna taking refuge in those regions for similar reasons as the humans. It is likely also that humans largely favoured the eastern half of Australia to settle for the same reasons that the megafauna had, because there was more fresh water there.

The onset of the last glacial event, with a drying palaeodrainage and lake system in Central Australia, was the last straw; most megafauna species had reached dangerously low numbers in some areas and local extinctions took place. Those that remained began to migrate out, moving south and southeast searching for food and water. In previous glacial events, these areas had been refuges and the pattern of megafaunal migration was to leave Central Australia at these times and trek south, east and north, or perhaps not north if the larger animals were not adapted to tropical environments. The original people had, till now, followed the megafauna, as their response to the drying. Now the slow increase in their numbers meant that they had spread to many parts of the centre, particularly the south. This may also have been a reaction to modern people arriving in the north, and so, for the first time, humans now followed the megafauna south, whereas previously they were left untouched, to recover their numbers and return to Central Australia when times were more plentiful.

To this point, the process of megafaunal extinction was not 'blitzkrieg' but a gradual and continuous series of events that, coupled with their vulnerability during glacial events, eventually led to a cascade of deleterious events inexorably leading to their demise over 50 ky or more. What was to follow could be termed a 'blitzkrieg' but in reality it was similar to the environmental changes and human pressures we have imposed on our modern fauna. Instead of the decades that our recent impact has taken to put our modern animals in danger, however, the final chapter marking the disappearance of Australia's huge marsupials and carnivores took thousands of years to complete. Neither was the final demise of these animals uniform. There is little doubt that extinction took place at different times in different parts of the continent, with those living in central parts of the continent disappearing first and the smaller populations in the western half of Australia next. One more thing, it is interesting to look at the demise of these animals from the perspective of them as marsupials. Had Australia not become an island continent 45 My earlier invasions by placental mammals might have replaced these animals, even before they had made much evolutionary progress towards the plethora of modern types. What we are seeing during the interglacial is testimony to the resilience of these animals, even in the face of invasion by the cleverest and most cunning of the placentals. In the end it was not a placental that brought the collapse but a series of events of which the placental was one.

By 55 ky the demise of the megafauna in Central Australia was almost complete. It marks a process that continues elsewhere as remaining animals are followed by Robust groups who, as usual, are looking for the same thing as they are – food and water. Pockets of animals accumulated in small refuges but humans are also attracted to them. It is likely that not all species of megafauna lived everywhere on the continent. Some may not have lived in the west or in particular environments such as rainforest or desert. Others may have been restricted to population patches in some areas and their numbers may not have been all that big. It is worth remembering that we have similar examples today of species separation between the western and eastern portions of the continent, e.g. numbats in the west and koalas in the east. Similar distributions could have existed among the megafauna.

By 55 ky Central Australia is virtually empty of large animals and humans. Only *Genyornis newtonii* hangs on as an element of the large creatures that roamed there before, but even it seems to have disappeared between 45 ky and 55 ky (Miller *et al.*, 1999). With the exception of this giant flightless bird, there is no evidence for the existence of megafauna in the Lake Eyre region after 55 ky (see Chapter 5). These animals either left the region, because of drying conditions, or became locally extinct for the same reasons. An alternative could be that already stressed populations were driven to extinction by anthropogenic activities, such as hunting and/or the use of fire. Perhaps the most parsimonious conclusion is that it was all of the above. Certainly humans are the most likely reason for tipping the balance this time, whereas before their arrival the animals survived poor conditions, probably by migrating out and returning when things improved. The demise of Central Australian megafaunal populations, therefore, looks as though it was instant. In reality, it was an accumulation of a lot of events and pressures over a long period of time: they did not become extinct there, they moved away to become extinct elsewhere. Just because they disappeared in that region there is no need to assume it was a blanket event, their story certainly continued for a little longer.

25–50 ky – the robust and AMH people

By 50 ky, Robusts had moved out of Central Australia because of degenerating environmental conditions and reduced animal numbers. Some may have perished as events overtook them and drought and famine were experienced. Those that survived headed towards resources in peripheral regions of the continent and they found these particularly in the southeast, a natural extension of a southerly trek out of the centre. This could have been a tactic employed previously but this time they found the area occupied by others, particularly along the Willandra, Darling and Murray river systems that flowed well during the glaciation. It is easy to see how the megafauna would have been trapped between the two meeting human groups with the new arrivals having already made an impact on these creatures as they worked around the coast and eastern river systems. The two groups begin to live with or close to each other, resulting in further impact on megafauna trying find increasingly rare refuge areas. The result was complete extinction of the giant faunal species sometime between 30 ky and 50 ky with, perhaps, some small populations hanging on in isolated pockets. So in a differential disappearing act that was anything but a uniform exit, those enormous creatures left this mortal coil in a pattern that continues to confound palaeontologists.

Sporadic and varying genetic mixing took place between old robust and modern gene pools, while small isolates of both remained in places away from initial point of mix. The resultant populations vary in their respective morphologies, which begins to produce a wide range of variation among late Pleistocene Australians from at least 25 ky to 35 ky onwards. Various proportions of genetic mixing occur across southeast Australia, in some areas modern gene frequencies are greater than those of robust groups and in a few places the condition is *vice versa*. The robust gene pool was, overall, smaller because of continuing migrations into Australia over the last 50 ky as well as the net loss of robust genes through a lack of replenishment from outside. Mixing continued and spread widely, but in varying proportions according to geographical area and where genetic integrity was greatest for both groups. On top of this, adaptation operating among each of the two groups favoured certain areas and a general adherence to those areas tended to retain morphological integrity for a while. With some minor exceptions, mixing favoured the modern group which had the larger population and constantly, albeit slowly, receives fresh genes through a series of migrations arriving throughout the glacial maximum and after, into the early Holocene. It is likely that there had been no fresh introduction of robust genes for 100 ky.

Prior adaptation to arid conditions favoured the robust to Central Australian conditions. After some mixing with modern groups people begin returning to central regions as the last Ice Age came to a close around 18 ky. We know, however, that some habitation of Central Australia did take place on an opportunistic basis even during the glacial maximum. Puritjarra in the Western Desert region was one of these with occupation at least 27 ky ago, Kulpi Mara south of the West MacDonnell Ranges in Central Australia is another at 29 ky and people were even visiting Lake Eyre at the height of the glaciation around 20 ky and at what we believe was a very dry, harsh and difficult time for human survival (Smith and Sharp, 1993; Thorley, 1998). Men over 190 cm tall were living and hunting in the Willandra at this time; far from Negrito-like, they displayed a robust build and the capability to run at 36 km per hour (Webb *et al.*, 2005). It was probably people like these that were making the opportunistic and sporadic incursions into Central Australia even during the Ice Age, probably when conditions occasionally ameliorated. Later population growth, particularly in the southeast as well as local, and the introduction of complexity in material culture did not affect the Central Australian populations as it did those living on around the edge of the continent. Instead the cultures of Central Australia became immersed in a very rich and complex ceremonial and religious life, the seeds of which had probably begun to emerge even before the two groups came together. After all a complex cerebral world is easier to carry around and store when living in the desert and moving great distances is a way of life.

9 *An incomplete jigsaw puzzle*

It makes no biological or demographic sense that the world's population grew to its present size in less than 10 ky. Nor does it account for the vast areas of the globe that were fully inhabited during the Middle and Upper Pleistocene. Comparatively large populations must have arisen during the latter, setting the stage for a Holocene population 'explosion'. Indeed, for the breadth of adaptation and production of modern people to occur around the world there had to be larger populations in the Upper Pleistocene than has been proposed. The 'explosion' itself is not so much an explosion but a time at which some sections of humanity began to change their lifestyles so that they became very visible. Escalation in population growth was taking place long before this.

It is also illogical to expect that *Homo erectus* lived in patches of splendid isolation around the world without some form of genetic interaction with neighbours. We are a group animal; gregarious, in need of company; we do not travel on our own; indeed we have to have others around us because we are too weak to survive when we try to face nature on an individual or small group basis. The erectine successes and the viable populations that arose from them can be traced from the Lower to the Upper Pleistocene and are testimony to that company. Nor does it make sense that all those descendent populations, begun by this hominid outside Africa, should suddenly capitulate and disappear before the 'onslaught' of anatomically modern humans out of Africa. It is even harder to believe that another *Homo erectus* living inside Africa (for which we have no firm fossil evidence) gave rise to those modern people, while those living outside Africa (also descendants of *Homo erectus* and who we *can* trace through the fossil record) could not. Nor, in my mind, is it possible for the latter erectine to give rise to a population so large that it could produce enough humans to populate Africa, then spill out across the world replacing all those living outside and, in particular, eliminate all females. This is counter to historical and anthropological fact that when wars are fought quite the opposite happens; females are taken and *men* are eliminated. Further, when the world is full of these anatomically modern humans, Africa then stops producing the 'human fountain' that had produced them in the first place. Then, assuming it is Africans leaving the continent, they have to change into all the different morphologically and adaptively different groupings we now find living across the planet in a short time. If these matters are not satisfactorily addressed, then the 'Out of Africa' idea begins to sound more like a new brand of Afrocentric 'Creationism,' at least to this author.

A more rational argument is one where Upper Pleistocene migration to different parts of the world was a product of broad world population growth and widespread gene flow between growing populations, albeit extremely slow initially. This was probably the case even during infinitely slow glacial/interglacial cycles when people adapted, not disappeared. Moreover, independent regional adaptations that took place over a very long time slowly provided both the biological and cultural wherewithall that armed these migrations with the ability and special requirements to eventually take them into high latitudes, allow them to live in marginal environments and arid regions, and migrate across large water gaps in their respective regions. This process occurred in many areas and increasingly facilitated and speeded up gene flow across the Old Word and eventually into the New World. I have proposed that two particular centres contributed to this process, China and the Indian subcontinent, and today they provide almost half the world's gene pool.

The ability of modern humans to live anywhere developed over a considerable period of time. Erectines were very capable creatures and not backward when it came to trying new environments and landscapes to live in, but they were not equipped for very high latitudes and probably not good at living in deserts, although adaptation to heat was no doubt one of their adaptive qualities. The adaptation and socio-cultural development of modern people could only come from a necessity to do so, derived from an increasingly strong driving force for them to alter and manipulate the world around them. That driving force was population growth and its commensurate genetic strengthening. From this emerged the increased capacity to develop broad biological, cultural and physical adaptations. Built upon a long adaptive lead up, these allowed us to migrate even farther. Adaptations emerged as traits, such as face, eye and body shape, skin colour and hair form, as well as physiological and immune responses and a range of nutritional, thermoregulatory and metabolic contrasts. These developed to fit the environmental circumstances in which humans found themselves during the long migratory processes. Our biological adaptations were then underpinned by culture and especially the slowly developing complexities of material culture, even though manufacturing prowess was limited during the early stages of migration and world settlement. For example, it is no accident or fluke that the oldest evidence for the boomerang found so far is approaching 20 ky and is found in the widely separated areas of Poland and Australia! People share and have always shared similar mental templates. Indeed, the 'wheel' has been reinvented several times because we humans carry similar mental templates for solving everyday problems.

Finally, we cannot ignore the fact that humans lived next door to Australia for nearly two million years before they arrived here. The fossil evidence indicates that they constituted a viable population living at the edge of the inhabited world, in Java and China. Their viability stretches beyond their fossil remains and into the realm of exploration as they explored beyond Java to Flores Island almost one million years ago, and that required a capability to cross at least two 22 km water gaps. The idea that this was an aberrant group and one that does not count in any assessment of regional human evolution is difficult to accept and, indeed, it does not take into consideration the will and determination of this erstwhile group of humans. Consider, if they made that inter-island crossing so long ago, what would have stopped them from continuing to the end of the archipelago and even further during the next 700 ky? These people or their descendants were the most obvious to be the first Australians, long before the 'advent' of anything called anatomically modern human. Why should we invoke a people from half way round the world to reach here first when this successful population thrived and explored their archipelago only a few hundred kilometres away from our continental shelf or little more than a hundred kilometres away during low sea levels? It is not possible that this persistent and successful group did not attempt to enter Australia and eventually succeed in doing so. They may have done this out of pure curiosity, probably tens, possibly hundreds of thousands of years before anybody else came anywhere near their new homeland. For me, the question is not *did* they arrive here, it is *when* did they arrive here. That is open for speculation and that is exactly what I have done here. My speculation, however, is conservative and based on evidence, not arm waving. While the earliest camp sites are probably under water and the early dates elude us for now, there can be no doubt that the robust morphology among early Australians is the banner of their arrival. It is the stark signal of a continuing process of exploration by people in the region that began almost 2 My ago. The distinct morphology of some Australian fossil crania cannot have an origin anywhere else other than Indonesia, it is not something developed within Australia, an aberration of the Ice Age, nor can it be one end of a wide range of anatomically modern human morphology that came 'Out of Africa'. Indeed, it was precisely that anatomically modern humans were *not* cranially robust that distinguishes them as such. Whatever the genetic evidence suggests, the only people around to convey such an outstanding morphology must have been *Homo soloensis* or its descendants. There is no other

Upper Pleistocene population from whom such a heavily built cranial structure could be inherited. I am not the first to invoke the Indonesian connection among Australia's fossil hominids and the fact that so many prominent physical anthropologists before me have come to the same conclusion only serves to underpin that drawn (once again) here, although arrived at independently.

The first and subsequent people to arrive had a variety of choices for exploring the continent. These varied according to glacial and interglacial conditions. Central Australia was not always hostile to exploring parties. For the vast majority of the last interglacial it could be accessed along large gallery forested river systems, flanked by wide savannahs; there was no particular need for people to cling to the coast although no doubt some did. When continental shelves disappeared under rising seas these corridors were open for travel and no doubt some took that opportunity.

Like everyone else, I have ended up by falling into the trap and talking about two foundation groups of people that eventually produced the modern Australian population. If only the process was as simple as that. Many migrations from many areas over many millennia contributed to modern Australian indigenous people, together with a steady adaptive evolutionary pathway on a continent that moved into, through and out of at least one Ice Age. The farther back our dates go, the more complicated the process was and, possibly, the more people contributed to that process and the longer they had to emerge as a unique culture, bearing an equally unique morphological appearance. Contrary to the inevitable conclusions many will draw from what I have said, my arguments concerning the time of the first arrivals and who they were have been drawn from the evidence at hand. I believe that the only logical conclusions to be drawn from the evidence at this time are those outlined here. I suppose it is a case that, while some would choose to dismiss certain evidence, I chose not to do so. The evidence may be sparse, but what there is of it cannot be dismissed so easily. Therefore, I cannot interpret it any other way than that which I have written.

Time and again it has been shown that we consistently underestimate the ability and guile of our earlier ancestors and are almost reluctant to update our perceptions of prehistory. For example, in 1960, it was firmly believed that people had been in Australia only 6 ky, 40 years later it is an order of magnitude longer. Sometimes we treat our data in a way we would not treat a piece from a jigsaw puzzle. It emerges, we try to fit it in, if it does not fit many would discard it instead of putting it aside for use later or speculate what the picture might look like using the characteristics of the piece. When one has enough pieces a broad picture emerges without necessarily having the pieces in between. I believe that is what we have in Australia, a small incomplete picture of a much vaster landscape, with many pieces of the jigsaw correct, but a far greater amount missing. Many would not wish to view the jigsaw till all the pieces are in place. We may never have all the pieces, so let us not ridicule or throw the pieces we have away because they cannot be logically fitted together. I have tried to make a picture from the ones we have.

The story of humans in Australia is similar to that of humans moving into the Americas. Australia has one advantage, however; we know people had to arrive by water craft. The American story has both sea and land to consider. But we both puzzle over the speed and direction taken by the first migrants as well as whether it was a coastal or interior colonisation. And then there is the timing. Ah, the timing. After many years of two views, one supporting a 12 ky Clovis entry and another clinging to limited evidence of a much older arrival, support is finally growing in American archaeology for the latter after the discoveries and 'vindication' of Monte Verde (Dillehay, 1989, 1997; Meltzer, 2004). We also both share tantalising but very limited evidence that could put human arrival much earlier than has been allowed to be discussed among consenting adults. There seems to be an innate dislike of any discussion that takes an order of magnitude view, for example one suggesting an arrival 100 ky or more that the existing evidence suggests. This is particularly puzzling when the universal lesson in archaeology has been time and again one of pushing back human existence or occupation farther and farther. I have always found this puzzling and cannot understand why, when most areas of science are willing to theorise, archaeology is generally unwilling to do so.

The topics covered in this book are written in a way I have not attempted before. The reason is to make people think again about the overzealous correctness about how we receive and what we do with data. Let's face it, whatever some may say, anthropology and archaeology are not exact sciences and never will be. We use science, much of which we do not fully understand, but bits of it help us forge our results into conclusions. The debate among various dating techniques on 'who's right' is a treat to behold. Nothing wrong with that, but we all deal with possibilities and predictions about what happened in the past, knowing we will never hear the whistle or chatter of the Upper Pleistocene stone tool napper or Middle Pleistocene raft makers. This book is, therefore, an appeal not to be too stuffy about the past, to keep one's mind open, even let it soar a little, and not close it on possibilities or conclusions which do not have 'enough data' or any data. Besides changing the name of the continent I have to agree with Meltzer (2004: 555) when in a review of the first North Americans he states ' In the meantime there is much to do. . . . At the most basic level, we need more archaeological data. Although we may never detect the very first archaeological "footprints" of people on the North American continent.' Be honest, we never have enough data and, I admit, this book is yet another example of that. There is always much more to do but my experience shows me that the more I find out, the less I know. So often we are confounded soon after 'the answer' has been provided to us by 'the data'. So a word of caution, those who would nail their colours too firmly to the mast of dogma can find the mast heavy when it falls, and the history of archaeological discovery shows that it falls all the time. In a continent like Australia, with so much yet to understand about its earliest inhabitants and when they arrived, I believe trying ideas on for size is a good thing. It opens up new avenues of direction and throws up further ideas. Many may be wrong, but so what! Those who would hide away from proposing controversial ideas will never help others, who never have an idea, participate in discovery. It is always worth bringing forth ideas and theories so that they can be tested, either in the short or long term.

If the propositions put above are not correct, then I challenge those who would doubt them to show me contrary evidence and particularly to hold high the ancestral form from which the early Australian robust morphology is derived. Moreover, we should keep an open mind to what sort of future archaeological and palaeontological discoveries could be made here in Australia and the region. The human story is far from told or being understood; *Homo floresiensis* is a good example. One thing is certain, however, we have yet to find the oldest archaeology there is to find on this ancient, very mysterious and little explored continent. The bottom line is that we need to keep looking because I believe Australia holds the key, not only for understanding salient migrational events in the Upper Pleistocene, but also unravelling the 'Out of Africa' ideas that have come to dominate world prehistory and the study of human evolution in recent years. Australian prehistory and palaeoanthropology has reached a stage when the fear of being criticised far outweighs the desire to propose new ideas and offer considered thoughts. It is time the pendulum swung the other way.

Appendix 1

Appendix 1. *Various permutations of Pleistocene population growth parameters*

Sv	c	Rg	Ro	c	Rg	Ro
$t = 10$ yrs, $r = 0.000005$				$t = 10$ yrs, $r = 0.000006$		
32%	6.404000	3.125152	1.000049	6.404073	3.125188	1.00006
35%	5.855094	2.857857	1.00005	5.855152	2.857314	1.00006
40%	5.123207	2.500125	1.00005	5.123258	2.500150	1.00006
42%	4.879230	2.381064	1.00005	4.879294	2.381095	1.00006
52%	3.940928	1.923173	1.00005	3.940967	1.923192	1.00006
$t = 15$ yrs, $r = 0.000005$				$t = 15$ yrs, $r = 0.000006$		
30%	6.831113	3.333583	1.000075	6.831216	3.333633	1.00009
32%	6.404169	3.125234	1.000075	6.404265	3.125281	1.00009
35%	5.855240	2.857357	1.000075	5.855328	2.857400	1.00009
40%	5.123335	2.500188	1.000075	5.123412	2.500225	1.00009
42%	4.879367	2.381131	1.000075	4.879440	2.381167	1.00009
50%	4.098668	2.000150	1.000075	4.098730	2.000180	1.00009
52%	3.941027	1.923221	1.000075	3.941086	1.923250	1.00009
$t = 10$ yrs, $r = 0.000007$				$t = 10$ yrs, $r = 0.000009$		
32%	6.404137	3.125219	1.00007	6.403746	3.122503	1.00009
35%	5.855211	2.857343	1.00007	5.854854	2.857169	1.00009
40%	5.123309	2.500175	1.00007	5.122999	2.500225	1.00009
42%	4.879035	2.380969	1.00007	4.879045	2.380974	1.00009
52%	3.940759	1.923090	1.00007	3.940767	1.923094	1.00009
$t = 15$ yrs, $r = 0.000007$				$t = 15$ yrs, $r = 0.000009$		
30%	6.831318	3.333683	1.000105	6.831523	3.333783	1.00135
32%	6.404361	3.125328	1.000105	6.404553	3.125422	1.00135
35%	5.855416	2.857443	1.000105	5.855591	2.857529	1.00135
40%	5.123489	2.500263	1.000105	5.123642	2.500338	1.00135
42%	4.880025	2.381452	1.000105	4.879659	2.381274	1.00135
50%	4.098791	2.000210	1.000105	4.098914	2.000270	1.00135

Appendix 1. (cont.)

Sv	c	Rg	Ro	c	Rg	Ro
52%	3.941145	1.923279	1.000105	3.941263	1.923367	1.00135
32%	6.403817	3.125063	1.00002	6.403945	3.125125	1.0004
35%	5.854918	2.857200	1.00002	5.855035	2.857257	1.0004
40%	5.123309	2.500175	1.00002	5.123156	2.500100	1.0004
42%	4.879035	2.380969	1.00002	4.879196	2.238105	1.0004
52%	3.940759	1.923090	1.00002	3.943096	1.923154	1.0004
$t = 15$ yrs, $r = 0.00002$				$t = 15$ yrs, $r = 0.00004$		

Sv	c	Rg	Ro	c	Rg	Ro
30%	6.832650	3.334333	1.0003	6.834700	3.335333	1.0006
32%	6.405610	3.129375	1.0003	6.407531	3.126875	1.0006
35%	5.856557	2.858000	1.0003	5.858314	2.858857	1.0006
40%	5.124488	2.500750	1.0003	5.126025	2.501500	1.0006
42%	4.879001	2.381452	1.0003	4.881928	2.382381	1.0006
50%	4.099590	2.000600	1.0003	4.100820	2.001200	1.0006
52%	3.941914	1.923654	1.0003	3.943096	1.924301	1.0006
$t = 10$ yrs, $r = 0.00005$				$t = 10$ yrs, $r = 0.00008$		

Sv	c	Rg	Ro	c	Rg	Ro
32%	6.404009	3.125156	1.00005	6.404200	3.125250	1.00008
35%	5.855094	2.857286	1.00005	5.855269	2.857371	1.00008
40%	5.123207	2.500125	1.00005	5.123361	2.500200	1.00008
42%	4.879245	2.381071	1.00005	4.879391	2.381143	1.00008
52%	3.940928	1.923173	1.00005	3.941047	1.923231	1.00008
$t = 15$ yrs, $r = 0.00005$				$t = 15$ yrs, $r = 0.00008$		

Sv	c	Rg	Ro	c	Rg	Ro
30%	6.832650	3.334333	1.00075	6.838798	3.337333	1.0012
32%	6.405610	3.129375	1.00075	6.411373	3.128750	1.0012
35%	5.856557	2.858000	1.00075	5.861827	2.860571	1.0012
40%	5.124488	2.500750	1.00075	5.129098	2.503000	1.0012
42%	4.879001	2.381452	1.00075	4.884856	2.383810	1.0012
50%	4.099590	2.000600	1.00075	4.103279	2.002400	1.0012
52%	3.941914	1.923654	1.00075	3.945460	1.925385	1.0012

Notes:
t – Female reproductive time span (years).
r – Rate of population growth (per generation).
Sv – Survivorship or the chances of reaching reproductive age.
c – Number of live births.
Rg – Gross reproductive rate.
Ro – Net reproductive rate.

Appendix 2

Appendix 2. *Cranial vault thickness (mm) by constituent bone type*

WLH no	Max width of ICT[a]	Max width of diploeic bone[a]	Maximum width of OCT[a]	Maximum cranial thickness[b]	Diploeic bone %[c]
1	1.5	4.0	2.5	8.0	50.0
2	3.5	2.0	4.0	8.0	26.6
3	1.0	8.0	1.5	11.0	76.2
9	1.0	–	1.0	5.0	–
10	<1.0	2.0	2.0	5.0	28.6
11	<1.0	6.5	1.0	7.5	76.5
12	<1.0	6.0	2.5	10.0	63.2
13	1.0	3.0	2.0	7.0	50.0
15	<1.0	6.0	1.0	7.0	75.0
16	2.0	4.0	2.0	10.0	50.0
17	1.5	4.0	2.0	7.5	53.3
18	1.0	10.0	2.5	13.5	74.1
19	2.0	7.5	2.5	15.0	62.5
20	2.0	7.0	3.0	11.5	58.3
21	1.0	3.5	2.0	7.5	53.9
22	2.0	8.0	3.0	12.0	61.5
23	1.0	4.5	1.0	10.0	69.2
24	<1.0	6.0	1.5	10.0	60.0
26	<1.0	7.0	1.5	9.0	73.7
27	1.5	10.0	2.0	14.0	74.1
28	2.0	6.0	4.0	12.0	50.0
29	<1.0	5.0	2.0	8.0	62.5
42	1.0	7.0	2.0	10.0	70.0
43	<1.0	6.5	<1.0	9.0	76.5
44	<1.0	5.0	<1.0	7.0	71.4
45	1.5	9.0	2.5	14.0	69.2
46	1.0	7.0	3.0	10.0	63.6
47	1.0	4.0	1.0	7.0	66.7
48	1.5	4.0	1.0	5.0	61.5
49	<1.0	6.0	2.0	8.0	66.7
50	1.0	14.0	1.0	19.0	87.5
51	<1.0	3.0	1.0	5.0	60.0
52	2.0	4.0	2.5	9.0	47.1
53	1.0	5.0	2.5	8.0	58.8
55	1.0	4.5	1.0	7.0	69.2

Appendix 2. (cont.)

WLH no	Max width of ICT[a]	Max width of diploeic bone[a]	Maximum width of OCT[a]	Maximum cranial thickness[b]	Diploeic bone %[c]
56	2.0	5.0	4.0	12.5	45.5
58	1.5	6.0	2.0	10.0	63.2
63	2.5	6.0	7.5	12.5	37.5
64	<1.0	6.0	1.0	9.0	75.0
67	<1.0	5.5	1.5	9.0	68.8
68	1.5	2.5	1.5	7.0	36.7
69	1.0	8.0	3.0	10.0	66.6
72	–	8.0	2.0	10.0	–
73	1.5	4.0	1.5	8.0	57.1
75	1.0	7.0	2.0	9.0	70.0
98	2.0	5.0	2.0	9.0	55.6
99	1.5	5.0	3.0	9.0	52.6
100	2.0	6.0	5.0	13.0	46.2
101	2.0	6.0	4.5	13.5	48.0
102	1.5	7.0	4.0	12.0	56.0
120	1.0	5.0	1.0	7.0	71.4
122	<1.0	3.0	2.0	6.0	50.0
123	1.0	3.0	1.0	7.0	60.0
124	2.0	6.0	4.0	11.5	50.0
125	1.5	6.0	2.5	9.0	60.0
127	1.0	6.5	2.0	10.0	68.4
128	1.0	3.0	2.0	6.0	50.0
129	1.0	6.0	2.0	11.5	66.7
130	1.5	5.0	1.5	9.0	62.5
134	<1.0	4.0	2.0	6.0	57.1

Notes:
[a] 1 – Measured anywhere on broken edges excluding asterion and occipital squama.
[b] 2 – Measured anywhere on vault excluding asterion and occipital squama.
[c] 3 – Column 2 as percentage of column 4.
ICT – Inner Cranial Table, OCT – Outer Cranial Table.

Appendix 3

Appendix 3A. *WLH Series sorted by maximum cranial vault thickness (mm)*

WLH No	Maximum width of ICT[a]	Maximum width of diploeic bone[a]	Maximum width of OCT[a]	Maximum cranial thickness[b]	Diploeic bone %[c]
9	1.0	–	1.0	5.0	–
10	<1.0	2.0	2.0	5.0	28.6
51	<1.0	3.0	1.0	5.0	60.0
48	1.5	4.0	1.0	5.0	61.5
128	1.0	3.0	2.0	6.0	50.0
122	<1.0	3.0	2.0	6.0	50.0
134	<1.0	4.0	2.0	6.0	57.1
68	1.5	2.5	1.5	7.0	36.7
13	1.0	3.0	2.0	7.0	50.0
123	1.0	3.0	1.0	7.0	60.0
47	1.0	4.0	1.0	7.0	66.7
55	1.0	4.5	1.0	7.0	69.2
44	<1.0	5.0	<1.0	7.0	71.4
120	1.0	5.0	1.0	7.0	71.4
15	<1.0	6.0	1.0	7.0	75.0
21	1.0	3.5	2.0	7.5	53.9
17	1.5	4.0	2.0	7.5	53.3
11	<1.0	6.5	1.0	7.5	76.5
2	3.5	2.0	4.0	8.0	26.6
1	1.5	4.0	2.5	8.0	50.0
73	1.5	4.0	1.5	8.0	57.1
29	<1.0	5.0	2.0	8.0	62.5
53	1.0	5.0	2.5	8.0	58.8
49	<1.0	6.0	2.0	8.0	66.7
52	2.0	4.0	2.5	9.0	47.1
98	2.0	5.0	2.0	9.0	55.6
99	1.5	5.0	3.0	9.0	52.6
130	1.5	5.0	1.5	9.0	62.5
67	<1.0	5.5	1.5	9.0	68.8
64	<1.0	6.0	1.0	9.0	75.0
125	1.5	6.0	2.5	9.0	60.0
43	<1.0	6.5	<1.0	9.0	76.5
26	<1.0	7.0	1.5	9.0	73.7
75	1.0	7.0	2.0	9.0	70.0
16	2.0	4.0	2.0	10.0	40.0

Appendix 3A. (*cont.*)

WLH No	Maximum width of ICT[a]	Maximum width of diploeic bone[a]	Maximum width of OCT[a]	Maximum cranial thickness[b]	Diploeic bone %[c]
23	1.0	4.5	1.0	10.0	69.2
12	<1.0	6.0	2.5	10.0	63.2
24	<1.0	6.0	1.5	10.0	70.6
58	1.5	6.0	2.0	10.0	63.2
127	1.0	6.5	2.0	10.0	68.4
42	1.0	7.0	2.0	10.0	70.0
46	1.0	7.0	3.0	10.0	63.6
69	1.0	8.0	3.0	10.0	66.6
72	–	8.0	2.0	10.0	–
3	1.0	8.0	1.5	11.0	76.2
124	2.0	6.0	4.0	11.5	50.0
129	1.0	6.0	2.0	11.5	66.7
20	2.0	7.0	3.0	11.5	58.3
28	2.0	6.0	4.0	12.0	50.0
102	1.5	7.0	4.0	12.0	56.0
22	2.0	8.0	3.0	12.0	61.5
56	2.0	5.0	4.0	12.5	45.5
63	2.5	6.0	7.5	12.5	37.5
100	2.0	6.0	5.0	13.0	46.2
101	2.0	6.0	4.5	13.5	48.0
18	1.0	10.0	2.5	13.5	74.1
45	1.5	9.0	2.5	14.0	69.2
27	1.5	10.0	2.0	14.0	74.1
19	2.0	7.5	2.5	15.0	62.5
50	1.0	14.0	1.0	19.0	87.5

Appendix 3B. *WLH Series sorted by maximum diploeic bone thickness (mm)*

WLH No.	Maximum width of ICT[a]	Maximum width of diploe bone[a]	Maximum width of OCT[a]	Maximum cranial thickness[b]	Diploeic bone %[c]
9	1.0	–	1.0	5.0	–
2	3.5	2.0	4.0	8.0	26.6
10	<1.0	2.0	2.0	5.0	28.6
68	1.5	2.5	1.5	7.0	36.7
13	1.0	3.0	2.0	7.0	50.0
123	1.0	3.0	1.0	7.0	60.0
128	1.0	3.0	2.0	6.0	50.0
122	<1.0	3.0	2.0	6.0	50.0
51	<1.0	3.0	1.0	5.0	60.0
21	1.0	3.5	2.0	7.5	53.9
16	2.0	4.0	2.0	10.0	40.0
52	2.0	4.0	2.5	9.0	47.1
1	1.5	4.0	2.5	8.0	50.0
73	1.5	4.0	1.5	8.0	57.1
17	1.5	4.0	2.0	7.5	53.3
47	1.0	4.0	1.0	7.0	66.7
134	<1.0	4.0	2.0	6.0	57.1
48	1.5	4.0	1.0	5.0	61.5
23	1.0	4.5	1.0	10.0	69.2
55	1.0	4.5	1.0	7.0	69.2
56	2.0	5.0	4.0	12.5	45.5
98	2.0	5.0	2.0	9.0	55.6
99	1.5	5.0	3.0	9.0	52.6
130	1.5	5.0	1.5	9.0	62.5
29	<1.0	5.0	2.0	8.0	62.5
53	1.0	5.0	2.5	8.0	58.8
44	<1.0	5.0	<1.0	7.0	71.4
120	1.0	5.0	1.0	7.0	71.4
67	<1.0	5.5	1.5	9.0	68.8
101	2.0	6.0	4.5	13.5	48.0
100	2.0	6.0	5.0	13.0	46.2
63	2.5	6.0	7.5	12.5	37.5
28	2.0	6.0	4.0	12.0	50.0
124	2.0	6.0	4.0	11.5	50.0
129	1.0	6.0	2.0	11.5	66.7
12	<1.0	6.0	2.5	10.0	63.2
24	<1.0	6.0	1.5	10.0	70.6
58	1.5	6.0	2.0	10.0	63.2
64	<1.0	6.0	1.0	9.0	75.0
125	1.5	6.0	2.5	9.0	60.0
49	<1.0	6.0	2.0	8.0	66.7
15	<1.0	6.0	1.0	7.0	75.0
127	1.0	6.5	2.0	10.0	68.4
43	<1.0	6.5	<1.0	9.0	76.5
11	<1.0	6.5	1.0	7.5	76.5

Appendix 3B. (*cont.*)

WLH No.	Maximum width of ICT[a]	Maximum width of diploe bone[a]	Maximum width of OCT[a]	Maximum cranial thickness[b]	Diploeic bone %[c]
102	1.5	7.0	4.0	12.0	56.0
20	2.0	7.0	3.0	11.5	58.3
42	1.0	7.0	2.0	10.0	70.0
46	1.0	7.0	3.0	10.0	63.6
26	<1.0	7.0	1.5	9.0	73.7
75	1.0	7.0	2.0	9.0	70.0
19	2.0	7.5	2.5	15.0	62.5
22	2.0	8.0	3.0	12.0	61.5
3	1.0	8.0	1.5	11.0	76.2
69	1.0	8.0	3.0	10.0	66.6
72	–	8.0	2.0	10.0	–
45	1.5	9.0	2.5	14.0	69.2
27	1.5	10.0	2.0	14.0	74.1
18	1.0	10.0	2.5	13.5	74.1
50	1.0	14.0	1.0	19.0	87.5

Appendix 3C. *Cranial vault thickness (mm) sorted by percentage of diploeic bone as a component of cranial thickness*

WLH No.	Maximum width of ICT[a]	Maximum width of diploe bone[a]	Maximum width of OCT[a]	Maximum cranial thickness[b]	Diploeic bone %[c]
72	–	8.0	2.0	10.0	–
2	3.5	2.0	4.0	8.0	26.6
10	<1.0	2.0	2.0	5.0	28.6
68	1.5	2.5	1.5	7.0	36.7
63	2.5	6.0	7.5	12.5	37.5
16	2.0	4.0	2.0	10.0	40.0
56	2.0	5.0	4.0	12.5	45.5
100	2.0	6.0	5.0	13.0	46.2
52	2.0	4.0	2.5	9.0	47.1
101	2.0	6.0	4.5	13.5	48.0
13	1.0	3.0	2.0	7.0	50.0
128	1.0	3.0	2.0	6.0	50.0
122	<1.0	3.0	2.0	6.0	50.0
1	1.5	4.0	2.5	8.0	50.0
28	2.0	6.0	4.0	12.0	50.0
124	2.0	6.0	4.0	11.5	50.0
99	1.5	5.0	3.0	9.0	52.6
17	1.5	4.0	2.0	7.5	53.3
21	1.0	3.5	2.0	7.5	53.9
98	2.0	5.0	2.0	9.0	55.6
102	1.5	7.0	4.0	12.0	56.0
73	1.5	4.0	1.5	8.0	57.1
134	<1.0	4.0	2.0	6.0	57.1
20	2.0	7.0	3.0	11.5	58.3
53	1.0	5.0	2.5	8.0	58.8
123	1.0	3.0	1.0	7.0	60.0
51	<1.0	3.0	1.0	5.0	60.0
125	1.5	6.0	2.5	9.0	60.0
48	1.5	4.0	1.0	5.0	61.5
22	2.0	8.0	3.0	12.0	61.5
130	1.5	5.0	1.5	9.0	62.5
29	<1.0	5.0	2.0	8.0	62.5
19	2.0	7.5	2.5	15.0	62.5
12	<1.0	6.0	2.5	10.0	63.2
58	1.5	6.0	2.0	10.0	63.2
46	1.0	7.0	3.0	10.0	63.6
69	1.0	8.0	3.0	10.0	66.6
47	1.0	4.0	1.0	7.0	66.7
129	1.0	6.0	2.0	11.5	66.7
49	<1.0	6.0	2.0	8.0	66.7
127	1.0	6.5	2.0	10.0	68.4
67	<1.0	5.5	1.5	9.0	68.8
23	1.0	4.5	1.0	10.0	69.2
55	1.0	4.5	1.0	7.0	69.2

Appendix 3C. (*cont.*)

WLH No.	Maximum width of ICT[a]	Maximum width of diploe bone[a]	Maximum width of OCT[a]	Maximum cranial thickness[b]	Diploeic bone %[c]
45	1.5	9.0	2.5	14.0	69.2
42	1.0	7.0	2.0	10.0	70.0
75	1.0	7.0	2.0	9.0	70.0
24	<1.0	6.0	1.5	10.0	70.6
44	<1.0	5.0	<1.0	7.0	71.4
120	1.0	5.0	1.0	7.0	71.4
26	<1.0	7.0	1.5	9.0	73.7
27	1.5	10.0	2.0	14.0	74.1
18	1.0	10.0	2.5	13.5	74.1
64	<1.0	6.0	1.0	9.0	75.0
15	<1.0	6.0	1.0	7.0	75.0
3	1.0	8.0	1.5	11.0	76.2
43	<1.0	6.5	<1.0	9.0	76.5
11	<1.0	6.5	1.0	7.5	76.5
50	1.0	14.0	1.0	19.0	87.5

References

Abbie, A. A. (1968) The Homogeneity of Australian Aborigines. *Archaeology and Physical Anthropology in Oceania* 3: 221–231.

(1976) Morphological variation in the adult Australian Aboriginal. In R. L. Kirk and A. G. Thorne (eds), *The Origin of the Australians*. Canberra AIAS.

Acsadi, G. Y. and J. Nemeskeri (1970) *History of Human Life Span and Mortality*. Budapest: Ackademia Kiado.

Aharon, P. and J. Chappell (1986) Oxygen isotopes, sea level changes and the temperate history of the coral reef environment in New Guinea over the last 10^5 years. *Palaeogeography, Palaeoclimatology, Palaeoecology* 56: 337–379.

Allen, H. R. (1972) Where the crow flies backwards. Unpublished Ph.D. thesis. Australian National University, Canberra.

(1974) The Bagunji of the Darling Basin: cereal gatherers in an uncertain environment. *World Archaeology* 5: 309–322.

Ammerman, A. J. (1975) Late pleistocene population alternatives. *Human Ecology* 3 (4): 219–223.

An, Z., W. Gao, Y. Zhu, X. Kan, J. Wang, J. Sun and M. Wei (1990) Magnetostratigraphic dates of Lantian *Homo erectus*. *Acta Anthropological Sinica* 9: 1–7.

Andel, T. H. van (1989) Late Quaternary sea-level changes and archaeology. *Antiquity* 63: 733–745.

Anderson, J. F., A. Hall-Martin and D. A. Russell (1985) Long-bone circumference and weight in mammals, birds and dinosaurs. *Journal of Zoology, London (A)* 207: 53–61.

Angel, J. L. (1966) Early skeletons from Tranquillity, California. *Smithsonian Contributions to Anthropology* 2: 1–15.

Antón, S. C. (2003) Natural history of *Homo erectus*. *Yearbook of Physical Anthropology* 46: 126–170.

Antón, S. C., F. Aziz and Y. Zaim (2001) Plio-Pleistocene Homo: patterns and determinants of dispersal. In *Colloquia in Human Biology and Palaeoanthropology: Humanity from African Naissance to Coming Millenia*. Florence: Florence University Press, pp. 97–108.

Antón, S. C., W. R. Leonard and M. L. Robertson (2003) An ecomorphological model of the initial hominid dispersal from Africa. *Journal of Human Evolution* 43: 773–785.

Archer, M. (1984) The Australian Marsupial Radiation. In M. Archer and G. Clayton (eds), *Vertebevate Zoogeography and Evolution in* Australia: Hesperin Press, pp. 633–808.

Arsuaga, J., I. Martinez, A. Garcia, J. Carretero, C. Lorenzo, N. Garcia and A. Ortega (1997) Sima de los Heusos (Sierra de Atapuerca, Spain). The Site. *Journal of Human Evolution* 33: 109–127.
Auffenberg, W. (1972) Komodo dragons. *Natural History* 81: 52–59.
Bakker, R. (1988) *The Dinosaur Heresies*. London: Penguin.
Barbetti, M. and H. R. Allen (1972) Prehistoric man at Lake Mungo, Australia, by 32 000 years BP. *Nature* 240: 46–48.
Barbetti, M. and H. Polach (1973) ANU radiocarbon date list V. *Radiocarbon* 15(2): 241–251.
Barrie, J. (1990) Skull elements and additional remains of the Pleistocene boid snake *Wonambi Naracoortensis*: *Memoirs of the Queensland Museum* 28: 139–151.
Basedow, H. (1914) Aboriginal rock carvings of great antiquity in South Australia. *Journal of the Royal Anthropological Institute* 44: 195–211.
Bellwood, P. (1978) *Man's Conquest of the Pacific*. London: Collins.
Bellwood, P. (1985) *Prehistory of the Indo-Malaysian Archipelgo*. New York: Academic Press.
Bergh, G. van den, P. Sondaar, J. de Vos, and A. Fachroel (1993) Chrono and biotratigraphy of the terrestrial Quaternary deposits in the Indonesian Archipelago and the impact of humans on the faunal evolution. Abstract, Inter-Inqua Conference April, Canberra, Australia (personal communication).
Bermudez de Castro, J., J. Arsuaga, E. Carbonell, A. Rosas, I. Martinez and M. Mosquera (1997) A Hominid from the Lower Pleistocene of Atapuerca, Spain: possible ancestor to Neanderthals and modern humans. *Science* 276: 1392–1395.
Bijlmer, H. T. (1939) Tapiro pygmies and Pania Mountain-Papuas. *Nova Guinea* (new series) 3: 113–184.
Birdsell, J. (1967) Preliminary data on the trihybrid origin of the Australian Aborigines. *Archaeology and Anthropology in Oceania* 2: 100–155.
(1972) *Human Evolution*. Chicago: Rand McNally.
(1977) The recalibration of a paradigm for the first peopling of Greater Australia. In J. Allen, J. Golson and R. Jones (eds), *Sunda and Sahul*. New York: Academic Press, pp. 113–167.
Birdsell, J. B. (1941) The Trihybrid origin of the Australian Aborigines. Ph.D. Thesis, Harvard University, Cambridge, MA.
(1957) Some population problems involving Pleistocene man. *Cold Spring Harbor Symposia on Quantitative Biology* 22: 47–70.
(1979) Physical anthropology in Australia today. *Annual Reviews in Anthropology* 8: 417–430.
Bohte, A. and P. Kershaw (1999) Taphonomic influences on the interpretation of the palaeoecological record from Lynch's Crater, northeastern Australia. *Quaternary International* 57(58): 49–59.
Bonte, M. and H. von Balen (1969) Prolonged lactation and family spacing in Rwanda. *Journal of Biosocial Science* 1: 97–100.
Borrie, W. D. (1970) *The Growth and Control of World Population*. London: Weidenfeld & Nicolson.

Bowdler, S. (1977) The coastal colonisation of Australia. In J. Allen, J. Golson and R. Jones (eds), *Sunda and Sahul*. New York: Academic Press, pp. 205–246.

Bowler, J. M. (1971) Pleistocene salinities and climatic change: evidence from lakes and lunettes in southeastern Australia. In D. J. Mulvaney and J. Golson (eds), Canberra: *Aboriginal Man and Environment in Australia*. Australian National University Press, pp. 47–65.

(1973) Clay dunes: their occurrence, formation and environmental significance. *Earth-Science Reviews* 9: 315–338.

(1975) Deglacial events in southern Australia: their age, nature, and palaeoclimatic significance. In R. P. Suggate and M. M. Cresswell (eds), *Quaternary Studies: Selected Papers from IX INQUA Congress*, Christchurch, New Zealand, 2–10 December, 1973, Wellington: Royal Society of New Zealand, Bulletin 13, pp. 75–82.

(1976a) Recent developments in reconstructing late Quaternary environments in Australia. In R. L. Kirk and A. G. Thorne (eds), *The Origin of the Australians*. Canberra: Australian Institute of Aboriginal Studies, pp. 55–77.

(1976b) Aridity in Australia: age, origins and expression in aeolian landforms and sediments. *Earth-Science Reviews* 12: 279–310.

(1980) Quaternary chronology and palaeohydrology in the evolution of Mallee landscapes. In R. R. Storrier and M. E. Stannard (eds), *Aeolian Landscapes in the Semi-Arid Zone of South Eastern Australia*. Proceedings of a conference held at Mildura, Victoria, Australia on 17–18 October 1979, Australian Society of Soil Science, Riverina Branch, Wagga Wagga (NSW), pp. 17–36.

(1983) Lunettes as indices of hydrologic change: a review of the Australian evidence. *Proceedings of the Royal Society of Victoria* 95(3): 147–168.

(1986) Quaternary landform evolution. In D. N. Jeans (ed.), *Australia: A Geography*. Sydney: Sydney University Press, pp. 117–147.

(1998) Willandra Lakes revisited: environmental framework for human occupation. In Willandra Lakes: People and Palaeoenvironments, *Archaeology in Oceania* 33 (3): 120–155.

Bowler, J. M. and J. W. Magee (2000) Redating Australia's oldest human remains: a sceptic's view. *Journal of Human Evolution* 38: 719–726.

Bowler, J. M. and A. G. Thorne (1976) Human remains from Lake Mungo: discovery and excavation of Lake Mungo III. In R. L. Kirk and A. G. Thorne (eds), *The Origin of the Australians*. Canberra: Australian Institute of Aboriginal Studies, pp. 127–138.

Bowler, J. and R. J. Wasson (1983) Glacial age environments of inland Australia. In J. C. Vogel (ed.), *Late Cainozoic Palaeoclimates of the Southern Hemisphere*. SASQUA International Symposium, Swaziland, 29 August–2 September 1983. A. A. Balkema, Rotterdam, pp. 183–208.

Bowler, J. M., R. Jones, H. Allen and A. G. Thorne (1970) Pleistocene human remains from Australia: a living site and human cremation from Lake Mungo, western New South Wales. *World Archaeology* 2: 39–60.

Bowler, J. M., A. G. Thorne and H. Polach (1972) Pleistocene man in Australia: age and significance of the Mungo skeleton. *Nature* 240: 48–50.

Bowler, J. M., G. A. T. Duller, N. Perret, J. R. Prescott and K. Wyroll (1998) Hydrologic changes in monsoonal climates of the last glacial cycle: stratigraphy and luminescence dating of Lake Woods, NT, Australia. *Palaeoclimates* 3(1–3): 179–207.

Bowler, J. M., H. Johnston, J. M. Olley, J. R. Prescott, R. G. Roberts, W. Shawcross and N. A. Spooner (2003) New ages for human occupation and climatic change at Lake Mungo, Australia. *Nature* 421: 837–840.

Bramble, D. M. and D. E. Leiberman (2004) Endurance running and the evolution of *Homo*. *Nature* 432: 345–352.

Brothwell, D. (1960) Upper Pleistocene human skull from Niah Caves, Sarawak. *Sarawak Museum Journal* 9: 323–349.

Brown, P. (1982) Coobool Creek: a prehistoric Australian hominid population. Unpublished Ph.D. Thesis, Australian National University.

(1987) Pleistocene homogeneity and Holocene size reduction: the Australian human skeletal evidence. *Achaeology in Oceania* 22: 41–71.

Brown, P., T. Sutinkna, M. J. Morwood, R. P. Soejono, E. Jatmiko, Wayhu Saptomo and Rokus Awe Due (2004) A new small-bodied hominin from the Late Pleistocene of Flores, Indonesia. *Nature* 431: 1055–1061.

Brown, T., S. K. Pinkerton, and W. Lambert (1979) Thickness of the cranial vault in Australian Aboriginals. *Archaeology and Physical Anthropology in Oceania* 14: 54–71.

Brues, A. M. (1977) *People and Races*. New York: Macmillan.

Brunet, M. F. Guy, D. Pilbeam, H. T. Mackaye, A. Likius, D. Ahounta, A. Beauvilain, C. Blondel, H. Bocherens, J.-R.Boisserie, L. De Bonis, Y. Coppens, J. Dejax, C. Denys, P. Duringer, V. Eisenmann, G. Fanone, P. Fronty, D. Geraads, T. Lehmann, F. Lihoreau, A. Louchart, A. Mahamat, G. Merceron, G. Mouchelin, O. Otero, P. P. Campomanese, M. P. De Leon, J.-C. Rage, M. Sapanet, M. Schuster, J. Sudre, P. Tassy, X. Valentin, P. Vignaud, L. Viriot, A. Zazzo and C. Zollikofer (2002) A new hominid from the Upper Miocene of Chad, Central Africa. *Nature* 418: 145–151.

Buikstra, J. E. and M. Swegle (1998) Bone modification due to burning: experimental evidence. In R. Bonnichsen and M. H. Song (eds), *Bone Modification*. Centre for the Study of the First Americans, Institute of Quaternary Studies University of Maine, Orono, Maine: Peopling of the Americas Publications.

Butlin, N. (1989) The palaeoeconomic history of Aboriginal migration. *Australian Economic History Review* 29(2): 3–57.

(1993) *Economics and the Dreamtime: A Hypothetical History*. Cambridge: Cambridge University Press.

Caddie, D. A., D. S. Hunter, P. J. Pomery and H. J. Hall (1987) The ageing chemist – can electron spin resonance (ESR) help. In W. R. Ambrose and J. M. J. Mummery (eds), *Archaeometry: Further Australasian Studies*. Canberra: Department of Prehistory, Australian National University, pp. 167–176.

Calaby, J. (1976) Some biogeographical factors relevant to the Pleistocene movement of Man in Australia. In R. L. Kirk and A. G. Thorne (eds), *The Origin of the Australians*. Canberra: Australian Institute of Aboriginal Studies, pp. 95–112.

Callan, R. A. and G. C. Nanson (1992) Discussion: formation and age of dunes in the Lake Eyre depocentre. *Geol. Rundsch.* 81(2): 589–593.
Campbell. A. H. (1981) Tooth avulsion in Victorian Aboriginal skulls. *Archaeology in Oceania* 16: 116–118.
Cappieri, M. (1974) *The Andamanese*. Field Research Projects, Miami.
Cavalli-Sforza, L. L., P. Manozzi, and A. Piazza (1993) Demic expansions and human evolution. *Science* 259: 639–646.
Chaloupka, G. (1993) *Journey in Time*. Sydney: Reed.
Chappell, J. and N. J. Shackleton (1986) Oxygen isotopes and sea level. *Nature* 324: 137–139.
Chivas, A. R., A. Garcia, S. van der Kaars, M. J. J. Coupel, S. Holt, J. M. Reeves, D. J. Wheeler, A. D. Switzer, C. V. Murray-Wallace, D. Banerjee, D. M. Price, S. X. Wang, G. Pearson, N. T. Edgar, L. Beaufort, P. De Dekker and C. B. Cecil (2001) Sea level and environmental changes since the last interglacial in the Gulf of Carpentaria, Australia: an overview. *Quaternary International* 83–85: 19–46.
Clark, P. (1987) Report on a plan of management for the Willandra Lakes region, western New South Wales. Unpublished report. New South Wales National Parks and Wildlife Service: Sydney.
Clark, P. and M. Barbetti (1982) Fires, hearths and palaeomagnetism. In W. R. Ambrose and P. Duerden (eds), *Archaeometry: An Australasian Perspective*. Canberra: Department of Prehistory, Australian National University, pp. 144–150.
Coale, A. J. (1974) The history of the human population. In *The Human Population*. Scientific American, San Francisco: W. H. Freeman & Co.
Colls, K. and R. Whitaker (1990) *The Australian Weather Book*. Sydney: Child & Associates.
Collings, H. D. (1938) Pleistocene site in the Malay Peninsula. *Nature* 142: 575–576.
Conroy, G. C., C. J. Jolly, D. Cramer, and J. E. Kalb (1978) Newly discovered fossil hominid skull from the Afar depression, Ethiopia. *Nature* 276: 67–70.
Cosgrove, R. (1989) Thirty thousand years of human colonization in Tasmania: new Pleistocene dates. *Science* 243: 1706–1708.
Curtis, G., C. Swisher and R. Lewin (2002) *Java Man*. London: Abacus.
Damon, A. (1977) *Human Biology and Ecology*. New York: W. W. Norton.
Day, M. H. (1971) Postcranial remains of *Homo erectus* from Bed IV, Olduvai Gorge, Tanzania. *Nature* 232: 383–387.
  (1984) The postcranial remains of *Homo erectus* from Africa, Asia and possibly Europe. *Courier Forschungsinstitut Senckenberg* 69: 113–121.
Dean, C., M. G. Leakey, D. Reid, F. Schreak, G. T. Schwartz, C. Stringer and A. Walker (2001) Growth processes in teeth distinguish modern humans from Homo erectus and earlier hominins. *Nature* 414: 628–631.
Deevey, E. S. Jr (1960) The human population. *Scientific American* 203(g): 195–204.
Delson, E. (1985) Palaeobiology and age of African *Homo erectus*. *Nature* 316: 762–763.
Derevianko, A. P. and V. T. Petrin (1992) Investigations of the layered Paleolithic open-type site of Kara-Bom. *Altaica* 1: 19–22.
Derevianko, A. P. and M. V. Shunkov (1992) Archaeological investigations in the valley of the Anui River. *Altaica* 1: 8–12.

Derevianko, A. P. and A. N. Zenin (1992) Archaeological study of the Strashnaya Cave. *Altaica* 1: 13–18.

Derevianko, A. P., M. V. Shunkov, D. T. Nash and L. Heong-Jong (1993) Paleolithic study at Denisova Cave. *Altaica* 2: 6–10.

Derevianko, A. P., J. W. Olsen, D. Tseveendorj, A. I. Krivoshapkin, V. T. Petrin, and P. J. Brantingham (2000) The stratified cave site of Tsagaan Agui in the Gobi Altai (Mongolia). *Archaeology Ethnology and Anthropology of Eurasia* 1: 23–36.

DeVogel, S. B., J. W. Magee, W. F. Manley and G. H. Miller (2004) A Gis-based reconstruction of late Quaternary paleohydrology: Lake Eyre, arid central Australia. *Palaeogeography, Palaeoeclimatology, Palaeoecology* 204: 1–13.

Diamond, J. M. (1991) Why are pygmies small? *Nature* 354: 111–112.

Dillehay, T. (1989) *Monte Verde: A Late Pleistocene Settlement in Chile, Volume 1: Palaeoenvironment and Site Context.* Washington DC: Smithsonian Institution Press.

(1997) *Monte Verde: A Late Pleistocene Settlement in Chile, Volume 2: The Archaeological Context and Interpretation* Washington DC: Smithsonian Institution Press.

Dongen, R. van (1963) The shoulder girdle and humerus of the Australian Aborigine. *American Journal of Physical Anthropology* 21: 469–488.

Dortch, C. E. and B. G. Muir (1980) Long range sightings of bushfires as possible incentive for Pleistocene voyagers to Greater Australia. *Western Australian Naturalist* 14: 194–198.

Dowling, P., G. Hamm, J. Klaver, J. Littleton, N. Sanderson and S. G. Webb (1985) Middle Willandra Creek archaeological site survey. Unpublished report to the New South Wales National Parks and Wildlife Service, Department of Prehistory and Anthropology, Australian National University, Canberra.

Elkin, A. P. (1938) Anthropological research in Australia, 1927–37, *Oceania* 8: 306–327.

English, P., N. A. Spooner, J. Chappell, D. G. Questiaux and N. G. Hill (2001) Lake Lewis basin, central Australia: environmental evolution and OSL chronology. *Quaternary International* 83–85: 81–101.

EPICA (2004) Eight glacial cycles from an Antarctic ice core. *Nature* 429: 623–628.

Finch, M. E. (1982) The discovery and interpretation of *Thylacoleo carnifex* (Thylacoleonidae, Marsupialia). In M. Archer (ed.), *Carnivorous Marsupials*. The Royal Zoological Society of New South Wales, pp. 537–551.

Flannery, T. F. (1990a) Who killed Kirlilpi? *Australian Natural History* 23: 234–241.

(1990b) Pleistocene faunal loss: implications of the aftershock for Australia's past and future. *Archaeology in Oceania* 25: 45–67.

(1994) *The Future Eaters*. Sydney: Reed.

Flood, J. (1999) *Archaeology of the Dreamtime*. revised edition, Sydney: Angus & Robertson.

Flower, W. H. (1880) On the osteology and affinities of the natives of the Andaman Islands. *Journal of the Anthropological Institute* 9: 108–133.

(1885) Additional observations on the osteology of the natives of the Andaman Islands. *Journal of the Anthropological Institute* 14: 115–121.

Francis, P. (1993) *Volcanoes: A Planetary Perspective*. London: Clarendon Press.

Gamble, C. (1994) *Time Walkers*. Cambridge, MA: Harvard University Press.
Gabunia, L. von, O. Joris, A. Justus, D. Lordkipanidze, A. Muschelisvili, M. Nioradze, C. Swisher III, A. Vekua, G. von Bosinski, R. Ferring, G. Majsuradze and M. Tvalcrelidze (1999) Neue Hominidenfunde Des Altpalaolithischen Fundplatzes Dmanisi (Georgien, Kaukasus) Im Kontext Aktueller Grabungsergebnisse. *Archaologisches Korrespondenzblatt* 29: 451–488.
Gibson, D. F. (1986) The Tanami Desert: research on Aboriginal land. *Australian Natural History* 21: 544–546.
Gillespie, R. (2002) Dating the first Australians. *Radiocarbon* 2: 455–472.
Gillespie, R. and R. G. Roberts (1999) On the reliability of age estimates for human remains at Lake Mungo. *Journal of Human Evolution* 38: 727–732.
Gillespie, R., D. R. Horton, P. Ladd, P. G. Macumber, T. H. Rich, R. Thorne and R. V. S. Wright (1978) Lancefield Swamp and the extinction of the Australian Megafauna. *Science* 200: 1044–1048.
Gregory, J. W. (1906) *The Dead Heart of Australia*. London: John Murray.
Groube, L. J., J. Chappell, J. Muke and D. Price (1986) A 40000-year-old human occupation site at Huon Peninsula, Papua New Guinea, *Nature* 324: 453–453.
Grun, R., N. A. Spooner, A. Thorne, G. Mortimer, J. J. Simpson, M. Mcculloch, L. Taylor and D. Curnoe (1999) Age of the Lake Mungo 3 skeleton, reply to Bowler and Magee and to Gillespie and Roberts. *Journal of Human Evolution* 38: 733–741.
Gunn, J. (1994) Global climatic and regional biocultural diversity. In C. Crumley (ed.), *Historical Ecology*. Santa Fe, New Mexico: School of American Research Press.
Harrison, G. A., J. S. Weiner, J. M. Tanner and N. A. Barnicot (1977) *Human Biology*. Oxford: Oxford University Press.
Harrisson, B. (1975) Tampan: Malaysia's Palaeolithic reconsidered. *Modern Quaternary Research in Southeast Asia* 1: 53–70.
Hassan, F. A. (1981) *Demographic Archaeology*. New York: Academic Press.
Hawks, J., S. Oh, K. Hunley, S. Dobson, G. Cabana, P. Dayalu and M. H. Wolpoff (2000) An Australian test of the recent African origin theory using the WLH50 calvarium. *Journal of Human Evolution* 39: 1–22.
Hecht, M. K. (1975) The morphology and relationships of the largest known terrestrial lizard, *Megalania prisca Owen*, from the Pleistocene of Australia. *Proceedings, Royal Society of Victoria* 87: 239–250.
Hope, G., A. P. Kershaw, S. van der Kaars, S. Xiangjun, L. Ping-Mei, L. E. Heusser, H. Takahara, M. McGlone, N. Miyoshi and P. T. Moss (2004) History of vegetation and habitat change in the Australia-Asian region. *Quaternary International* 118–119: 103–126.
Hope, J. (1978) Pleistocene mammal extinctions: the problem of Mungo and Menindee. *Alcheringa* 2: 65–82.
Horton, D. R. (1979) The great megafaunal extinction debate: 1879–1979. *Artefact* 4: 11–25.
  (1980) A review of the extinction question: man, climate and megafauna. *Archaeology and Physical Anthropology in Oceania* 15: 86–97.
  (1981) Water and woodland: the peopling of Australia. *Australian Institute of Aboriginal Studies, Newsletter* 16: 21–27.

Horton, D. R. and R. V. S. Wright (1981) Cuts on Lancefield bones: carnivorous *Thylacoleo*, not humans, the cause. *Archeaology in Oceania* 16: 73–80.
Howells, W. W. (1970) Anthropometric grouping analysis of Pacific peoples. *Archaeology and Physical Anthropology in Oceania* 5: 192–217.
  (1973) *The Pacific Islanders*. London: Weidenfeld & Nicolson.
Hublin, J. J. (1985) Human fossils from the North African middle Pleistocene and the origin of *Homo sapiens*. In Delson E. Liss (ed), *Ancestors: The Hard Evidence*, New York, pp. 283–288.
Ivanhoe, F. (1979) Direct correlation of human skull vault thickness with geomagnetic intensity in some Northern Hemisphere populations. *Journal of Human Evolution* 8: 433–444.
Jablonski, N. G. and G. Chaplin (2000) The evolution of human skin coloration. *Journal of Human Evolution* 39: 57–106.
Jacob, T. (1966) The sixth skull cap of *Pithecanthropus erectus*. *American Journal of Physical Anthropology* 25: 243–259.
Jacob, T. K. (ed.) (1991) *In the Beginning: A Perspective on Traditional Aboriginal Societies*. Western Australia: Ministry of Education.
Johnson, A. W. and T. Earle (1987) *The Evolution of Human Societies*. Stanford, CA: Stanford University Press.
Johnson, B. J., G. H. Miller, M. L. Fogel, J. W Magee, M. K. Gagan and A. R. Chivas (1999) 65 000 years of vegetation change in Central Australia and the Australian summer monsoon. *Science* 284: 1150–1152.
Johnston, H. and P. Clark (1998) Willandra Lakes archaeological investigations 1968–98. *Archaeology in Oceania* 33(3): 105–119.
Jones, R. M. (1973) Emerging picture of Pleistocene Australians. *Nature* 246: 275–281.
  (1979) The fifth continent: problems concerning the human colonization of Australia. *Annual Reviews in Anthropology* 8: 445–466.
  (1987) Pleistocene life in the dead heart of Australia. *Nature* 328: 666.
Jurmain, R., H. Nelson and W. A. Turnbaugh (1990) *Understanding Physical Anthropology and Archeology*. 4th edition, New York: West Publishing Co.
Keesing, F. M. (1950) Some notes on early migrations in the southwest Pacific area. southwestern. *Journal of Anthropology* 6: 101–119.
Kennedy, K. A. R. and S. U. Deraniyagala (1989) Fossil remains of 28 000-year-old hominids from Sri Lanka. *Current Anthropology* 30: 394–398.
Kershaw, A. P. (1985) An extended late Quaternary vegetation record from northeastern Queensland and its implications for the seasonal tropics of Australia. *Proceedings of the Ecological Society of Australia* 13: 179–189.
  (1986) The last two glacial-interglacial cycles from northern Australia: implications for climate change and Aboriginal burning. *Nature* 322: 47–49.
Kershaw, A. P. and C. Whitlock (2000) Palaeoecological records of the last glacial-interglacial cycle: patterns and causes of change. *Palaeogeography, Palaeoclimatology, Palaeoecology* 155: 1–5.
Kershaw, A. P., J. G. Baird, D. M. D'Costa, P. A. Edney, J. A. Peterson and K. M. Strickland (1991) A comparison of long pollen records from the Atherton and Western Plains volcanic provinces, Australia. In M. A. J. Williams, P. DeDeckker and A. P. Kershaw (eds), *The Cainozoic in Australia: A*

Re-appraisal of the Evidence. Geological Society of Australia, Special Publication No. 18, pp. 288–301.

Kershaw, A. P., G. M. McKenzie and A. McMinn (1993) A quaternary vegetation history of northeastern Queensland from pollen analysis of ODP site 820. *Proceedings Ocean Drilling Program, Scientific Results.* 133: 107–114.

Kershaw, P., P. Moss and S. van der Kaars (2003) Causes and consequences of long-term climatic variability on the Australian continent. *Freshwater Biology* 48: 1274–1283.

Kimber, R. (1983) Black lightning: Aborigines and fire in Central Australia and the Western Desert. *Archaeology in Oceania* 18: 38–45.

Lambeck, K. and J. Chappell (2001) Sea level change through the last Glacial Cycle. *Science* 292: 679–686.

Lampert, R., S. G. Webb, J. Magee, J. Luly, J. Ash, D. Fergie, P. Clarke and J. Klaver. (1989) The Lake Eyre Basin project: Archaeology and Palaeoenvironments. Unpublished report, ANU, Canberra.

Larick, R., R. L. Ciochon, Y. Zaim, S. Sudijono, Y. Rizal, F. Aziz, M. Reagan and M. Heizler (2001) Early Pleistocene ^{40}Ar/ ^{39}Ar ages for Bapang Formation hominins, Central Java, Indonesia. *Proceedings of the National Academy of Science.* 98(9): 4866–4871.

Larnach, S. L. and N. W. G. Macintosh (1966) The craniology of the Aborigines of Coastal New South Wales. *Oceania Monographs* No.13, University of Sydney: Sydney.

(1970) The craniology of the aborigines of Queensland. *Oceania Monographs* No.15, University of Sydney: Sydney.

Latz, P. (1995) *Bushfires and Bushtucker.* Alice Springs: IAD Press.

Leigh, S. R. (1992) Cranial capacity evolution in *Homo erectus* and early *Homo sapiens. American Journal of Physical Anthropology* 87: 1–13.

Lindsay, H. A. (1954) Australia's first human population. *Walkabout*, 1 January, pp. 34–35.

Lumley, H. de and A. Sonakia (1985a) Context Stratigraphique et Archeologique de L'Homme de la Narmada, Hathnora, Madhya Pradesh, Inde. *L'Anthropologie* (Paris) 89(1): 3–12.

Lumley, M.-A. de and A. Sonakia (1985b) Premier Decouverte D'un *Homo erectus* sur le Continent Indien a Hathnora, dans la Moyenne Vallee de la Narmada. *L'Anthropologie* (Paris) 89: 13–61.

Lunney, D. (1990) Size of animal populations. In H. F. Recher, D. Lunney and I. Dunn (eds), *A Natural Legacy: Ecology in Australia.* Second edition, Oxford: Pergamon Press, pp. 189–217.

MacIntosh, N. W. G. and S. Larnach (1976) Aboriginal affinities looked at in world context. In R. L. Kirk and A. G. Thorne (eds), *The Origin of the Australians*, Canberra, pp. 113–126.

Magee, J. W. (1997) Late quaternary environments and palaeohydrology of Lake Eyre, arid central Australia. Ph. D. Thesis, Australian National University.

Magee, J. W. and G. H. Miller (1998) Lake Eyre palaeohydrology from 60ka to the present: beach ridges and glacial maximum aridity. *Paleogeography, Palaeoclimatology, Palaeoecology* 144: 307–329.

Malthus, T. (1798) *Essay on the Principles of Population*. London.
Manzi, G., F. Mallegni and A. Ascenzi (2001) A cranium for the earliest Europeans: Phylogentic position of the hominid from Ceprano, Italy. *Proceedings of the National Acasdemy of Science* 98(17): 10011–10016.
Maringer, J. and Th. Verhoeven (1970a) Die oberflachenfunde aus dem Fossilgebeit von Mengeruda und Olabula auf Flores. *Indonesien: Anthropos* 65: 530–546.
Maringer, J. and Th. Verhoeven (1970b) Mengeruda auf Flores. *Indonesien: Anthropos* 65: 229–247.
Martin, R. (1928) *Lehrbuch der Anthropologie*, Germany: Jena.
McEvedy, C. and R. Jones (1978) *Atlas of World Population History*. Harmondsworth: Penguin.
Meltzer, D. J. (2002) What do you do when no one's been there before? Thoughts on the exploration and colonization of new lands. *Memoirs of the California Academy of Sciences* 27: 25–56.
  (2003) Lessons in landscape learning. In M. Rockman and J. Steele (eds), *Colonization of Unfamiliar Landscapes: The Archaeology of Adaptation*. London: Routledge.
  (2004) Peopling of North America. *Developments in Quaternary Science* 1: 539–563.
Miller, G. H., J. W. Magee, B. J. Johnson, M. L. Fogel, N. A. Spooner, M. T. McCulloch and L. K. Ayliffe (1999) Pleistocene extinction of *Genyornis Newtoni*: human impact on Australian Megafauna. *Science* 283: 205–208.
Molnar, R. E. (1990) New cranial elements of a giant varanid from Queensland. *Memoirs of the Queensland Museum* 29: 437–444.
Molnar, S. (1985) *Human Variation*. 2nd edition, New Jersey: Prentice-Hall.
Morant, G. M. (1938) The form of the Swanscombe skull. *Journal of the Royal Anthropological Institute* 68: 67–96.
Morton, S. R. (1990) Land of uncertainty: the Australian arid zone. In H. F, Recher, D. Lunney and I. Dunn (eds), *A Natural Legacy: Ecology in Australia*. 2nd edition, Oxford: Pergamon Press, pp. 122–144.
Morwood, M. (2002) *Visions from the Past*. Sydney: Allen & Unwin.
Morwood, M., R. P. Soejono, R. G. Roberts, T. Sutinka, C. S. M. Turney, K. E. Westaway, W. J. Rink, J.-X. Zhao, G. D. Van Den Bergh, Rokus Awe Due, D. R. Hobbs, M. W. Moore, M. I. Bird and L. K. Fifield (2004) Archaeology and age of a new hominin from Flores in eastern Indonesia. *Nature* 431: 1087–1091.
Morwood, M. J., P. B. O'Sullivan, F. Aziz and A. Raza (1998) Fission-track ages of stone tools and fossils on the east Indonesian island of Flores. *Nature* 392: 173–176.
Morwood, M. J., F. Aziz, P. B. O'Sullivan, Nasruddin, D. R. Hobbs P. B. and A. Raza (1999) Archaeological and palaeotological reseach in central Flores, east Indonesia: results of fieldwork, 1997–98. *Antiquity* 73: 273–286.
Moss, P. and A. P. Kershaw (1999) Evidence from marine ODP site 820 of fire/vegetation/climate patterns in the humid tropics of Australia over the last 250 000 years. Proceedings, Australian Bushfire Conference, Albury, Australia.
  (2000) The last glacial cycle from the humid tropics of northeastern Australia: comparison of a terrestrial and marine record. *Palaeogeography, Palaeoclimatology, Palaeoecology* 155: 155–176.

Mountford, C. P. (1979) *Nomads of the Australian Desert.* Sydney: Risby.
Mulvaney, D. J. (1961) The Stone Age of Australia. *Proceedings of the Prehistoric Society* 27: 56–107.
  (1973) Summary report on first Mungo project season: 17 August–1 September. *Australian Institute of Aboriginal Studies Newsletter* 1: 21–22.
  (1975) *The Prehistory of Australia.* Penguin: Ringwood (Vic.)
Mulvaney, D. J. and J. Kamminga (1999) *Prehistory of Australia.* Sydney: Allen & Unwin.
Murray, P. (1984) Extinctions downunder: a bestiary of extinct Australian monotremes and marsupials. In P. S. Martin and R. G. Klein (eds), *Quaternary Extinctions: a Prehistoric Revolution.* Tucson: University of Arizona Press, pp. 600–628.
Nag, M. (1962) Factors affecting fertility in non-industrial societies: a cross-cultural study. *Yale University Publications in Anthropology* No. 66. New Haven, Connecticut: Human Relations Area Files Press.
Nanson, G. C., R. W. Young, D. M. Price and B. R. Rust (1988) Stratigraphy, sedimentology and late Quaternary chronology of the Channel Country of southwest Queensland. In R. F. Warner (ed.), *Fluvial Geomorphology of Australia.* Sydney: Academic Press, pp. 151–175.
Nanson, G. C., D. M. Price and S. A. Short (1992) Wetting and drying of Australia over the past 300ka. *Geology* 20: 791–794.
Nanson, G. C., R. A. Callen and D. M. Price (1998) Hydroclimatic interpretation of Quaternary shorelines on South Australian Playas. *Palaeogeography, Palaeoclimatology, Palaeoecology* 144: 281–305.
Ninkovich, D., R. S. J. Sparks and M. T. Ledbetter (1978) The exceptional magnitude and intensity of the Toba Eruption, Sumatra: an example of the use of deep sea Tephra Layers as a Geological Tool. *Bulletin of Volcanology* 41–3: 286–298.
Nobbs, M. and R. I. Dorn (1988) Age determination for rock varnish formation within petroglyphs: catio–catio dating of 24 motifs from the Olary region, South Australia. *Rock Art Research* 5: 108–146.
Oakley, K. P. (1969) Analytical methods of dating bones. In *Science in Archaeology.* D. Brothwell, E. Higgs and G. Clark (eds), London Thames & Hudson.
O'Connell, J. F. and J. Allen (2004) Dating the colonization of Sahul (Pleistocene Australia-New Guinea): a review of recent research. *Journal of Archaeological Science* 31: 835–853.
O'Connor, S. (1995) Carpenter's Gap rokshelter 1: 40 000 years of Aboriginal occupation in the Napier Ranges, Kimberley, WA. *Australian Archaeology* 40: 58–59.
Omoto, K. (1985) The Negritos: genetic origins and microevolution. In R. Kirk and E. Szathmary, *Out of Asia: Peopling the Americas and the Pacific* Published by the *Journal of Pacific History*, Canberra: Australian National University Press.
Ortner, D. J. (1968) Description and classification of degenerative bone changes in the distal joint surfaces of the humerus. *American Journal of Physical Anthropology* 28: 139–156.
Pavlov, P., J. I. Svendsen and S. Indrelld (2001) Human presence in the European Arctic nearly 40 000 years ago. *Nature* 413: 64–67.

Pearce, R. H. and M. Barbetti (1981) A 38 000 year old archaeological site at Upper Swan, Western Australia. *Archaeology in Oceania* 16: 173–179.
Perrens, S. (1982) Australia's water resources. In W. Hanley and M. Cooper (eds), *Man and the Australian Environment* Sydney: McGraw-Hill, pp. 24–36.
Pfeiffer, J. E. (1977) *The Emergence of Society*. New York: McGraw-Hill.
Pienaar, U. de, (1969) Predator-prey relations amongst the larger mammals of the Kruger National Park. *Koedoe* 12: 108–176.
Quammen, D. (1996) *The Song of the Dodo*. London: Random House.
Rampino, M. R. S. Self and R. B. Stothers (1988) Volcanic winters. *Annual Review of Earth Planetary Science* 16: 73–99.
Renfrew, C. (1987) *Archaeology and Language*. London: Jonathan Cape.
Reuther, J. G. (n.d.) The Reuther manuscript and map, translated by P. Scherer, Tanunda, SA 1975. Microfiche edition, AIAIS.
Revelle, R. and W. W. Howells (1977) Changing numbers of mankind. In A. W. Damon, *Human Biology and Ecology*. New York: Norton & Co., pp. 304–325.
Rich, P. V. (1985) *Genyornis newtoni*: A Mihirung. In Rich, P. V. and G. F. van Tets (eds), *Kadimakara: Extinct Vertebrates of Australia*. Victoria: Pioneer Design Studio, pp. 188–194.
Rich, P. V., G. F. van Tets and F. Knight (1985) *Kadimakara: Extinct Vertebrates of Australia*. Victoria: Pioneer Design Studio.
Rightmire, G. P. (1990) *The Evolution of Homo Erectus*. Cambridge: Cambridge University Press.
Roberts, R. G., R. Jones and M. A. Smith (1990) Thermoluminescence dating of a 50 000-year-old human occupation site in northern Australia. *Nature* 345: 153–156.
Roberts, R. G., R. Jones, N. A. Spooner,, M. J. Head, A. S. Murray, and M. A. Smith (1994) The human colonisation of Australia: optical dates of 53 000 and 60 000 years bracket human arrival at Deaf Adder Gorge, Northern Territory. *Quaternary Science Reviews* 13: 575–583.
Roberts, R. G., H. Yoshida, R. Galbraith, G. Laslett, R. Jones and M. Smith (1998) Single-aliquote and single-grain optical dating confirm thermoluminescence age estimates at Malakunanja II rock shelter in northern Australia. *Ancient TL* 16: 19–24.
Roberts, R. G., T. F. Flannery, L. K. Ayliffe, H. Yoshida, J. M. Olley, G. J. Prideaux, G. M. Laslett, A. Baynes, M. A. Smith, R. Jones and B. L. Smith (2001) New ages for the last Australian megafauna: continent-wide extinction about 46 000 years ago. *Science* 292: 1888–1892.
Roebroeks, W., N. C. Conard and T. van Kolfschoten (1992) Dense forests, cold Steppes, and the palaeolithic settlement of Northern Europe. *Current Anthropology* 33: 551–586.
Santa Luca, A. P. (1980) *The Ngandong Fossil Hominids*. Pub. Department of Anthropology, Yale University.
Schaller, G. (1974) *The Serengeti Lion*. Chicago: University of Chicago Press.
Schebesta, P. (1952) *Die Negrito Asiens*. Vienna: St Gabriel Verlag.
Sekiya, A. (2000) History of Palaeolithic studies in Japan: for the 50th anniversary of Iwajuku excavations. *Archaeology Ethnology and Anthropology of Eurasia* 3(3): 13–17.

Shackleton, N. (1987) Oxygen isotopes, ice volume and sea level. *Quaternary Science Reviews* 6: 183–190.
Shaffer, M. L. (1981) Minimum population sizes for species conservation. *BioScience* 31(2):
  (1987) Minimum viable populations: coping with uncertainty. In M. E. Soulé (ed.), *Viable Populations for Conservation* (The Blue Book). London: Cambridge University Press.
  (1990) Population viability analysis. *Conservation Biology* 4(1): 22–30.
Shawcross, F. W. (1975) Thirty thousand years and more. *Hemisphere* 19: 26–31.
Shawcross, F. W. and M. Kay (1980) Australian archaeology: implications of current interdisciplinary research. *Interdisciplinary Science Reviews* 5(2): 112–128.
Shipman, P. (1992) Human ancestor's early steps out of Africa. *New Scientist* 133 (1806): 16.
Shipman, P., G. Foster and M. Schoeninger (1984) Burnt-bones and teeth: an experimental study of colour, morphology, crystal structure and shrinkage. *Journal of Archaeologicl Science* 11: 307–325.
Simmons, T. and F. H. Smith (1991) Human population relationships in the late pleistocene. *Current Anthropology* 32: 623–627.
Simpson, J. J. and R. Grun (1998) Non-destructive gamma spectrometric U-series dating. *Quaternary Geochronology* 17: 1009–1022.
Singh, G. E. and A. Geissler (1985) Late Cainozoic vegetation of vegetation, fire, lake levels and climate, at Lake George, New South Wales, Australia. *Philosophical Transactions of the Royal Society of London* Series B 311: 379–447.
Slome, D. (1929) The osteology of a Bushman tribe. *Annals of the South African Museum* 24: 33–87.
Smith, H. D. (1912) A study of pygmy crania, based on skulls found in Egypt. *Biometrika* 8: 262–66.
Smith, M. A. and N. Sharp (1993) Pleistocene sites in Australia, New Guinea and island Melanesia: geographic and temporal structure of the archaeological record. In M. A. Smith, M. Spriggs and B. Fankhauser (eds), *Sahul in Review: Pleistocene Archaeology in Australia, New Guinea and Island Melanesia*. Occasional Papers in Prehistory 24, Department of Prehistory, Research School of Pacific Studies, Australian National University, Canberra, pp. 37–59.
Sobbe, I. H. (1990) Devils on the Darling Downs – the tooth mark record. *Memoirs of the Queensland Museum* 28: 299–322.
Sonakia, A. (1985a) Skull cap of an early man from the Narmada Valley Alluvium (Pleistocene) of Central India. *American Anthropologist* 87: 612–616.
  (1985b) Early *Homo* from Narmada Valley. In E. Delson (ed.), *Ancestors: The Hard Evidence*. New York: Alan R. Liss, pp. 334–338.
Spencer, B. and F. G. S. Walcott (1911) The origin of cuts on bones of australian extinct marsupials. *Proceedings of the Royal Society of Victoria* 24: 92–123.
Steegman, A. T. (1977) Human adaptation to cold. In A. Damon (ed.), *Physiological Anthropology*. Second Printing, Oxford: Oxford University Press, pp. 130–166.
Stirling, E. C. (1896) The newly discovered extinct, gigantic bird of South Australia. *Ibis* 7(11): 593.

(1900) Physical features of Lake Callabonna. *Memoirs, Royal Society of South Australia* 1: i–xv.

(1913) Fossil remains of Lake Callabonna. Part IV. 1. Description of some further remains of *Genyornis newtoni*, Stirling and Zeitz. 2. On the identity of *Phascolomys* (*Phascolonus*) *gigas*, Owen, and *Sceparnodon ramsayi*, Owen, with a description of some remains. *Memoirs, Royal Society of South Australia* 1: 111–178.

Stirling, E. C. and A. H. C. Zeitz (1896) Preliminary notes on *Genyornis newtoni*; a new genus and species of fossil struthious bird found at Lake Callabonna, South Australia. *Transactions, Royal Society of South Australia* 20: 171–190.

(1899) Fossil remains of Lake Callabonna. Part 1: Description of the manus and pes of *Diprotodon australis*, Owen. *Memoirs, Royal Society of South Australia* 1: 1–40.

(1900) Fossil remains of Lake Callabonna. 1. *Genyornis newtoni*. A new genus and species of fossil struthious bird. *Memoirs, Royal Society of South Australia* 1(2): 41–80.

Stirton, R. A. (1967) The Diprotodontidae from the Ngapakaldi fauna, South Australia. *Bureau Mineral Research, Bulletin* 85: 1–44.

Stirton, R. A., R. H. Tedford and A. H. Miller (1961) Cenozoic stratigraphy and vertebrate palaeontology of the Tirari Desert, South Australia. *Records of the South Australian Museum* 14: 19–61.

Stirton, R. A., R. H. Tedford and M. O. Woodburne (1967) A new Tertiary formation and fauna from the Tirari Desert, South Australia. *Records of the South Australian Museum* 15: 427–462.

Stirton, R. A., M. O. Woodburne and M. Plane (1967) A phylogeny of the Tertiary Diprotodontidae and its significance in correlation. *Bureau of Mineral Resources, Bulletin* 85: 149–160.

Strehlow, C. (1907) *Die Aranda – und Loritja Stamme in Zentral Australien*. Frankfurt am Main.

Stringer, C. and P. Andrews (1988) Genetic and fossil evidence for the origin of modern humans. *Science* 239: 1263–1268.

Stringer, C. and C. Gamble (1993) *In Search of the Neanderthals*. London: Thames & Hudson.

Swisher III, C. C., G. H. Curtis, T. Jacob, A. G. Gatty and A. Suprijo Widiasmoro (1994) Age of the earliest known Hominids in Java, Indonesia. *Science* 263: 1118–1121.

Tchernov (1987) The age of the 'Ubeidya Formation, an Early Pleistocene Hominid Site in the Jordan Valley, Israel. *Israel Journal of Earth Sciences* 36: 3–30.

Tedford, R. H. and R. T. Wells (1990) Pleistocene deposits and fossil vertebrates from the dead heart of Australia. *Memoirs of the Queensland Museum* 28: 263–284.

Tedford, R. H., R. T. Wells and F. Barghoorn (1992) Tirari Formation and contained faunas, Pliocene of the Lake Eyre basin, South Australia. *The Beagle, Records of the Northern Territory Museum of Arts and Sciences* 9: 173–194.

Theil, B. (1987) Early settlement of the Philippines, Eastern Indonesia, and Australia-New Guinea: a new hypothesis. *Current Anthropology* 28: 236–241.

Thorley, P. B. (1998) Pleistocene settlement in the Australian arid zone: occupation of an inland riverine landscape in the central Australian ranges. *Antiquity* 72: 34–45.

Thorne, A. G. (1975) Kow Swamp and Lake Mungo: towards an osteology of early man in Australia. Unpublished Ph.D. Thesis, Australian National University.

(1976) Morphological contrasts in Pleistocene Australians. In R. L. Kirk and A. G. Thorne (eds), *The Origin of the Australians*, Australian Institute of Aboriginal Studies, Canberra, pp. 95–112.

Thorne, A. (1977) Separation or reconciliation? Biological clues to the development of Australian society. In Allen, J., J. Golson and R. Jones (eds), *Sunda and Sahul*, New York: Academic Press. pp. 187–204.

Thorne, A. G. and M. H. Wolpoff (1981) Regional continuity in Australian Pleistocene hominid evolution. *American Journal of Physical Anthropology* 55: 337–349.

Thorne, A., R. Grun, G. Mortimer, N. A. Spooner, J. J. Simpson, M. McCulloch, L. Taylor and D. Curnoe (1999) Australia's oldest human remains: age of Lake Mungo 3 skeleton. *Journal of Human Evolution* 36: 591–612.

Tindale, N. B. (1974) *Aboriginal Tribes of Australia*. 2 vols, Canberra: Australian National University Press.

Torgersen, T., M. F. Hutchinson, D. E. Searle and H. A. Nix (1983) General Bathymetry of the Gulf of Carpentaria and the Quaternary Physiogeography of Lake Carpentaria. *Palaeogeography, Palaeoclimatology, Palaeoecology* 41: 207–225.

Torgersen, T., J. Luly, P. De Dekker, M. R. Jones, D. E. Searle, A. R. Chivas and W. J. Ullman (1988) Late Quaternary environments of the Carpentaria Basin, Australia. *Palaeogeography, Palaeoclimatology, Palaeoecology* 67: 245–261.

Trinkaus, E. (1984) Western Asia. In F. H. Smith and F. Spencer (eds), *The Origins of Modern Humans*, New York: Alan R. Liss, pp. 251–293.

Turney, C. S. M., A. P. Kershaw, P. Moss, L. K. Fifield, R. G. Cresswell, G. M. Santos, M. L. Di Tada, P. A. Hausladen and Y. Zhou (2001a) Redating the onset of burning at Lynch's Crater (North Queensland): implications for human settlement in Australia. *Journal of Quaternary Science* 16(8): 767–771.

Turney, C. S. M., M. I. Bird, L. K. Fifield, R. G. Roberts, M. A. Smith, C. E. Dortch, R. Grün, E. Lawson, L. K. Ayliffe, G. H. Miller, J. Dortch and R. G. Cresswell (2001b) Early human occupation at Devil's Lair, southwestern Australia 50 000 years ago. *Quaternary Research* 55: 3–13.

Valde-Nowak, P., A. Nadachowski and M. Wolsan (1987) Upper Palaeolithic boomerang made of a mammoth tusk in south Poland. *Nature* 329: 436–438.

Vignaud, P., P. Duringer, H. T. Mackaye, A. Likius, C. Blondel, J.-R. Boisserie, L. De Bonis, V. Eisenmann, M.-E. Etienne, D. Geraads, F. Guy, T. Lehmann, F. Lihoreau, N. Lopez-Martinez, C. Mouer-Chauvire, O. Otero, J.-C. Rage, M. Schuster, L. Viriot, A. Zazzo and M. Brunet (2002) Geology and palaeontology of the Upper Miocene Toros-Menalla hominid locality. *Nature* 418: 152–155.

Voris, H. K. (2000) Maps of Pleistocene sea levels in Southeast Asia: shorelines, river systems and time durations. *Journal of Biogeography* 27: 1153–1167.

Walker, A. and R. Leakey (1993) *The Nariokotome Homo Erectus Skeleton*. Cambridge, MA: Harvard University Press.

Walker, D. and A. de G. Sieveking (1962) The Palaeolithic industry of Kota Tampan, Perak, Malaya. *Proceedings of the Prehistoric Society* 28: 103–139.
Walsh, G. L., and M. Morwood (1999) Spear and spear thrower evolution in the Kimberley region, NW Australia: evidence from rock art. *Archaeology in Oceania* 34: 45–58.
Walsh, G. L. and M. J. Morwood (1999) Spear and spearthrower evolution in the Kimberley region, N. W. Australia: evidence from rock art. *Archaeology in Oceania* 34: 45–58.
Wang, X., S. van der Kaars, P. Kershaw, M. Bird. and F. Jansen (1999) A record of fire, vegetation and climate through the last three glacial cycles from Lombok Ridge core G6–4, eastern Indian Ocean, Indonesia. *Palaeogeography, Palaeoclimatology, Palaeoecology* 147: 241–256.
Warner, R. F. (1986) Hydrology. In D. N. Jeans (ed.), *Australia: A Geography, Vol. 1, The Natural Environment*. Sydney: University of Sydney Press, pp. 49–79.
Webb, S., M. L. Cupper and R. Robins (In press) Pleistocene human footprints from the Willandra Lakes, southeastern Australia, *Journal of Human Evolution* 50.
Webb, G. and C. Manolis (1989) *Crocodiles of Australia*. Sydney: Reed Books.
Webb, S. G. (1987) A palaeodemographic model of late Holocene Central Murray Aboriginal Society, Australia. *Human Evolution* 2(5): 385–406.
  (1989) *The Willandra Lakes Hominids*. Canberra: Panther Press.
  (1995) *Palaeopathology of Australian Aborigines*. Cambridge: Cambridge University Press.
Webb, S. G., M. Cupper, H. Johnson, and R. Robbins (2005) Fossil human footprints from the Willandra Lakes, Southeastern Australia, *J. Hum. Evol.* 50.
Weidenreich, F. (1943) The skull of *Sinanthropus pekinensis*; a comparative study on a primitive hominid skull. *Palaeontologia Sinica*, New Series No. 10. Lancaster, PA: Lancaster Press.
  (1951) Morphology of solo man. *Anthropological Papers of the American Museum of Natural History, New York* 43(3): 205–290.
Wells, R. T. (1985) *Thylacoleo carnifex* A Marsupial Lion. In P. V. Rich, G. F. van Tets and F. Knight (eds), *Kadimakara: Extinct Vertebrates of Australia*. Australia: Pioneer design Studio, pp. 225–229.
Wells, R. T. and R. A. Callan (eds) (1986) The Lake Eyre Basin – Cainozoic sediments, fossil vertebrates and plants, landforms, silcretes and climatic implications. *Australasian Sedimentologists Group Field Guide Series No. 4*. Sydney: Geological Society of Australia.
White, J. P. and J. F. O'Connell (1982) *A Prehistory of Australia, New Guinea and Sahul*. New York: Academic Press.
Williams, M. A. J., D. L. Dunkerley, P. De Dekker, A. P. Kershaw and T. Stokes (1993) *Quaternary Environments*. London: Edward Arnold.
Wolpoff, M. H. (1980) *Paleoanthropology*. New York: Alfred A. Knopf.
  (1999) *Paleoanthropology*. 2nd edition, Boston: McGraw Hill.
Wolpoff, M. H., W. X. Zhi and A. G. Thorne (1984) Modern *Homo sapiens* origins: a general theory of hominid evolution involving the fossil evidence from East Asia. In F. H. Smith and F. Spencer (eds), *The Origins of Modern*

Humans: A World Survey of the Fossil Evidence. New York: Alan R. Liss, pp. 411–483.
Wolpoff, M. H., A. G. Thorne, J. Jelinek and Z. Yinyun (1994) The case for sinking *Homo erectus*: 100 years of *Pithecanthropus* is enough! *Cour Forsch Inst Senckenburg* 171: 341–361.
Woo, Ju-Kang (1966) The Skull of Lantian Man. *Current Anthropology* 7: 83–86.
Wood, B. (1991) *Koobi Fora Research Project, Volume 4. Hominid Cranial Remains*. Oxford: Clarendon Press.
Worms, E. A. (1955) Contemporary and prehistoric rock paintings in central and northern North Kimberley. *Anthropos* 50: 546–566.
Wroe, S. (1999) Killer kangaroos and other murderous marsupials. *Scientific American* 280(5): 58–64.
  (2002) A review of terrestrial mammalian and reptilian carnivore ecology in Australian fossil faunas, and factors influencing their diversity: the myth of reptilian domination and its broader ramifications. *Australian Journal of Zoology* 50: 1–24.
Wu, Xhinzhi (1981) A well-preserved cranium of an archaic type of early *Homo sapiens* from Dali, China. *Scienta Sinica* 24(4): 530–539.
Wu, Xinzhi and F. E. Poiroer (1995) *Human Evolution in China. Oxford: Oxford University Press.*
Yamei H., R. Potts, Y. Baoyin, G. Zhengtang, A. Deino, W. Wei, J. Clark, X. Guangmao and H. Weiwen (2000) Mid-Pleistocene Acheulean-like stone technology of the Bose Basin, South China. *Science* 287 (5458): 1622.
Yokoyama, Y., A. Purcell, K. Lambeck and P. Johnston (2001) Shore-line reconstruction around Australia during the Last Glacial Maximum and Late Glacial Stage. *Quaternary International* 83–85: 9–18.
Young, J. Z. (1979) *An Introduction to the Study of Man*. Oxford: Oxford University Press.
Zhu, R. X., K. A. Hoffman, R. Potts, C. L. Deng, X. Y. Pan, B. Guo, C. D. Shi, Z. T. Guo, B. Y. Yuan, Y. M. Hou and W. W. Huangs (2001) Earliest presence of humans in northeast Asia. *Nature* 413: 413–417.

Index

Aboriginal communities 189
Aboriginal people 95, 100, 113, 174, 249, 251
Aboriginal population 131
Abraded meander channels 139
Acheulian hand-axe 39
Acheulian tradition/tools 13, 30, 57
adaptation 66, 101, 242
adaptive conditions 199
adaptive lifestyle 231
adaptive qualities and capabilities 24, 28, 50
adaptive social system 230–31
Adelaide River 257
aeolian sediments 187
Aeta 238
Afghanistan 45
Africa 2, 5–8, 10, 21, 22, 23, 30, 39, 45, 49, 58, 72, 79, 94, 247, 271
African bushmen 79
African origins 18
African savannas
agrarian revolution 44
Agta 238
alchemy 114
allometry 234
Alor 34, 35
Altai 76, 79
alveolar bone 226
ambient temperatures 18
Ambrona 30
amenorrhoea 50
Americas 41, 58
amino acid racemisation dating 137
anaemia 43
anatomically modern humans (AMH) 59, 94, 107, 173, 184, 238, 247, 268, 271, 273
ancestors 6, 72
Ancestral Beings 112
Andaman Islands 238
Andaman Sea 84
animal behaviour 15
animal domestication 43
animal herds 23
animal populations 12
ankylosing 229

Antarctica 7
anthropogenic burning 117, 135, 173
anthropogenic pressures 180
anthropoids 5, 6
anthropologist 37
Arabian peninsula 17
Arafura Plain 96, 107, 120, 252, 256
Arafura Sea 97, 156
Arafuran palaeocoast 104, 119, 121, 123
archaeological discovery 48
archaeological evidence 5, 39, 69, 71
archaeological investigation 33
archaeological philosophy 38
archaeological record 66, 181
archaeologist 37, 38, 39
archaic brows 203
archaic *sapiens* 17, 95, 240
Arctic 32
Arctic Circle 10, 79
area-per-person 46
Arnhem Land 94, 107, 120, 132, 257
arrival of humans in Australia 117
artefacts 72
arthritic conditions 43
Aru 98
Aru Hills 119
Aru Islands 117, 124
ash layer 89
ash-fall 91
Ashmore reef 35
Asia 7, 21, 23, 46, 77, 79, 100
Asia Minor 69
Asian fossil evidence 38
Asian population 58
Assam 67
Asteraceae 139
Atapuerca 9
Atherton Tableland 257
Ati 238
Atlantic coast 9
Atlantic Ocean 7
atlatl elbow 229
atmospheric aerosols 92
atmospheric cooling 92
attrition – dental (see dental wear)

Index

Australasian region 41
Australia 1, 32, 35, 37, 41, 71, 72, 76, 80, 81, 82, 86, 89, 95, 96, 186, 237, 238
Australia – colonisation of 128, 134
Australia – human entry 85, 173
Australia's landmass 70
Australian continental shelf 173
Australian megafauna (see megafauna)
Australoids 183
australopithecines 6, 8
avalanche 54
average life expectancy 50

backshore 187
backup groups 64
bags 229
Balabac Straits 85
Bali 34, 85, 86
bamboo 96
bamboo knives 99
Banda Sea 82
bandicoots (Peramelinae) 140
bands 10, 19, 76, 124, 230
Bangladesh 22
Barkly-Selwyn Tablelands 124
Barkly Tableland 133, 134, 257, 259
barrier pool 66
Barrier Ranges 225
Basilian Islands 85
Bassian Bridge 252
Batadomba Lena 236
Batak 238
bathometric topography 34, 84, 97
Bay of Bengal 19, 89, 147
Beagle Channel 7
bear 29
behavioural modification 80
Belgium 91
belief systems 50
Bering Bridge 31, 76
Berkley River 123
bifacial hand-axe 33
bifacial stone tools 79
big cats 13
bindings 230
bio-cultural adaptation 78
biogeography 6
biological adaptation 10, 13, 80
biological anthropologist 37
biological population 53
biological variability 80
biology 7
biomass 12, 74, 87
biotic change 117
Birdsell, J. 20, 85, 128, 130, 184, 235

bird's head 119
birth interval 6
birth spacing 50
bison 18, 29
Black Death 54
blitzkrieg 161, 180
boar 29
boat 35
boat people 2, 94
body decoration 230
body fluids 26
body hair 27, 28
Bonaparte Bay 122
Bonaparte Depression 122
bone cracking 222
bone infection 229
bone smashing 223, 225
boomerang 230
Borneo 82, 102, 103
Bose basin 30, 39
bottle necks – population 88
Bowdler, S. 264
Bowler, J. 185
Boxgrove 10, 30
Bradshaw figures (see Gwion Gwion) 229
brain size 23–24
Bramaputra River 67
breast feeding 50
Britain 10, 48
Brothwell, D. 236
brow ridges 214, 240, 256
browsers 153
bubonic plague 54
budding groups 108
budding mechanism 17
budding-off 5, 17, 127, 128, 129
burial practices 95, 230, 234
Burma – see Myanmar
bush fires 121, 174
butchering 30
Byzovaya 79

calcination 221
caldera 89
Callitris 120, 139
calvaria 199
Cambodia 67
campsites 72, 254
Cape Londonderry 122
Cape Mangkalihat 85
Cape York 120, 124, 132, 257
capturing 75
carbohydrate 74
carnassial complex 144
carnivores 9, 88, 145, 146
carrying capacity 49, 69, 70

Casuarina 120
Casuarinacaea 139
Caucasians 67
cave dwelling 29
Celebes Sea 85
census – earliest 47
central Africa 238
central Asia 39
Central Australia 3, 131, 133, 135, 155, 174, 177, 242, 247, 248, 255, 259, 260, 267
central Malaya 91
central Russia 76
central Siberia 76, 78, 79
Ceprano 9
Ceram 98
ceremonies 95, 229
Chad 70
channel systems 140
charcoal particles 117
Charnley River 123
charring 221
Chelodina sp. 160
Chenopodiaceae 139
child spacing 50
China 1, 8, 10, 17, 21, 37, 40, 41, 45, 58, 65, 67, 78, 84, 86, 101, 103, 184, 247
Chinese migrations 84
chopping tools 13, 33
Christ 48
Christian era 60
cleaning country 175
cleavers 30
Cleland Hills 115
climatic extremes 14
clinal effect 66, 77
clothing 79
Clovis 275
coastal adaptation 82
coastal and riverine navigation 82
coastal contour 83
coastal economy 231, 266
coastal environments 70, 72
coastal length 84
coastal migration 86
coastal travel 82
coastal voyaging 265
Coburg Peninsula 113
cognitive abilities 33
Cohuna 245
cold climates 79
cold effects 28
colonisation 3, 97
colonising group 132
colour changes 223
complex language 31, 76

complex ritual 230
complex social relationships 50
complex tool kits 19
conception frequencies 50
continental shelves 12, 18, 34, 81, 87, 130, 253
continuous movement 132
contraceptive effects 49–53
Coobool Creek 204, 245
Coobool females 217
cool winters 11
cooperative hunting 50
Cooper Creek 134, 135
cortical bone 219
Cossack 245
cradle of humanity 58
cranial buttressing 204
cranial capacity 23, 57
cranial features – robust 256
cranial morphology 26, 67
cranial sutures 233
cranial thickening 193, 194, 195, 239
craniometric data 215
cremation – human 170, 189, 193, 219
crocodile (*Crocodylus porosus*) salt water 1, 32, 34, 115, 116, 131, 142, 147
cross-currents 35
crouch burial 225
crowd diseases 43
crustaceans – freshwater 140
cultural adaptation 24, 79
cultural baggage 81, 95
cultural complexity 226
cultural debris 71
cultural development 38, 62, 80
cultural fluctuation 101
cultural indicators 219
cultural modification 73
cunning 30
curiosity 15
currents 35
customs 50
cut marks 79

Daly River 123, 257
Darling Downs 149, 156, 158
Darling River 268
dating 37, 114–116
dating problems 10
Deaf Adder Gorge 179
debitage 71
deer 18, 29
Deevey, E.S. 46
demographic development 62
demographic history 73
demographic parameters 52, 54

demography 6
density differential 73
dental wear 228, 229
dermal layers 27
desert landscapes 11
deserts 12
Devil's Lair 179
Dhaky 22
Diamantina River 135
diaspora 2
Dieri Tribe 134
diet 79
dietary change 23
dingo (*Canis familiaris*) 145
diploeic bone 195, 200
Diprotodon minor 140
Diprotodon optatum 115, 140, 142, 144, 147, 160, 181
Diprotodontids 142, 146, 160
Diring-Ur'akh 9, 79
dispersal patterns 9
distance travelled 15
Djankawu 112
Dmanisi 8, 9, 16
DNA 248
domestication 43
doubling time – population 54
Dreaming 112, 251
Dreaming stories 112
Dreamtime 112
dried food 96
drifting 35
droughts 54
dry summers 11
Drysdale River 122
dugong (*Dugong dugon*) 124
Dumagat 238
dune building 137, 187
Durack River 123
dwindling animal populations 179

early seafarers 118
earthquakes 158
East Africa 17, 37
East Alligator River 257
east coast of Sunda 84
eastern Asia 18
eastern Siberia 78, 79
East Sunda River System 85, 103
ecological adaptation 246
eco-niches 12
economic change 41
economic diversification 73
economic intensification 43, 78
ecosystems 12
ecosystems – marginal 13

Efe people 238
electrolytes 26
elephants 30, 116
embryonic diapause 152
empty country 132
Emydura sp. 160
endemic malaria 28
endemic malarial areas 199
endogamous system 19
endurance running 12, 27
England 10, 30
environmental barriers 12
epicanthic fold 28, 79, 237, 250
equator 23, 79, 85, 87
equatorial 12, 26, 35
equatorial rainforests 11
erectine 6, 8, 23, 34, 35, 50, 59, 79, 95, 207, 271, 272
erectine bands 19, 22
erectine crania 15
erectine females 50, 51
erectine height 26
erectine migration 12, 29, 99
erectine palaeodemography 21, 37, 50–53
erectine skills and capabilities 12, 23
erectines – Chinese 30, 32, 36, 272
erectines – Dmanisi 27
erectines – east Asian 24
erectines – southeast Asian 27
erectines Indonesian 26
eruption column 89
estuarine environments 70
Etadunna Formation 137
ethnographic observation 175
ethnohistoric 175
Eurasia 18, 30
Europe 7, 9–10, 23, 29, 45, 62, 69, 78, 229
Europe – eastern 76
European landscape 10
Europeans 7
Euryzygoma dunense 140
Eve hypothesis 59
evolution 6
evolution of human populations 67
excessive exercise 50
exploitative strategies 57
exploration 131
explorative behaviour 80
exploratory groups 108
exponential population growth 42, 43
extinction 15, 108
extinction spiral 159
extractive behaviour 56
extractive potential 69, 75
extrinsic population growth 130

facial sinuses 28
family groups 230
Far East 9, 58
faunal assemblages 10
faunal populations 11
faunal remains 29
feast and famine diet 238
feeding strategies 29
female exchange 19
female losses 88
female replacement 51
female survivorship 51
fertility rates of 6
fire 29, 57, 79, 81
fire prone plant species 117
firing 177, 258, 264, 266
first arrivals 173
first Australians 1, 114, 116, 183, 184
fish 140
fishing spears 265
fish traps 265
fission 127, 130
fission model 132
Fitzmaurice River 123
Fitzroy River 123, 257
flaking techniques 30
flexed 225
floatation 96
floating log 32
floods 54, 71
floral populations 11
Flores 33, 35, 39, 94, 95, 104, 107
flotillas 130
fluctuating sea levels 11
fluorine 162, 163
fluvial activity 137
fluvial sediments 142
Fly-Strickland River system 121, 124, 257
fodder collapse 159
forward movement 15
fossae 215
fossil assemblage 23
fossil faunal populations 145
fossil sites 141
founder effect 12, 66, 77, 101, 104, 242
founder frequency 66
founder population 41, 52, 54, 62
founding world populations 53
freshwater highways 124, 132, 266
freshwater lakes 181
frontal bone – flat 239, 256
frontal pneumatisation 193
frontal sinus 212
fuel build up 175
funerary rites 225
Furneaux Island 100

gait 26
gallery forest 87, 260
Garawa 238
gene drainage 65
gene exchange 19
gene flow 12, 40
gene frequencies 77
gene migration 104
gene penetration 66
gene pool – 'daughter' 107
gene pool – 'mother' 107
gene pools – small 249
gene pools 12, 77
general intelligence 35
generalist hunter-gatherer 126
generations 262
genes 6
genetic affiliation 102
genetic assortment 108
genetic barrier 64, 65
genetic diversity 81
genetic drift 66, 77, 101, 158, 242, 248
genetic expansion 12
genetic intermixing 87
genetic load 158
genetic loss 65
genetic mixing 265
genetic mutation 101
genetic paths 68
genetic pool 64, 101
genetic pressure gradients 62, 75
genetic selection – general 101
genetic selection – local 101
genetic tunnel 64, 66
gene transfer 80, 101
Genyornis egg shell 161
Genyornis newtonii 115, 137, 140, 161, 264, 268
Geochelone sp. 33
geographical barriers 6
geographical isolation 81
geo-political borders 17
Georgia 8, 9, 39, 59
giant bears 13
Gibraltar 9
Gibson Desert 255
glabella 212
glacial conditions 9, 15
glacial cycles 87, 254
glacial events 69
glacial-interglacial cycle 17, 252
glacial maxima 33, 35, 269
glacial peaks 82
glacial sea levels 34

Index

glaciation 22
glaciers 10, 12
Glasgow University 136
glenoid fossa 193, 214
glenoid length 214
Gobi Desert 116
Goliath people 238
Gongwangling (Lantian) 8
gracile 251
gracile fossils 184
gracile morphology 184
gracile origins 236
gracile people 234
gracile skeleton 234
Grand Dolina 9
grandmothering 19
Great Australian Bight 256
Great Divide – the 256
Greater Australia 246
Great Indian Desert 116
Great Sandy Desert 122
Great Victoria Desert 155, 260
Green Gully 245
Gregory, J. W. 136
gross reproduction rates 51
group cooperation 30
group extinction 108
group organisation 30
growth potential 51
growth rates 41, 56
growth spurt 26
Gulf of Carpentaria 97, 156
Gulf of Malacca 84
Gulf of Martaban 83, 84
Gulf of Thailand 82, 84, 85, 87
Gwion Gwion (Bradshaw figures) 229

habitable land 12
hafted axes 230
Halmahera Islands 98, 103
hand axes 30, 81
Hassan, F. A. 20
Hathnora cranium 17
HbA 28
HbE 199
HbS 199
Hb variants 28
hearing 36
hearths 187, 191
herbivore 146
herd animals 15
herd trailing 15
high humidity 27
high latitudes 12, 28, 78, 79
high metabolic rates 28
high mortality rates 48, 51

high pressure centres 64
Himalayan Mountains 45, 67
hippopotamus 116
history 81
Holocene 40, 41, 42, 44, 58, 251, 269
Holocene population explosion 41
hominin bands 50
hominin brain size 23
hominin females 52
hominin fossils 199–202
hominin migration 11, 36
hominins 5, 6, 7, 8, 18, 31, 32, 36, 95
Homo 6
Homo antecessor 6
Homo erectus 1, 2, 6, 13, 18, 21, 24, 26, 30, 33, 36, 50, 51, 73, 91, 94, 98, 173, 199, 237, 240, 271
Homo erectus – survivorship 51
Homo ergaster 6
Homo floresiensis 39, 93, 95, 99, 100, 106, 236, 239, 256
Homo sapien bands 31
Homo sapiens 7, 14, 23, 37, 49, 72, 80, 81, 100
Homo soloensis 95, 102, 106, 107, 125, 173, 186, 199, 240, 246, 249, 256, 261
horse 18, 30
Huang (Yellow) River 74
human demography 21
human-empty areas 45
human evolution 37
human expansion 81
human genome 81
human health 43
human migration 2, 134
human palaeodemography 9, 40
human populations – regional 87
human remains 187
human stature 43
human variation 101
humeral length 217
hunter-gatherer lifestyle 132
hunter-gatherers 6, 46, 50, 123
hunter-scavengers 6
hunting bands 10
hunting behaviour 19
hunting capabilities 56, 57
hunting-gathering-scavenging 16
hunting skills 13
hunting techniques 57
hyenas 13
Hyperplatycnemia 219

ice 12
Ice Age 18, 78, 116, 159, 252, 264, 274
Ice Age - Australia 135
ice-bergs 232

ice build up 11
ice cores 69
ice-free corridors 18
ice sheets 15
implement manufacturing 73
incipient herding 74
incisors 226
increased brain size 23, 56, 57
India 17, 30, 40, 41, 58, 65, 86, 101
Indian Ocean 83, 84, 89
Indian subcontinent 18, 58, 76
Indonesia 10, 28, 32, 96
Indonesian archipelago 95, 173, 256
Indonesian (Javan) hominids 95, 240
infant mortality 108
infectious disease 43, 157
infertility 108
initiation ceremony 227
inner cranial tables 195, 198
inundation 88
intellectual development 80
intelligence increase 23
interglacial faunal ecology 141
interglacial humans 170
interglacial palaeoenvironments 125
interglacials 15, 79, 181
inter-island crossings 35, 95–96
inter-lake areas 139
inter-membral indices 144
interorbital region 212
inter-riverine 139
inter-Wallacian islands 97
intrinsic population growth 60, 128, 130, 131, 251
Inuit 79
Iran 17
Irian Jaya 97, 110
Irian Jaya–New Guinea cordillera 119
Irrawaddy delta 83
Isampur 30
Isdell River 123
Isernia la Pineta 9
island hopping 9, 33, 73
island populations 99
island southeast Asia 76
isobaric patterns 65
isolated islands 12
isolation 19, 66, 82, 131, 242
Italy 9

Japan 9, 45
Java 8, 17, 21, 59, 82, 102, 104, 184, 199
Javans 103
Java Sea 82, 87
Johore River 85

joint destruction 229
Jolo 85
Jordan River 8
Joseph Bonaparte Plain 123
jungle 82

Kadimakara 134, 135
Kalahari 238, 239
Kallakoopah Creek 137, 169
Kambing Island 34
Kamitakamori 9
Kampar River 85
Katherine River 123
Katipiri assemblages 139
Katipiri Formation 137, 161, 169
Keilor 245
Kelapa 34
Kenniff Cave 185
Kenya 146
Kimberley 107, 121, 131, 156, 252, 257
Kimberley palaeocoast 125
Kimberley plateau 122
Kimberley shelf 104
King Edward River 122
King George River 122, 123
KNM-ER1481 8
KNM-ER3228 8
KNM-WT15000 26
koala (*Phascolarctos* sp.) 139, 268
Komodo dragon (*Varanus komodoensis*) 33, 98, 142, 143, 147
Komodo Island 34
Kow Swamp 184, 185, 200, 204, 245
Krakatoa 89
Kuala Lumpur 83
Kulpi Mara 269
Kutjitara Formation 137

Lachlan River 187
La Gravette 229
Lake Ara 121
Lake Carpentaria 97, 118, 119, 120, 124, 134, 143, 256, 257, 258, 261
Lake Carpentarian river system 104, 120
Lake Eyre 139–140, 147, 156, 181, 259, 268
Lake Eyre basin 120, 133, 134, 159, 255
Lake Eyre catchment 120, 253, 261
Lake Eyre drainage system 160
Lake Eyre extinctions 161
Lake Eyre – megalake phases 143, 259
Lake Eyre region 176
Lake Fura 121
Lake Garnpung 187, 189
Lake Leaghur 189
Lake Londonderry 122

Index

Lake Menindee 265
Lake Mulurulu 187
Lake Mungo 94, 95, 161, 184, 247
Lake Mungo sedimentary sequence
 – arumpo 190
Lake Mungo sedimentary sequence – Gol
 Gol 191
Lake Mungo sedimentary sequence – lower
 Mungo 191
Lake Mungo sedimentary sequence
 – Mulurulu 190
Lake Mungo sedimentary sequence – upper
 Mungo 191
Lake Mungo sedimentary sequence
 – Zanci 190
Lake Mungo sedimentary sequences 190
Lake Tandou 265
Lake Turkana 26
Lake Victoria 256
land-based migration 64
land bridge 9, 10, 17, 81
landscapes 10
language 31, 81
language development 31, 57
Lantian 20
Laos 67
last glacial maximum 137
last glaciation 78
last Ice Age 180
last interglacial 78
late Pleistocene skeletal morphology
 – Australia 184
late Quaternary 137
laterally expanding groups 65
lateral restrictors to migration 64
Leigh, S.R. 24
Lena River 9, 79
length and robusticity modules 208
leopard 29
Lesser Sunda archipelago 84
Lesser Sunda chain 33, 36
Levant 9
lexicon 31
LGM 243
life expectancy 6
limited tool kit 35
Linea aspera 219
Liran 35
littoral environments 27
littoral hunting 36
Liujiang 247
live births 52
living areas 71
local adaptation 239
local extinctions 175

local ingenuity 75
logs 35
Lombok 34
Lorentz River 119
low birth rates 108
lower latitudes 11
lowering seas 12
low latitudes 28
low reproduction rates 49
lunettes 187
Luzon 238

Mackay 147
Macropods 142, 146
Macropus spp. 140
magic 230
magma 92
major glaciations 11
Malacca 102
Malaccan coast 83
Malacca Strait River System 84, 85
Malacca–Sumatra bridge 84
Malakunanja 94, 179, 246
Malar 204
Malar fossa 193
malaria 28
Malar morphology 208
Malaysian Peninsula 18, 67, 76, 81, 84,
 91, 238
Malita shelf 35
Malita Valley 122, 123
Malthus, T. R. 43
mammoth 18
mammoth tusk 79, 230
Mammotovaya Kurya 79
manufacturing skills 35
Mapa cranium 78, 106, 247
marginal environments 57
marginal niches 74
marginal regions 73
marginal territory 79
mark of ancient Java 184
marsupial wolf (*thylacinus cynocephalus*) 145
Mary River 257
mass movement 14
mastoid pneumatisation 193, 213
mastoid processes 214
material culture 227
mating rules 19, 128
matrilineal descent 19
maximum cranial thickness 205
May River 123
McHenry, H. 24
measles 54
medullary cavity 217

Index

megafauna 1, 140, 154, 174, 254, 258, 265, 266
megafauna collapse 154, 160
megafaunal extinctions 172, 267
megafaunal populations 268
megafaunal species 115
megalake phase 265
Megalania prisca 98, 116, 132, 142, 143, 145, 147
Melanesia 81, 110
melanisation 27
Meltzer, D. 275
menstrual cycle 51
metacarpal 167
micro-evolutionary mechanisms 110
Middle Ages 37, 54
Middle East 7, 19, 37, 39, 45, 59, 69, 76, 106, 237, 264
migrant gradient pressures 66
migrant populations 101
migrant settlement centres 65
migrants – 'single line' 19
migrating groups 238
migration 88, 239, 242
migration – 'accidental' 14
migration – 'background' 14
migration corridors 80, 123
migration – 'illegal' 14
migration – deliberate 15
migration – earliest 38
migration – human 14
migration – inter-island 39
migration – longitudinal 100
migration – natural barriers to 14
migration – oceanic 37
migration – purposeful 15
migration routes 80
migration routes – optimum 12
migrations 85
migration strategies 125
migration – slow 14
migration – trans-world 7
migratory bands 111
migratory gene pool 66, 104
migratory patterns 15
migratory pool 66
Mincopies 238
Mindanao 85
Mindel ice age 29, 37
Mindoro Strait 85
minimum numbers 48
Miocene 137
Mitchell Plateau 1
Mitchell River 123
mitochondrial DNA 59
mobile colonist 75

modern drainage system 139
modern fisherman 35
modern human crania 24
modern humans 50, 79
Modjokerto (see also Perning) 8, 20, 32
Momats River 119
Mongolia 9
Mongoloids 67, 68
monsoon – southerly incursion 139, 181, 253
monsoon 261, 263
Monte Verde 275
Moon 35
Mootwingee 226
Moroccan Neanderthal 94
Morocco 9
morphological change 76
morphological contrast 183, 186
morphological differences 235
morphological variation 38, 101, 187, 193, 236, 242, 244
mortality 6
mortality rates 41, 128
mosaic burning 175, 176
mosquitoes 1
Mossgiel 245
mountain chains 12
Mount St Helens 89
Moyle River 123
Multiregional Continuity hypothesis 58
Mulvaney, D. J. 125
Mungo Lady 185, 233
Murray–Darling Basin 187
Murray River 184, 247, 268
muscle markings 234
muscular attachment areas 240
musk ox 18
mutation events – random 66, 242
mutation rates – local 77
mutations – adaptive 80
Myanmar (Burma) 67, 83, 102

Namarrgon 113
Nandong 212
Nanwoon 230
Narmada Valley 17, 52
natality 6
natural catastrophe 157
natural disaster 15
navigation skills 82
navigators 94
Neanderthals 69, 78, 79, 94, 207, 218, 219
Negrito migration 239
Negritos 236, 238
Negros 238
net construction 230
net reproduction rate 51

nets 230
neural networks 57
neurological complexity 23
neurological organisation 23
New Guinea – see Papua New Guinea
New South Wales 186, 187, 252
new species 12
New World 72, 73
New Zealand 7, 48, 77
Ngandong 32, 53, 71, 241
Ngandong XI 202
Niah 236, 237
Nihewan Basin 8, 9, 29, 39, 79
Nitchie 245
nomadic littoralists 82
nomadism 74
North Africa 9, 69, 76
North America 32
northern Australia 147
northern Europe 10
northern palaeocoast 131
northern Sumatra 83
northern territory 122, 179
North Sunda River System 85
northwest Africa 10
Nototherium sp. 140
nuchal area 204
nuchal crest 240
nuclear winter 92
Nullarbor Plain 156, 158, 260
numbat (*Myrmecobius fasciatus*) 268
nutritional resources 12

occipital bone 231
occipital torus 256
occipitomastoid border 213
oceanic distance 34
oceanic tephra 89
oceanic ventures 36
oceanic water 18
oceanic water gaps 33
ochre 189, 225, 230, 234
ODP 820 core 172, 173
Old World 8, 39, 45, 72, 95
Ombai Strait 34
omnivorous diet 13
Onge 238
Oogeroo Nunuccal 112
open forest 11
open ocean 12
open sites 10
opportunistic breeders 154
Ord River 123, 257
origin of modern humans 41
osteoarthritis 43, 229
outer cranial table 198

outer tables 195
Out of Africa hypothesis 58, 59, 95, 184, 207, 237, 251
oxygen isotope stages 255

Pacific islands 48
Pacific voyages 7
Pacitanian 33
palaeoanthropologist 37, 38
palaeobeach 187
palaeobiology 24
palaeochannels 137, 142, 261
palaeocoasts 82, 117, 254
palaeodemographic trends 41
palaeodune cores 138
palaeoecological conditions 146
palaeoenvironmental change 128
palaeoenvironmental reconstruction 135
palaeolithic 51
palaeoshoreline 256
palaeovalleys 259
Palawan 36, 85, 238
Palimnarchos pollens 116, 131, 142
Palorchestes sp. 140
palynology 117, 172
Panaramitee art 115
Papua New Guinea 32, 85, 86, 96, 107, 110, 237, 238, 256, 257, 261
parallel evolution 79
parasites 157
parasitism 43
parietal boss 194
parietal notch 214
patchy populations 179
pathological revolution 43
pathology 198, 219
patrilineal descent 19
pebble tools 33
Penida 34
Pentecost River 257
people generator 59
people movement 180
Perak Valley 91
permafrost 10
Perning 8, 32
petroglyphic art 95
Phascolonus gigas 140
Phascolonus medius 140
phenotypic characteristics 237
Philippines 36, 85, 86, 102
photosynthesis 93
physical barriers to migration 64, 107
physiological adaptation 15, 49
phytoliths 228
Pienaar, U. de 146
pigs 116

pirates 2
Pitcairn Island 100
Pithecanthropus 212
plant fibre 230
playa lakes 141
Pleistocene 2, 54, 86
Pleistocene Australian populations 207
Pleistocene – Late 47, 58, 247
Pleistocene – Lower 6, 7, 8, 22, 56
Pleistocene – Middle 6, 10, 13, 22, 24, 37, 42, 48, 49, 56, 79, 80, 91, 95, 236, 271
Pleistocene migrations 80
Pleistocene – palaeodemography 1, 38
Pleistocene population growth 130
Pleistocene – populations 41
Pleistocene – Upper 24, 42, 48, 56, 67, 74, 78, 80, 87, 91, 97, 100, 183, 230, 241, 259, 271
Plio-Pleistocene 116
pneumatisation 212, 240
Poaceae 139
Poland 230
polymorphisms 199
poor nutrition 43
population 'chain' 21
population bottle neck 93
population boundaries 238
population centres 75
population chains 22
population decline 54
population doubling time 51, 52
population explosions 12
population generators 58, 111
population growth 39, 40, 48, 51, 53, 105, 131
population growth centres 79
population pressure 62, 74
populations – high latitude 29
population spread 22
population valves 76
population viability 51
possums (*Trichosurus vulpecula*) 139, 140
post-cranial measurements 49
post-partum 50
posture 26
pouch young 152
pre-adaptation 126
pre-agricultural 44
pre-arranged traps 30
precursor migratory population 21
pre-existing gene pool 107
prehistoric world population growth 41
pressure flaking 230
prey populations 145
primate populations 28
Prince Regent River 123
Procoptodon sp. 140, 152

progenitor populations 54
prominent muscle markings 256
protein biomass 22
protein diet 74
Protemnodon spp. 140, 152
proto-hominins 6
Proto-Malays 238
pseudo-opposable thumb 144
puberty 50
Pulau River 119
Puritjarra 269
pygmy elephant (see Stegodon)
pyroclastic flows 91
pyroclastic rock 89
Pyrophytes 117

Qafzeh 106, 237
quartz grains 228
Quaternary 12, 88, 93, 117
Quaternary sea levels 34
Queensland 252

raft-bands 128
raft building 99
raft flotillas 99
rafts 35
rainforest 12, 27
rainforest environments 238, 239
rainforest fauna 173
rainforest species 12
rapid movement 14
rate of natural increase 45
red kangaroo (*Macropus rufus*) 152, 153
refugees 267
regional continuity 183, 241
regional human evolution 273
regional morphological variability 81
regional Pleistocene migration 80
regional population growth 80
regional variation 40
reindeer 18
relative age 10
Renzidong 8, 79
replacement theory 59
reproductive biology 6, 41
reproductive cycle 50
reproductive life 50
reproductive potential of early people 41
reproductive span 50, 51, 52
resident group 17
retouched flakes 33
rhinoceros 30
rhizomorphs 139
rhyolytic ignumbrite 89
Rintja 34
rites of passage 227

ritual and ceremony 50, 227, 230, 234
riverine ecosystem 230
riverine environments 27, 70
riverine system 255
river refuges 160
robust groups 268
robusticity module 209
robust morphology 71, 183, 184, 191, 213, 239, 242, 247, 251, 268
robust origins 241
robust populations 235, 244
rock art sites 226
rock shelters 29, 231
rocky landscapes 29
Roe River 123
Roman Empire 47
Rufus Hare-wallaby (*Lagorchestes hirsutus*) 174
Russian Arctic 79

Sahara Desert 255
Sahul 35, 37, 81, 86, 95, 96, 107, 244, 252, 258, 263, 264
Sahul banks 98
Sahulian migrations 105
Sahul palaeodemography 109
Sahul rise 98
Sahul shelf 96, 122, 256
sailing skills 95
Sambungmacan 32
sand flies 1
Sangiran 32
Sangiran fossils 8
Santa Luca, A. P. 53
Santorini 89
Sarawak (Sabah) 236
satellite imagery 123
savanna 5, 119, 147, 254, 255
Scandinavian ice sheets 47
scavengers 27
scavenger targets 12
scavenging 29, 149
scrapers 13, 30
sea crossings 37
sea currents 34
sea levels 12, 70, 80, 87, 96, 122, 254
seasonal hunting 15
secondary burial 225
sedentism 78, 131
Selwyn Range 133
Semang 238
semi-arid environments 73
semi-arid regions 70
semi-permanent settlement 84
Senoi 238
Serengeti National Park 146

Sewing 230
sexual dimorphism 245
sharks 34
shellfish 140
shell middens 187, 191
shields 230
shortest route 85, 96, 97, 121
Siam Plain 85
Siam River System 85, 103
Siberia 18, 31, 39, 45, 68, 78
Siberian Far East 76
Sibutu Passage 85
sickle cell anaemia 28, 199
side scrapers 79
Sidi Abderrahman 9
sight 36
Simosthenurus spp. 140, 152
Simpson Desert 137, 169, 181, 255
Simpson Savanna 133, 146, 153, 154, 181, 259
Sinai 20
Sinai Peninsula 7, 9
Sinanthropus 212
site interpretation 10
size-length module 208, 209
skeletal biology 110
skeletal morphology 81, 107
skeletal record 66, 109
skills – behavioural 35, 99
skills – linguistic 35
skills – technological 35, 69, 94, 99
skin colouration 27
skins and fur 29
smallpox 54
smell 36
snake bite 28
social cohesion 234
social structure 230
societal development 70
socio-cultural change 80
socio-cultural development 78
Soleilhac 9
Solo 33, 53, 241
Solo River 22, 256
Somalia 70
sorcery 230
South Alligator River 257
South Australia 134, 259
South China Sea 82, 84, 89
south coast of Sumatra 83
Southeast Asia 264
southern hemisphere 252
Southern Ocean 7
Southern Siberia 79
Southern Sunda 84
south Javan coast 84
Southwest Tasmania 252

Southwestern Australia 252
space travel 81
Spain 9, 30
sparse populations 65
spears 57, 230
spear thrower 229, 230, 266
spheroid pounders 13
Spinifex (*Triodia* sp.) 175
spirit beings 112
spotting danger 36
Sri Lanka 45, 89, 236
stable population 17
starvation 50
stature – small 234, 236, 250
steerable craft 35
steering boards 96
Stegodon (*Stegodon trigonocephalus florensis*) 33, 34
steppe 10, 12
Sthenurus spp. 140, 152
stochasticity – demographic 157
stochasticity – environmental 157
stochasticity – genetic 157
stone artefacts 79
stone flakes 71
stone points 230
stone tools 91
stone tools 10
Straits of Hormuz 17
Straits of Malacca 83
strategy building 73
stratigraphic dating 37
stratospheric dust 92
string 230
sub-alpine 11
sub-arctic 78
subequatorial 12, 26
sub-glacial environments 10
Suez Canal 22
Sula 104
Sulawesi 85, 103
Sulawesi-Ceram corridor 118
sulphur 92
sulphuric acid 92
Sulu Archipelago 36, 85
Sulu Sea 85
Sumatra 22, 81, 84, 86, 89, 102, 238
Sumatran coast 84
Sumbawa 33, 34
sun 18
Sunda 76, 81, 83, 86, 88, 95, 102, 116, 264,
Sunda archipelago 35
Sunda coastline 82
Sunda – eastern 86
Sunda shelf 11, 70, 85, 105

Sunda – south 85
Sunda subcontinent 18, 37
Sunda to Sahul 37
Sunda – west coast 83, 84
superciliary ridges 204
supernatural 91
supramastoid crest 214, 256
supraorbital development 193, 202
supraorbital module 202, 207
survival 61
survivorship 51, 145
swamps 122
Swan River 252
sweat glands 26, 27
sweating 27
sweat mechanism 26
swimming 34
synchondrosis – suboccipital 233

T. rex 98
taboos 50
Tadjikistan 78
Tampanian industry 91
Tanami Desert 123
Tanimbar 98, 119, 124
Tanzania 256
taphonomic processes 219, 223
Tapiro 238
target areas 64
target destinations 15
Tasmania 3, 48, 81, 116, 231, 252, 258
Tasmanian Aboriginal people 250
Tasmanian devil (*Sarcophilus laniarius*) 145
Tasman Sea 7
taste 36
technological change 43, 75
technological development 80
technological innovation 75
tectonic activity 54, 71, 84, 88
teeth – avulsion 227, 230
teeth – canine 226
teeth – molars 230
teeth – premolars 226
temporal lines 256
terrestrial biosphere 12
territoriality 21
Teshik Tash 78
Thai coast (peninsula) 83, 84
Thailand 11, 67, 102
thalassaemia 28, 199
the crowded planet 43
thermoregulation 272
Thomas' Quarry 9
Thorne A. G. 184
Thurrawah family 113
Thylacoleo carnifex 144

Tierra del Fuego 48
time 37
Timor 33, 34, 36, 95, 96, 98, 104, 105
Timor Sea 143
Tirari Formation 137, 162, 163
Toba Lake 89
Toba Mount 86, 88–93, 263, 264
tool kit 33
tool manufacturing 57
toral arch 204
Torralba 30
Torres Strait Bridge 124
tortoises 160
total fertility 51
trabecular bone 219
trading 76, 226
transcontinental crossings 266
transgressive events 97, 253, 257
transoceanic travel 80
transverse cracking 221
trans-world migration 7
treadmill 43, 44
trigonic area 204
tri-hybrid theory 184
Trinil 32
Tristan da Cunha 100
trophic pyramid 181
trophic 'shape' 145
tropical environments 18
tropical regions 12
tropical storms 54
Tsagaan Agui 9
tsunami 54, 84
tuff layer 89, 91
tundra 11, 12
Turkana 'boy' 26
Typha 120

Ubeidya 8
uninhabited landscapes 101
Unio sp. 229
univariate comparison 205
Upper Miocene 5
Upper Pleistocene 215
Upper Pleistocene migration 126
UV radiation 27

Vallonet Cave 9
varanid lizard 33
vasoconstriction 28
vault pathology 207
vault thickness 197
vegetation change 117
verbal communication 31
viability 109
viable population 51, 96, 110

Victoria 184
Victoria River 123
Victoria River (WA) 257
Vietnam 67
Vietnamese coast 84
Viking voyages 53
vitamin D_3 28
vocal coordination 31
volcanic activity 54, 88
volcanic eruption 91, 158
volcano 86
Vostok ice core 89
voyaging 80

Wallace Line 116
Wallacia 81, 85, 86, 96
Warburton River 135, 160
warfare 54, 75
water craft 9, 32–37, 94, 96, 100, 124
water gaps 36
water intake 27
water lilies 120
weaning 50
weaponry 30, 234
weapons 230
weaving 230
Weber Deeps 96
Weidenreich, F. 52, 184, 199, 212, 240
Western Australia 121
Western Desert 158, 174
western Sumatra coast 84
West MacDonnell Ranges 269
West Papua 238
Wetar 35
wet-dry episodes 259
wetlands 122
wet periods 137
wet tropics 11
whirlpools 35
Willandra Creek 187
Willandra Lakes collection 184
Willandran robusticity 219
Willandra region 242, 247, 248, 265, 268
WLH1 170, 185, 186, 189, 195, 202, 204, 206, 217, 221, 225, 233, 234, 239, 245
WLH10 222, 234
WLH100 194, 206, 214, 239
WLH101 206, 214, 239
WLH102 195
WLH107 219, 239
WLH11 202, 204, 206, 217, 234, 235
WLH110 217, 239
WLH115 222, 234
WLH120 223
WLH121 219

WLH122 191, 222, 234
WLH123 222
WLH124 206, 214
WLH126 223
WLH130 234
WLH132 222
WLH134 206, 235
WLH135 225, 234
WLH152 202, 206, 229
WLH154 186, 225, 239
WLH16 239
WLH18 195, 202, 206, 210, 239, 245
WLH19 194, 198, 202, 206, 210, 216, 239, 245
WLH2 189, 195, 222, 225, 245
WLH20 217
WLH22 194, 217, 223, 227, 239
WLH23 191
WLH24 191, 223
WLH27 194, 210, 239
WLH28 194, 210, 239
WLH29 234
WLH3 94, 185, 186, 202, 204, 206, 217, 219, 225, 227, 228, 233, 234, 245
WLH44 191, 214, 234
WLH45 194, 204, 217, 239, 245
WLH50 184, 194, 202, 206, 210, 239, 245
WLH51 202, 204, 206, 235
WLH52 191
WLH56 195, 222, 239
WLH6 223, 234
WLH63 195, 223, 239
WLH67 195, 206, 219
WLH68 195, 202, 204, 206, 222, 234, 235, 245
WLH69 202, 206, 239
WLH72 206, 214
WLH73 202, 206, 239
WLH9 191, 222, 234
WLH93 223
wolf 13
wombats (Vombatidae) 140
Wonambia naracoortensis 132, 144
world population 39, 46, 53, 54, 58, 81, 88
world population end-Holocene 44

Yali, 238
yams (*Dioscorea* sp.) 120
Yangtze River 74
Yuanmou 8

zero population growth 42, 61
Zhoukoutien 22, 28, 30, 32, 38, 52, 199, 212
Zhoukoutien – humerus II 218
Ziyang 247
zoonotic 43
zygomatic arch 212, 214
zygomatic trigones 234, 240
Zygomaturus trilobus 140